6th ITB International Geothermal Workshop (IIGW 2017)

IOP Conference Series: Earth and Environmental Science
Volume 103

Bandung, Indonesia
22 – 23 March 2017

ISBN: 978-1-5108-5554-0
ISSN: 1755-1307

Printed from e-media with permission by:

Curran Associates, Inc.
57 Morehouse Lane
Red Hook, NY 12571

Printed by Curran Associates, Inc. (2018)

For permission requests, please contact the Institute of Physics at the address below.

Institute of Physics
Dirac House, Temple Back
Bristol BS1 6BE UK

Phone: 44 1 17 929 7481
Fax: 44 1 17 920 0979

techtracking@iop.org

Additional copies of this publication are available from:

Curran Associates, Inc.
57 Morehouse Lane
Red Hook, NY 12571 USA
Phone: 845-758-0400
Fax: 845-758-2633
Email: curran@proceedings.com
Web: www.proceedings.com

Table of contents

Volume 103

6th ITB International Geothermal Workshop (IIGW2017)

22–23 March 2017, Bandung, Indonesia

Accepted papers received: 18 December 2017
Published online: 10 January 2018

Preface

Papers

Engineering

Exploration

IOP Conf. Series: Earth and Environmental Science **103** (2017) 011001 doi:10.1088/1755-1315/103/1/011001

PREFACE

ITB International Geothermal Workshop (IIGW) is an annual event which is organized by Prodi Teknik Panas Bumi, Fakultas Teknik Pertambangan dan Perminyakan (FTTM), ITB (Geothermal Study Program, Faculty of Mining and Petroleum). The 6th IIGW was held on March 20- 24th, 2017. The workshop has become a special event in supporting geothermal development acceleration program in Indonesia, particularly in capacity building, research and professional interest.

In the past year, Indonesian Government has made significant reforms that directly and indirectly affect a delay in geothermal development in Indonesia. Nevertheless we still have the faith that Indonesia geothermal will someday be developed. The journey of ITB Geothermal Workshop up to now in the sixth year has proven that the faith is worthy of care. Until this year we are implementing the sixth workshop as proof of support and strong commitment from Program Studi Teknik Panas Bumi FTTM ITB for the future development of geothermal in Indonesia in particular and worldwide. These brought to the theme of this year workshop that is ***THE RISE OF GEOTHERMAL DEVELOPMENT IN INDONESIA; Resource Transformation Strategy towards 7000 MWe Capacity***. We understand that we have to be more creative and innovative in scientific, engineering and financial approach to reach Indonesia Geothermal target. Therefore, the content of Plenary Session in the workshop present some breakthrough and innovation in the current geothermal activities in order to open the opportunity for further geothermal development in a less supportive atmosphere of geothermal development.

We hope the workshop will create a knowledges sharing forum among the key stakeholders in energy and extractive resource sectors. We hope participants take benefit the exchange of views, knowledges and experiences on latest technologies and researches by bringing together scientists, engineers, academicians, experts, and industrial stakeholders involved in geothermal studies and developments.

As part of this year commitment, outstanding papers are published as an open access volume of IOP Conference Series: Earth and Environmental Science. Other papers are published in the conference proceeding in print version. We hope wider geothermal communities will gain the same benefits as our conference attendee. *Rise Geothermal!*

Sincerely,

Suryantini, Dr. Eng.

Chairman of the ITB International Geothermal Workshop (IIGW) 2017

IOP Conf. Series: Earth and Environmental Science **103** (2017) 011001 doi:10.1088/1755-1315/103/1/011001

LIST OF SPONSORSHIP 6th ITB INTERNATIONAL GEOTHERMAL WORKSHOP 2017

Sponsored by

Supported by

Endorsed by

Media Partner

Organized by

IOP Conf. Series: Earth and Environmental Science **103** (2017) 011001 doi:10.1088/1755-1315/103/1/011001

LIST OF SPONSORSHIP 6th ITB INTERNATIONAL GEOTHERMAL WORKSHOP 2017

1. Chevron
2. PT. Pertamina Geothermal Energy
3. PT. Star Energy
4. PT. Bank Negara Indonesia
5. PT. Indonesia Power
6. PT. LAPI ITB
7. Yayasan LAPI ITB
8. Leapfrog
9. Thermochem Indonesia
10. PT. Geo Dipa Energi
11. Indonesia Geothermal Association
12. Ministry of Energy and Mineral Resources of Republic Indonesia
13. GEOCAP
14. Think Geoenergy
15. Eksplorasi
16. Good News From Indonesia
17. HMGZine
18. Pers Mahasiswa
19. Petro Energy
20. Institut Teknologi Bandung
21. Geothermal Technology Study Program, Institut Teknologi Bandung

IOP Conf. Series: Earth and Environmental Science 103 (2017) 011001 doi:10.1088/1755-1315/103/1/01

LIST OF COMMITTEES

Steering Committee	Sutopo and staff of Geothermal Technology Study Program, FTTM – ITB
Chairman	Suryantini
Vice Chairman	Teguh Rahat Prabowo, Muhammad Firdaus Al Hakim, Ichwan El Fajrie Agusta
General Secretary	Yuniar Zhafira Abdillah, Nanda Hanyfa Maulida, Yutty Hendrawati
Treasurer	Lestari Apriani, Almas Ghasani SFU, Qanitah
Sponsorship	Riviani Kusumawardhani, Mohamad Imbar Fitriadi, Bramono Prabowo, Yuniar Zhafira Abdillah, Jonathan Sharon Widiatmo
Creative, Publication and Media	Hendro Wibowo, Reza Syahputra Mulyana, Indra Agoes Nugroho, Nanda Hanyfa Maulida, Rhamadhana Sultan, Citra Aulian Chalik, Immanuel Lumban Gaol, Riviani Kusumawardhani, Ribka Firtania Asokawaty
Pre-Workshop Courses	Immanuel Lumban Gaol, Ribka Firtania Asokawaty
Technical and Scientific Session	Jonathan Sharon Widiatmo, Ribka Firtania Asokawaty, Beta Kurniawahidayati, Indra Agoes Nugroho, Muhammad Hafidz, Rhamadhana Sultan, Lestari Apriani, Bramono Prabowo, Milki Fabian
Field Trip	Faisal Ahmad, Muhamad Imbar Fitriadi, Immanuel Lumban Gaol
Field Camp	Joshua Satriana, Muhammad Hafidz
Logistic and Accommodation	Indra Agoes Nugroho, Citra Aulian Chalik, Joshua Satriana, Muhammad Hafidz, Milki Fabian, Faisal Ahmad, Muhamad Imbar Fitriadi, Bramono Prabowo, Jonathan Sharon Widiatmo
Exhibition	Beta Kurniawahidayati, Almas Ghasani SFU, Faisal Ahmad, Reza Syahputra Mulyana, Qanitah
Editorial	Asep Saepuloh (ITB), Suryantini (ITB), Hendro Wibowo (ITB), Fiorenza Deon (GFZ Germany), Chris Hecker (ITC University of Twente), Antonie de Wilde (SATE PTE LTD - Singapore), Jonathan Sharon Widiatmo (ITB), Ribka Firtania Asokawaty (ITB), Joshua Satriana (ITB)
Reviewer Team	Suryantini (ITB), Hendro Wibowo (ITB), Sutopo (ITB), Jooened Hendrasakti (ITB), Asep Saepuloh (ITB), Fiorenza Deon (GFZ Germany), Chris Hecker (ITC University of Twente), Antonie de

Wilde (SATE PTE LTD - Singapore), Hendra Grandis (ITB), Nenny Miryani Saptadji (ITB), Sutopo (ITB), Dimas Taha (ITB), Heru Berian Pratama (ITB), Nurita Putri Hardiani (ITB), Mirzam Abdurrahman (ITB), Soedarto (ITB), Irwan Iskandar (ITB), Ali Ashat (ITB), Bambang Priadi (ITB), Sutikno Bronto (Geological Agency Indonesia), Yudi Indra Kusuma (Star Energy)

IOP Conf. Series: Earth and Environmental Science **103** (2017) 011001 doi:10.1088/1755-1315/103/1/011001

EDITORIAL COMMITTEE

Antonie de Wilde
adewilde@gmail.com
SATE PTE LTD
30 CECIL STREET
#19-08
PRUDENTIAL TOWER
SINGAPORE (049712)

Jonathan Sharon Widiatmo, S. T.
Geothermal Technology Master Program, Faculty of Mining and Petroleum Engineering
Bandung Institute of Technology (ITB)
Jl. Ganesha No. 10, Bandung, West Java, Indonesia. 40132

Ribka Firtania Asokawaty, S. T.
Geothermal Technology Master Program, Faculty of Mining and Petroleum Engineering
Bandung Institute of Technology (ITB)
Jl. Ganesha No. 10, Bandung, West Java, Indonesia. 40132

Joshua Satriana, S. T.
Geothermal Technology Master Program, Faculty of Mining and Petroleum Engineering
Bandung Institute of Technology (ITB)
Jl. Ganesha No. 10, Bandung, West Java, Indonesia. 40132

Peer review statement

All papers published in this volume of *IOP Conference Series: Earth and Environmental Science* have been peer reviewed through processes administered by the proceedings Editors. Reviews were conducted by expert referees to the professional and scientific standards expected of a proceedings journal published by IOP Publishing.

Slim hole drilling and testing strategies

Dennis L. Nielson[1], Sabodh K. Garg[1], Colin Goranson[1]

[1]DOSECC Exploration Services, LLC, Salt Lake City, Utah USA

Email: dnielson@dosecc.com

Abstract. The financial and geologic advantages of drilling slim holes instead of large production wells in the early stages of geothermal reservoir assessment has been understood for many years. However, the practice has not been fully embraced by geothermal developers. We believe that the reason for this is that there is a poor understanding of testing and reservoir analysis that can be conducted in slim holes. In addition to reservoir engineering information, coring through the cap rock and into the reservoir provides important data for designing subsequent production well drilling and completion. Core drilling requires significantly less mud volume than conventional rotary drilling, and it is typically not necessary to cure lost circulation zones (LCZ). LCZs should be tested by either production or injection methods as they are encountered. The testing methodologies are similar to those conducted on large-diameter wells; although produced and/or injected fluid volumes are much less. Pressure, temperature and spinner (PTS) surveys in slim holes under static conditions can used to characterize temperature and pressure distribution in the geothermal reservoir. In many cases it is possible to discharge slim holes and obtain fluid samples to delineate the geochemical properties of the reservoir fluid. Also in the latter case, drawdown and buildup data obtained using a downhole pressure tool can be employed to determine formation transmissivity and well properties. Even if it proves difficult to discharge a slim hole, an injection test can be performed to obtain formation transmissivity. Given the discharge (or injection) data from a slimhole, discharge properties of a large-diameter well can be inferred using wellbore modeling. Finally, slim hole data (pressure, temperature, transmissivity, fluid properties) together with reservoir simulation can help predict the ability of the geothermal reservoir to sustain power production.

1. INTRODUCTION

Slim-hole drilling, and specifically coring, has long been used in the geothermal industry. The Geothermal Resources Council and Sandia National Laboratories presented a comprehensive "Slimhole Technology Workshop" in 1996 that summarized experience up to that time. Much of the slim-hole activity in the geothermal industry has focused on temperature gradient drilling, but we are promoting a comprehensive reservoir assessment approach. Coring has been used for reservoir assessment at several sites in Japan [1], Steamboat Springs in Nevada [2], The Geysers in California [3] and Tiwi in the Philippines [4]. In Indonesia, coring has been used at Awibengkok [5], Karaha [6], Wayang Windu and Darajat [7].

Unocal used a hybrid drilling technique to core numerous exploration holes in the evaluation of the Sarulla Block in North Sumatra [8]. Hybrid technology has also been used to drill several scientific holes, including the Long Valley Scientific Drilling Program [9]. The hybrid approach uses a conventional rotary rig to drill and emplace large diameter casing to the reservoir cap. Then a high-speed coring head is attached to the rotary rig and continuous core is collected through the cap and into the reservoir. This phase may involve multiple directional core holes to provide a greater understanding of the reservoir character. Following the completion of the coring, the coring head is removed and the well can be completed at production size using the rotary rig. The disadvantage of this scenario is that the operator is essentially paying for two drilling systems during the coring phase. With improvements in coring equipment in the last decade, it is possible to core and complete deep slim holes without using hybrid rigs.

The high cost of discovering and assessing geothermal resources remains an impediment to the expansion of the geothermal industry. We have recently analyzed the use of slim-hole drilling and testing of production zones and concluded that geological and reservoir data equivalent to that from large-diameter production wells can be collected in slim holes [10]. In that paper, we advocate for the use of slim holes to provide a comprehensive assessment of a geothermal reservoir and illustrate not only cost savings, but also risk reduction.

2. SLIM HOLE DRILLING METHODOLOGY

Slim holes can be drilled using different techniques: rotary, diamond coring or mud motors. Rotary drilling uses high weight on bit and low rpm to crush rock, and the only recovered samples are rock fragments (cuttings). Once major lost circulation zones (LCZ) are encountered, it may be impossible to recover cuttings and the well is drilled without obtaining subsurface geologic data. In addition, the loss of circulation increases mud costs and can lead to the loss of the well due to plugging of permeable formations. Diamond coring uses high rpm and low weight on bit to cut rock and recover a core sample. Lost circulation generally does not impact core collection, and because of the small cutting size along with small mud circulation volumes, plugging of productive formations and possible loss of the well is avoided. Rotary drilling is normally faster whereas diamond coring produces a straighter hole, superior lithologic samples and less thermal and chemical disturbance.

Slim-hole drilling has been applied to geothermal directly from the mining industry, and there has been little effort to optimize the technology for geothermal applications. Most core rigs drill to depths of 1000 m. Rigs that will core to 2000 m or more are not as common, but they are available and can be adapted to geothermal service. Several rig manufactures offer multipurpose rigs that can drill the upper part of the hole with air hammers or rotary tri-cone drill bits, and following emplacement of casing, complete the hole using core drilling. The geothermal industry needs multipurpose slim hole rigs with a minimum drilling capacity of 31,750 kg (70,000 pounds). This provides the capacity to establish casing above the reservoir for pressure control, and will allow the drilling of HQ core into the reservoir to nominal depths of 2000 m. This size hole give us the option of reducing to NQ should hole conditions dictate. Since most geothermal fluid production is found in steeply inclined faults and fractures, it is often advantageous to drill holes perpendicular to the trend of the fractures. Core rigs commonly drill angle holes by tilting the derrick. Mud motors can also be used in slim holes and can directionally drill from a vertical casing or perform multilateral completions. Mud motors can also be used to drill ahead without core sampling above the cap and then transition to coring once the reservoir is intersected.

A significant advantage of slim hole drilling equipment over production well equipment is its smaller size and greater mobility. This translates into smaller well pads and less environmental disturbance as well as lower daily costs for crews, fuel and consumables. For remote locations, the savings in road construction and transportation costs can also be significant.

2.1 Fluid Circulation

The fluid circulation system pumps drilling fluid (mud) down the hole through the center of the drill pipe or drill rod, past the cutting surface of the bit and returns the mud and entrained cuttings to the surface through the annular area. The purpose is to cool and lubricate the bit, remove cuttings and stabilize the hole. The cost of mud can be a significant component of the overall drilling costs [11]. Diamond coring typically circulates at a rate of about 40 liters per minute. This contrasts with production well (rotary) methods that circulate more than 3000 liters per minute. Another important point is that rotary drilling crushes rock and produces centimeter-size cuttings while core drilling cuts the rock, produces millimeter-size cuttings and essentially mines about 50% of the rock as core. Along with the smaller hole size, this results in much less formation damage compared to rotary drilling.

Recent mud technology developments include centrifuge systems that effectively clean mud for recirculation. This decreases mud cost and is advantageous in many parts of the world because it also decreases water consumption. The oil & gas industry has pioneered underbalanced drilling methods that control well pressure to less than the hydrostatic reducing the likelihood of lost circulation and formation damage.

To-date, geothermal mud-loss data have not been employed to make quantitative determinations of the permeability of natural fractures encountered during drilling. In petroleum drilling, mud losses as small as 3000 liters have been measured and used as permeability indicators [12]. Also, petroleum industry has developed mathematical models [13,14] for computing permeability of natural fractures using mud-loss data. We have proposed using this approach to quantify permeability in real time as the drill penetrates a target reservoir.

2.2 Tubulars

Drilling rods used in coring operations have flush joints in contrast to the large tool joints on drill pipe used in production well drilling. Standard rod sizes are P (4.5" x 4"), H (3.5" x 3.063") and N (2.75"x 2.375"). In the event of hole problems, drill rod can be left in the hole as casing and the next smaller drill string can be used to continue drilling. The annular space between the rod and the hole is small, typically 63 to 127 mm (0.25 to 0.5 inches). This is adequate to clear the fine-grained cuttings and also helps to prevent the buildup of mud rings that can cause plugging of the annular area. The inner diameter of the drill rods is also flush, and the bits are open at the end. This allows the driller to use the larger size drill rod as casing with the next smaller size drill rod used to drill out the end of the assembly, without pulling it back to the surface, and continue coring operations.

2.3 Hole Design

Slim holes should be designed differently than production wells. We often see "slim hole" drilling plans that are similar to production wells, except for casing strings that are one or two sizes smaller. These designs still require large drilling rigs and multiple casing strings that eliminate cost advantages, and drilling using rotary techniques reduces sample quality.

A reference design for a 2000 m slim hole is shown in Figure 1. A 12-1/4 inch diameter hole is drilled to a depth of 30 m and a 9-5/8 inch conductor is cemented in place. Following that, an 8-1/2 inch hole is drilled to 350 m and a 7 inch surface casing is cemented in place and pressure tested. Depending on rig availability, it may be beneficial to drill the upper parts of the hole with small mud rotary or air hammer rigs. Regulatory agencies typically require the surface casing to be 10% of the total depth of the hole. An annular blowout preventer, in contrast to a full BOP stack, is often acceptable. But provision should be made for a full BOP stack. From 350 m to the top of the reservoir, that we are assuming to be at 750 m depth, the hole can be cored PQ or rotary at 6-1/4 inch. Either way, 4-1/2 inch casing is cemented to the top of the reservoir. From that point, HQ core can be collected in the reservoir to a depth of 2000 m. NQ can be used as a backup should hole conditions dictate. Delahunty *et al.* [15] describe drilling experience using a similar, but somewhat smaller hole design. Our reference 2000 m slim hole has a cost in the US of about $1.5 million.

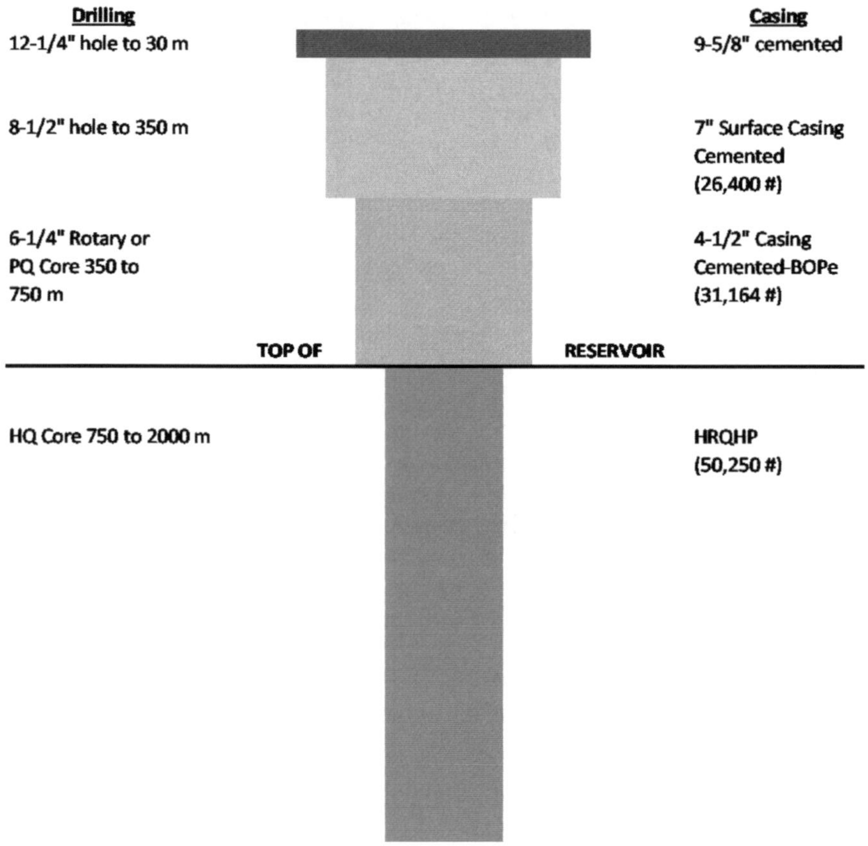

Figure 1: Reference design for a 2000 m geothermal assessment slim hole.

Slim holes are intended for reservoir assessment, investigating the permeability structure and pressures and temperatures as a function of depth in the reservoir. An additional benefit is that this information allows us to better design and budget for the subsequent drilling of production wells. Core material obtained from the slim hole can also be used to determine rock properties of both the geothermal reservoir and the caprock system [16]. Permeability tests and clay analyses can be performed to better understand reservoir properties. Clay swelling tests can also be performed to determine mud properties needed to reduce clay swelling that can lead to stuck drill pipe and formation damage.

3. TESTING & ANALYSIS

3.1 Testing

Slim holes are not normally intended for geothermal production. Production and reservoir engineering data from slim holes are usually obtained through injection testing. However, proper design of the slim hole completion will allow for discharge testing. Standard downhole pressure, temperature and spinner (PTS) survey data can be obtained during both discharge and injection testing. In addition, slim holes can be used in reservoir pressure interference testing [17] and used for long-term monitoring of subsurface reservoir pressure and temperature of the geothermal system during long term production and injection operations.

The techniques for slim hole testing and scaling the results to large-diameter holes have been known for many years [18,17,1,19,20]. Our approach to reservoir testing is to characterize permeable zones as drilling advances. This is different than most production well drilling plans that drill the well to a predetermined total depth before performing reservoir tests. We believe that this destroys shallow permeability that may be the most productive component of the geothermal system.

In order to use slim holes for reservoir assessment, it is necessary to be able to predict the discharge characteristics of large-diameter wells based on discharge and/or injection tests on small-diameter boreholes. To compute the probable discharge characteristics of a large-diameter well, a relationship between the injectivity and/or productivity of slim holes and large-diameter production and/or injection wells is required. Garg and Combs [1] examined production and injection data from slim holes and large-diameter wells at five geothermal fields (Oguni, Sumikawa, Takigami, and Kirishima, Japan; Steamboat Hills, U.S.A.) in order to establish relationships (1) between productivity and injectivity indices, and (2) between discharge capacity of slim holes and large diameter wells. For boreholes with liquid feed zones, the productivity and injectivity indices are more or less equal, and the productivity (or injectivity) index is independent of borehole diameter. It therefore follows that the productivity index of large-diameter production wells can be estimated based on testing of slim holes. If the productivity index is sufficiently large, the discharge rate of the well will not be limited by the formation; in this case, a multi-phase wellbore simulator can be employed to investigate the production characteristics of a well regardless of the borehole diameter. It may also be feasible to predict the performance of wells with two-phase feed zones [19]. Garg and Combs [20] indicate the importance of data collection for thermal modeling associated with early stages of exploration.

When permeable zones are encountered during drilling, they will be identified by fluid losses during coring. Once there is total mud loss, the drilling will stop to allow testing. Figure 2 shows the surface testing equipment needed to collect flow data from slim holes. In the caprock or other zones with permeability insufficient to maintain continuous fluid discharge, injection tests can be performed. Permeability testing of the caprock, or other low permeability zones, is useful in creating an accurate 3D numerical model of the reservoir system. Production testing is the preferred method in that additional data on fluid and gas geochemistry is also obtained.

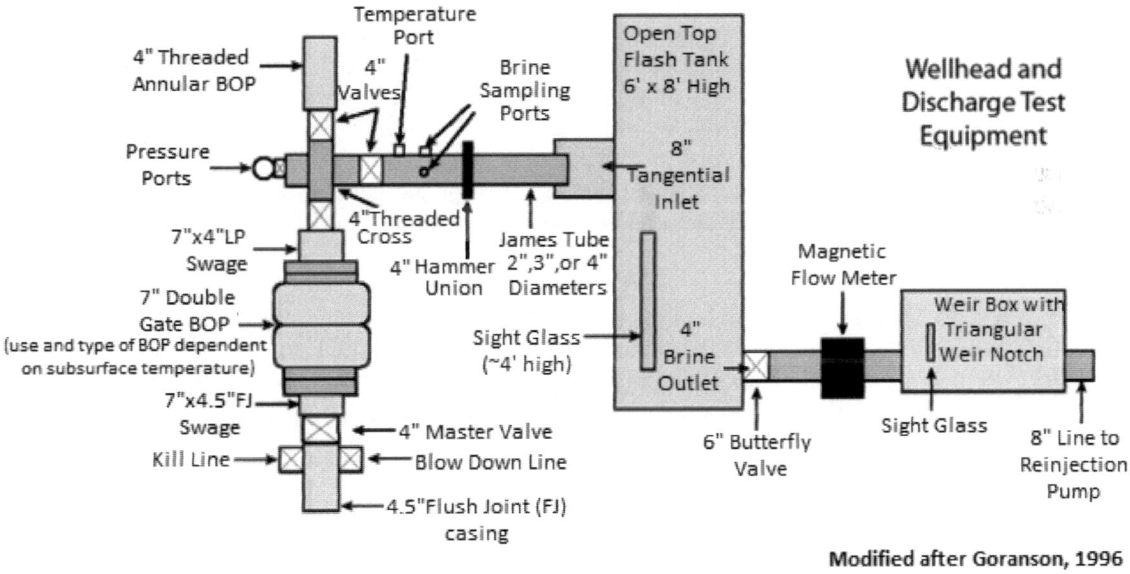

Figure 2. Surface testing equipment for geothermal slim hole assessment.

Below the caprock, production and injection testing can be performed as follows.

1) For flow testing, set up wellhead system, surface flow test equipment and surface measurement equipment (WHP, WHT, James Tube Pressure, Orifice plate pressures, geochemical sample point equipment, etc).
2) Set up test monitor equipment for injection tests.

3) Allow well to heat up for 12-24 hours.
4) Run static pressure-temperature surveys
5) Attempt to flow the well. Pressure up the well with air if it does not self-flow.
6) If the well flows, run the PTS tool and determine the influx zone. Flow the well for 12-24 hours. Run periodic (determined in the field) PTS surveys to determine subsurface discharge conditions.
7) Set the PTS tool at the main fluid entry zone, shut in the well, measure downhole pressure build-up (or pressure fall-off in the case of injection testing) for 12-24 hours (the actual time will depend on subsurface conditions).
8) If the well does not flow, inject fresh water while monitoring the loss zone using the PTS tool.
9) Inject fresh water for ~12 hours. Run PTS surveys during injection.
10) Identify injection zone from PTS surveys
11) Set PTS tool at main injection zone.
12) Shut off injection and monitor downhole PT for +/- 24 hours.

3.2 Development of 3D Numerical Model

In addition to the borehole and well testing data obtained from coring operations, a review of the geology, geophysics, geochemistry, hydrogeology and topography of the surrounding region will provide a basis for establishing geographic relationships of reservoir properties. A preliminary conceptual model including a description of the reservoir volume, heat source, permeability structure, fluid source, *etc.* of the geothermal reservoir is then developed.

A suitable grid for the numerical model in the reservoir simulation software program is developed to encompass the complete geothermal system. The grid should include the entire known permeable region of the geothermal reservoir and lateral boundaries of the system. Based on available data and analyses, preliminary values are assigned for the required formation properties (porosity, density, heat capacity, thermal conductivity, permeability, relative permeability for liquid and vapor phases, *etc.*). Available heat flow data is used to specify preliminary boundary conditions (*i.e.* heat and mass flux) along the bottom boundary of the numerical grid. A natural state model of the geothermal system is created by varying the permeability distribution and boundary conditions. The model is constrained by matching the known information (*e.g.* pressure and temperature distribution, heat flux at the surface, *etc.*) about the geothermal system prior to the initiation of production. The constrained numerical model can then be used to forecast the response of the geothermal reservoir under future production and injection scenarios developed in consultation with the client.

4. CONCLUSIONS

This paper describes a comprehensive process for the assessment of geothermal resources starting with slim hole drilling and extending through the testing and initiation of numerical modeling. As previously discussed [10], slim hole drilling reduces risk and cost in geothermal exploration and development. We believe that this approach has not been more widely applied because developers are unaware of the advantages and history of the approach.

REFERENCES

[1] Garg, S. K. and Combs, J. (1997) Use of slim holes with liquid feedzones for geothermal reservoir assessment: *Geothermics*, **26**, 153-178.

[2] Combs, J. and Goranson, C. (1994), Use of Slim Holes for Reservoir Evaluation at the Steamboat Hills Geothermal Field, Nevada, U.S.A. : *Proceed. 19th Workshop on Geothermal Reservoir Engineering*, Stanford University, 131-138.

[3] Hulen, J. B., Koenig, B. and Nielson, D. L. (1994) The Geysers coring project - a cooperative investigation of reservoir controls in a vapor-dominated geothermal system: *Geothermal Resource Council Trans.*, **18**, 317-323.

[4] Nielson, D. L., Clemente, W. C., Moore, J. N. and Powell, T. S. (1996) Fracture permeability in the Mataliong-25 corehole, Tiwi geothermal field, Philippines: *Proceed. 21st Workshop on Geothermal*

Reservoir Engineering, Stanford University, 64-70.

[5] Stimac, J. and Sugiaman, F. (2000) The AWI 1-2 core research program: Part 1, geologic overview of the Awibengkok geothermal field, Indonesia: *Proceed. World Geothermal Congress 2000.* 2221-2226.

[6] Moore, J., Lutz, S., Renner, J., McCulloch, J. and Petty, S. (2000) Evolution of a volcanic-hosted vapor-dominated system: petrologic and geochemical data from corehole T-8, Karaha-Telaga Bodas, Indonesia: *Geothermal Resources Council Trans.*, **23**, 259-263.

[7] White, P., MacKenzie, K., Verghese, K. and Hickson, C. (2010) Deep slimhole drilling for geothermal exploration: *Geothermal Resources Council Trans.*, **34**, 269-272.

[8] Gunderson, R., Ganefianto, N., Riedel, K., Sirad-Azwar, L. and Suleiman, S. (2000) Exploration results in the Sarulla Block, North Sumatra, Indonesia: *Proceed. World Geothermal Congress*, 1183-1188.

[9] Nielson, D. L. (2001) Deep wireline coring in geothermal reservoirs: *Geothermal Resources Council Trans.* **25**, 115-118.

[10] Nielson, D. L. and Garg, S. K. (2016) Slim hole reservoir characterization for risk reduction: *Proceed. 41st Workshop on Geothermal Reservoir Engineering*, Stanford University, 113-120.

[11] Silverman, R. L., Lukawski, M. J. and Tester, J. W. (2014) Uncertainty analysis of geothermal well drilling and completion costs: Geothermal Resources Council *Trans.*, **38**, 419-422.

[12] Dyke, C.G., Wu, B., and Milton-Tayler, D. (1995) Advances in characterizing natural-fracture permeability from mud-log data: *SPE Formation Evaluation*, **10**,160-166.

[13] Huang, J., Griffiths, D.V, and Wong, S.-W. (2011) Characterizing natural-fracture permeability from mud-loss data: *SPE Journal*, **16**, 111-114.

[14] Akin, S. (2013) Estimating natural fracture permeability from mud-loss data: *Proceed. 38th Workshop on Geothermal Reservoir Engineering*, Stanford University

[15] Delahunty, C., Nielson, D. L. and Shervais, J. W. (2012) Deep core drilling of three slim geothermal holes, Snake River Plain, Idaho: *Geothermal Resources Council Trans.* **36**, 641-648.

[16] Boitnott, G. N. (2002) Core analysis for the development and constraint of physical models of geothermal reservoirs: *Geothermal Resources Council Trans.* **26**, 387-392.

[17] Combs, J. and Goranson, C. (1995) Reservoir evaluation using discharge and injection data from slim holes and large-diameter production wells at the Steamboat Hills geothermal field, Nevada, USA: *Proceed. World Geothermal Congress*, Florence, Italy, **3**, 1517-1524.

[18] Pritchett, J. W. (1993) Preliminary study of discharge characteristics of slim holes compared to production wells in liquid-dominated geothermal reservoirs: *Proceed. 18th Workshop on Geothermal Reservoir Engineering*, Stanford University, 181-187.

[19] Garg, S. K. and Combs, J. (2000) Geothermal reservoir assessment using data from slim holes: *Geothermal Resources Council Trans.*, **24**, 581- 587.

[20] Garg, S. K. and Combs, J. (2015) A reformulation of the USGS volumetric "heat in place" resource estimation method: *Geothermics*, **55**, 150-158.

IOP Conf. Series: Earth and Environmental Science **103** (2017) 012002 doi:10.1088/1755-1315/103/1/012002

Geothermal reservoir simulation of hot sedimentary aquifer system using FEFLOW®

Hardi Nur Hidayat and Maximillian Gala Permana

Department of Geothermal Engineering, Institute of Petroleum Engineering, TU Clausthal, 38678 Clausthal-Zellerfeld,Germany

Email : hardi.nur.hidayat@tu-clausthal.de maximillian.gala.permana@tu-clausthal.de

Abstract. The study presents the simulation of hot sedimentary aquifer for geothermal utilization. Hot sedimentary aquifer (HSA) is a conduction-dominated hydrothermal play type utilizing deep aquifer, which is heated by near normal heat flow. One of the examples of HSA is Bavarian Molasse Basin in South Germany. This system typically uses doublet wells: an injection and production well. The simulation was run for 3650 days of simulation time. The technical feasibility and performance are analysed in regards to the extracted energy from this concept. Several parameters are compared to determine the model performance. Parameters such as reservoir characteristics, temperature information and well information are defined. Several assumptions are also defined to simplify the simulation process. The main results of the simulation are heat period budget or total extracted heat energy, and heat rate budget or heat production rate. Qualitative approaches for sensitivity analysis are conducted by using five parameters in which assigned lower and higher value scenarios.

1. Introduction

1.1. Background

As geothermal industry, especially the EGS technologies', growing in the last couple of decades, low enthalpy geothermal system is gained more attention. One of the plays in the low enthalpy geothermal system is hot sedimentary aquifer (HSA). Moeck [1] classified the geothermal plays by its geological setting. HSA is a conduction-dominated hydrothermal play type utilizing deep aquifer, which is heated by near normal heat flow.

HSA play type typically is found in passive plate tectonic where tectonic activity is relatively rare and no near active igneous body acted as a heat source. HSA play type is mainly located in intracratonic basin, orogenic belts and associated foreland basins [1]. The target reservoirs are typically high porosity/high permeability or high porosity/low permeability formations that need to be enhanced or stimulated using EGS to gain high productivity. Figure 1 shows the typical lithologies or formations which is targeted in HSA. With temperature gradient 32˚C / km as an assumption, category A is less than 3 km depth geothermal play and mostly used for district heating; while category B geothermal plays that located at more than 3 km deep and is suitable for heating and electricity, and category C is potential HDR system. Reservoir fluid of HSA is rarely coming from meteoric water, instead the water is the result of deep circulation system within the basin or major crustal faults.

Typically, doublet well is used in order to utilize HSA system. Doublet well is an open loop system, which uses an injection and production well to circulate water to and from the reservoir. One of the successful HSA geothermal plays is Bavarian Molasse Basin of the Alps, which used as heating and electricity generation in Southern Germany. Bavarian Molasse Basin is a foreland basin associated with orogeny belt. As the sedimentary sequences of foreland basin are subsiding towards the orogen due to the weight of the formations, the basin formed wedge shape and normal faults regime, thus caused the positive geothermal gradient within the basin.

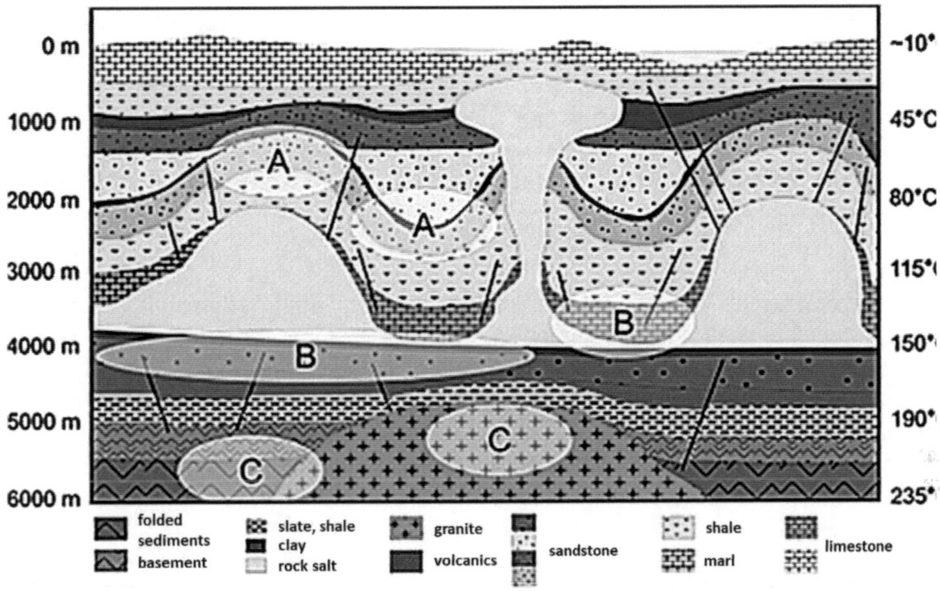

Figure 1. Schematic of Various Geothermal Play Types within Intracratonic Sedimentary Basin. [1]

1.2. *Objective*

The purpose of this assignment are to generate a transient HSA model using FEFLOW software package and analyse technical feasibility and the performance regarding the extracted energy from this concept. The procedures of the 3D modelling is briefly described in the next chapter. The input data for the base and modified model are also represented in the next chapter. Furthermore, the heat recovery from different scenarios (base and modified model) is compared in order to obtain the significant influences of given parameters.

2. Modelling

2.1. *FeFlow*

The HSA modeling in this project assignment is performed by FEFLOW, a numerical and computation software. FEFLOW is a commercial simulation software for groundwater flow and heat transfer which is fully associated by density dependent and finite element [2]. The related processes in this software are operated in one software environment, one user interface, and one simulation model.

As an intuitive program, FEFLOW is a relatively user-friendly simulation which is concurrently powerful and comprehensive for complicated subsurface modelling. The stages of FEFLOW-modelling consist of preprocessing, simulation, and post-processing [3]. Complex fluid flow and geometry are also possible to be handled as an excellence in this software.

The example of utilizations related to groundwater application are mine water management, tunneling, agriculture, and geothermal energy [3]. FEFLOW provides several built-in modules those quite applicable for Geothermal modelling such as Borehole Exchanger and Multilayer well. The approaching method for this HSA modelling is using a built-in module or a special program interface that included multilayer well modelling [4].

2.2. Procedure
In the FEFLOW software, defining reservoir area by plotting the coordinates is the first stage of the modelling. Afterwards, the well locations are defined by plotting points in the reservoir area. After generating and smoothing the mesh, defining the hydraulic head and conductivity, 3D layer configuration is created in order to define the properties of each layer. Subsequently, the transient state is chosen for this modelling instead of steady-state condition. Moreover, the injection and production are created in certain coordinates those have been determined at the previous stage. Before running the simulation, the temperature difference and simulation time have to be defined by using open loop plug-in and simulation time control respectively.

2.3. Input Data
Four basic information of reservoir, aquifer, temperature, and well are involved as significant parameters in developing this HSA model. The basic reservoir information offers four components such as area, hydraulic head, hydraulic conductivity, and porosity as represented in table 1. The values of aquifer are also represented in this table; thereby, the properties of aquifer in fourth layer can be described with these parameters.

The temperature characteristic is organized into two sections for whole layers. The temperature of reservoir section is determined by temperature gradient and surface temperature as represented in table 1. Otherwise, the aquifer section is fixed by 95°C. The different temperature of injection- production well is furthermore specified by the open loop. The procedure of multilayer well implementation is described in previous chapter.

The values of longitudinal dispersivity, volumetric heat capacity of fluid, thermal conductivity of fluid are inputted identically as the FEFLOW's default values. Moreover, the values of radius, well depth, well distance, and flow rate have to be defined respectively as shown in the table 1. These data are further simulated and used as basic model, which is compared to other models with modified parameter. In order to simplify the modelling process, following assumptions were made such as: 1) reservoir is considered to be homogenous and isotropic; 2) the inflow temperature at a certain depth is assumed to be equal to the corresponding initial temperature; 3) only the heat transport is simulated and the mass transport is considered in steady state; 4) the flow rate of well is pumped constantly for 10 simulation years.

Table 1. Parameter Information for Base Model

Parameter	Unit	Value
Basic Reservoir Information		
Reservoir area	m^2	$2x10^6$
Hydraulic head (east border)	m	29
Hydraulic head (west border)	m	30
Hydraulic conductivity	m/day	1
Porosity	-	0.3
Basic Aquifer Information		
Aquifer location	m	1500-2000
Hydraulic conductivity	m/day	1
Porosity	-	0.3

Basic Temperature Information		
Surface temperature	°C	15
Geothermal gradient	°C/100m	3
Aquifer temperature	°C	95
Injection-Production temp. difference	°C	-60
Basic Well Information		
Injector-producer distance	m	450
Well depth	m	2000
Flow rate (Injection)	m^3/day	-60
Flow rate (Production)	m^3/day	60
Well radius	Inch	8.5
Longitudinal dispersivity	m	5
Volumetric heat capacity of fluid	$MJ/m^3/K$	4.2
Thermal conductivity of fluid	J/m/s/K	0.65

Table 2. Parameter Information for modified Model

Parameter	Unit	Low	Base	High
Depth	m	1500	2000	2500
Porosity	%	0.2	0.3	0.4
Flow rate	m^3/day	40	60	80
Hydraulic conductivity	m^3/day	1	2	3
Thermal conductivity of solid	J/(m.s.K)	2	3	4

3. Simulation Result

3.1. Heat Period Budget

Heat period budget represents produced thermal energy of HSA system after 10 simulation years. Negative values of heat period budget are obtained after running the simulation, since FEFLOW defines the extracted energy as negative value and the storage energy as positive value. Otherwise, these values were expressed in Joule, which were inappropriate due to the huge numbers. Thus, the authors decided to convert the Joule to Gigawatthour unit. The heat period budget for base model showed approximately 15.5 GWh after 10 years as represented in figure 2.

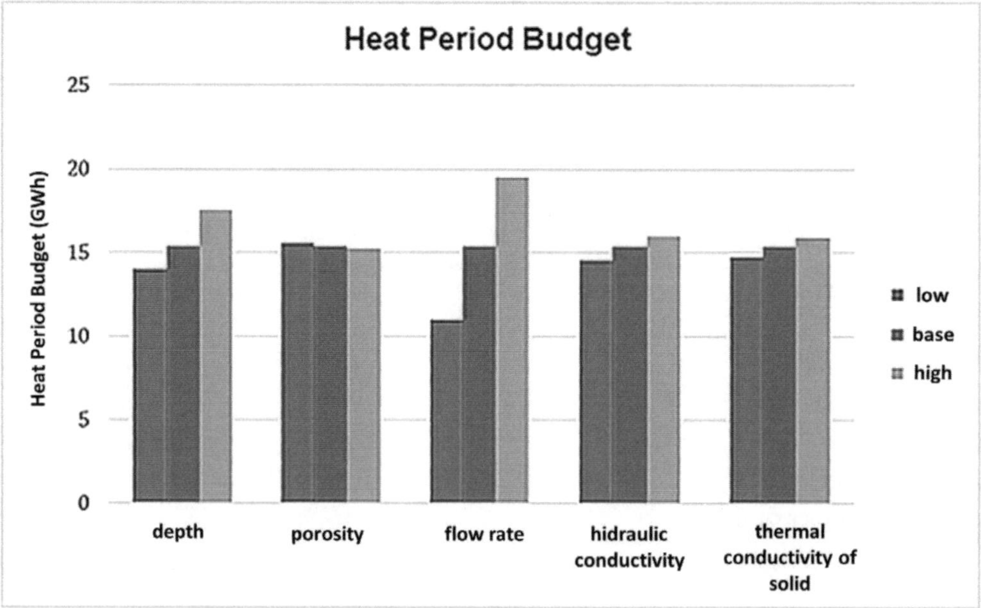

Figure 2. Heat Period Budget after 10 Simulation Years

Flow rate was the most significant parameter, that influenced the produced heat period budget. The heat period budget reached almost 20 GWh, when the high scenario was applied in this HSA system. Inversely, low scenario of flow rate for this system caused a less heat period budget. Therefore, the flow rate was direct proportional to the heat period budget. The depth played also a significant role for this HSA system. Based on lower and higher scenario in figure 2, it could be concluded that the deeper the aquifer depth could generate more thermal energy of HSA system. The hydraulic conductivity and thermal conductivity of solid were also direct proportional to the heat period budget. The relationship of those two parameter is represented also in figure 3. In contrast, the heat period budget was inverse proportional to the porosity. By comparing between the low and high scenario to the base model, the highest heat period budget was generated by the low scenario.

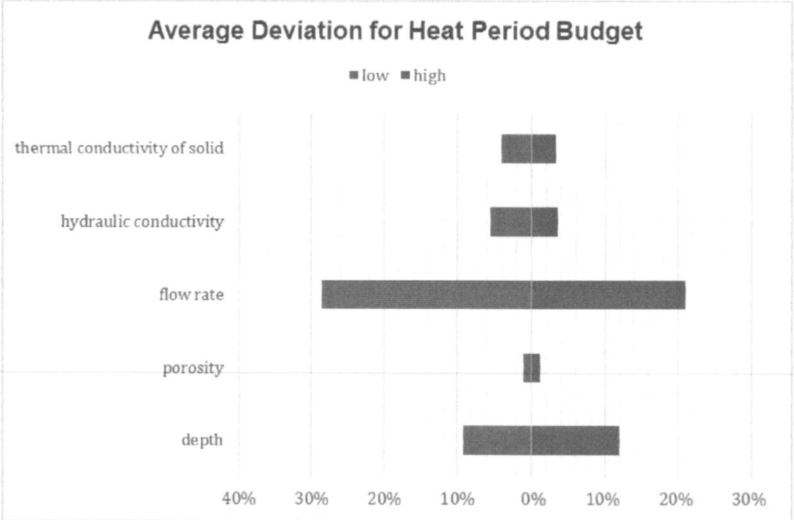

Figure 3. Average Deviation for Heat Period Budget.

3.2. Heat Rate Budget
One of the results obtained from the simulation is heat rate budget. Heat rate budget is the amount of thermal power extracted by the production wells. FEFLOW plotted the result in a graph of thermal power against the simulation time. As mentioned in previous chapter, this study simulated five different parameters by three different scenarios, the heat rate budget of each parameter are described as follows. Graphs of the comparisons are attached in the appendix of the paper.

3.2.1. Comparison by The Depth
The first parameter that was investigated is the depth of aquifer. Appendix 1 shows the comparison between three depths of aquifer scenarios. The base model scenario was aquifer with thickness of 500 m and bottom depth at 2000 m, the second scenario was 1500 m bottom depth with 500 m thick aquifer layer, and the last scenario was model with bottom depth of aquifer at 2500 m. At 10 years of simulation time, the production well at base model scenario performed at around 170 kW, while the low scenario extracted 154 kW thermal power and the high scenario obtained higher value at 196 kW. The high scenario with 0.0034 kW / day was the smallest decline rate of thermal power from the production well than other scenarios.

3.2.2. Comparison by Porosity
Porosity is the second parameter that was analysed. Appendix 2 shows the comparison between three scenarios of porosity of the aquifer. The porosity in the aquifer was applied throughout the layer since the aquifer layer considered as homogeneous and isotropic. The base model scenario was aquifer with porosity of 0.3, the value of porosity in the low scenario was 0.2 and the last scenario was the high scenario with porosity at 0.4. There were only small differences, around 2 kW, between one scenario and the other. After 10 years, the thermal power produced at low, base and high scenarios were 173, 171 and 169 kW, respectively. The production well performed relatively similar with decline rate difference of 0.03 x 10-5 kW / day to each scenario.

3.2.3. Comparison by Flow Rate
The next parameter that was examined in this study is flow rate of the well. As mentioned in modelling section, the flow rates on the injection well and production well were applied at constant and same value for 10 years of simulation time but at different direction. Appendix 3 shows the comparison between flow rates of the production well at three different scenarios. The base model scenario was 60 m3 / day, the low scenario was 40 m3 / day, and the high scenario was 80 m3 / day. After simulated for 10 years of simulation time, the production well at base model scenario extracted 171 kW, while for the low scenario was 123 kW and the high scenario obtained thermal power at 216 kW. However, the high scenario performed relatively high decline rate at 2.563 x 10^{-5} kW / day, meanwhile the low scenario had decline rate of 1.934 x 10^{-5} kW / day and the base scenario continuously declined at 2.303 x 10^{-5} kW / day.

3.2.4. Comparison by Hydraulic Conductivity
The fourth parameter which was focused on in this study is the hydraulic conductivity of the aquifer. Since the model was simplified and the layers considered as homogeneous and isotropic, the hydraulic conductivity of the aquifer had the same value for all axis. Appendix 4 shows the comparison between hydraulic conductivity of the aquifer at three different scenarios. The base model scenario was 2 m / day, the low scenario was 1 m / day, and the high scenario was 3 m / day. After simulated for 10 years of simulation time, the production well at base model scenario obtained 171 kW thermal power, while for the low scenario and the high scenario were 161 kW and 178 kW, respectively. Even though the decline rate of all the scenarios had only small differences, around 0.03 x 10^{-5} kW / day, the high scenario performed relatively small decline rate at 2.067 x 10^{-5} kW / day, meanwhile the low scenario and the base scenario had decline rate of 2.623 x 10^{-5} kW / day and 2.303 x 10^{-5} kW / day, subsequently.

3.2.5. Comparison by Thermal Conductivity of Solid

The last parameter that was examined in this study is the thermal conductivity of solid of the aquifer. This parameter was applied homogeneously on the aquifer layer to simplify the model. Appendix 5 shows the comparison between thermal conductivity of solid of the aquifer at three different scenarios. The base model scenario was 3 J / (m.s.K), the low scenario was 2 J / (m.s.K), and the high scenario was 4 J / (m.s.K). After simulated for 3650 days of simulation time, the production well at base model scenario obtained 171 kW thermal power, the low scenario gained 164 kW, and the high scenario extracted 177 kW thermal power. The production well thermal power declined at a small difference rates with base scenario at 2.303×10^{-5} kW / day, low scenario at 2.446×10^{-5} kW / day and high scenario at 2.164×10^{-5} kW / day.

3.3. Temperature Distribution

One of the results of the simulation is the temperature distribution within the model and the wells. As discussed on modelling section, the temperature boundary condition and initial temperature inside the model was defined before the model was simulated. After 3650 days of simulation, the temperature distribution within the model was altered, especially near the wellbore region. Only temperature distribution of base model is presented in this study since there are only small differences of temperature distribution between all of the scenarios that were simulated.

Figure 4 is 2D view of the model at 2000 m depth slice. This figure shows temperature distribution of the model at the deepest part of the model. After 10 years of simulation time, there were only small area, around maximum radius of 75 m from the center of the wells, was affected by the injection wells. The lowest temperature was at the very center of the well at 8.8°C. Meanwhile, the lowest temperature on the production well was around 80 °C and did not affected the near well bore region of the well.

Figure 5 shows the 3D view of the temperature distribution of the base model. This figure represents that after 3650 days of simulation time, the injection wells caused more disturbances in temperature in the model than the production well. However, the disturbance developed only on the near region of the well while the rest of the model was at steady state condition.

Figure 6 shows the base model's average temperature of production and injection well. After 10 years of simulation time, the average temperature of production well was 69 °C and the injection well average temperature was 9°C. The production well was able to maintain high enough temperature to be extracted for longer time.

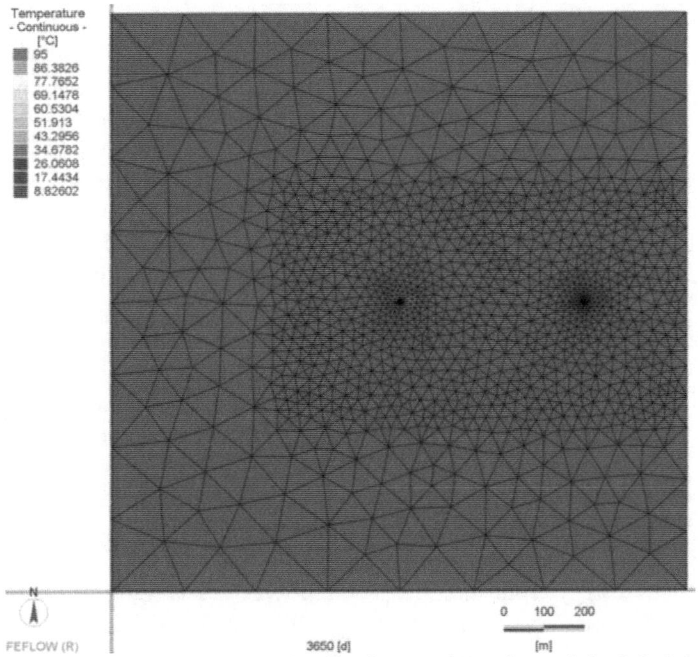

Figure 4. Temperature Distribution on 2D View of the Model at 2000m Depth Slice

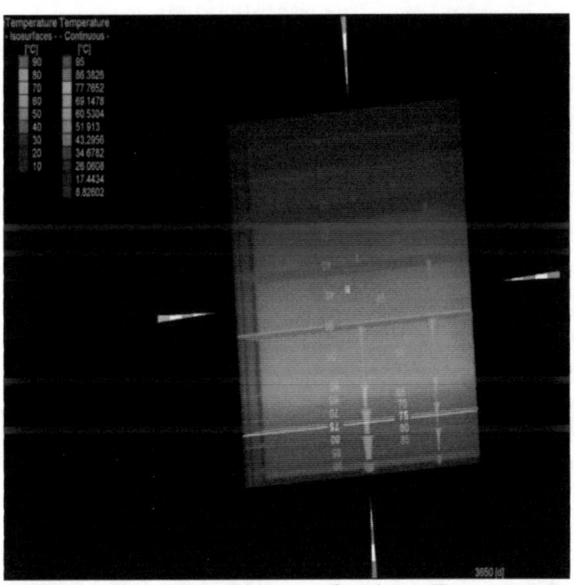

Figure 5. 3D View of the Temperature Distribution of the Base Model

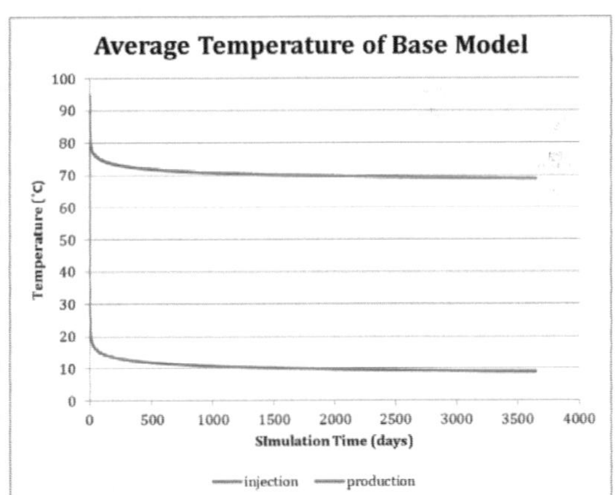

Figure 6. Average Temperature of Base Model

4. Discussion

As represented in previous chapter, the greater depth of the aquifer showed higher heat production rate with lower decline rate than the base scenario, which had a direct correlation with thermal gradient. Higher flow rate also generated a higher heat production rate similar to the depth correlation. On that note, the decline rate had to be considered, since the higher flow rate showed a higher decline rate too.

In contrast, the porosity expressed a different proportionality to the heat production rate. The lower porosity showed higher heat production rate with lower decline rate. This behaviour could be occurred due to low thermal conductivity of the fluid compared to thermal conductivity of solid in the aquifer. Thus, conduction had higher effect on heat transfer from the solid in the aquifer than the convection of the fluid.

Thermal conductivity of solid and hydraulic conductivity was indeed dependent on the lithology of the aquifer. The higher value of both parameters could conduct more thermal energy to the surface, thus higher conductivity could generate higher heat production rate. The sensitivity analysis was performed only to identify the qualitative relationship between the heat production rate and those parameters.

The temperature distribution in the reservoir is only disturbed for a short distance, around 75 meters, after 10 simulation years. The production well could still generate for longer time until this well had to be shut off, in order to replenish the heat in the aquifer.

The generated thermal energy from this HSA system was not able to be converted into electrical energy, since the average temperature was only circa 70°C. The binary power plant cycle requires the minimum temperature of 85°C at the surface in order to generate electrical energy. If the average temperature exceeded 85°C [5] , the extracted energy was also not economically beneficial, since the thermal energy and the efficiency of binary cycle were to low (ca. 10-13%) [3]. Therefore, this HSA system is more suitable to utilize for heating system, such as district heating. However, the surface temperature from production well cannot be defined clearly. Hence, the extracted heat power from this system cannot be estimated accurately.

The mass balance of this fluid is not considered in this simulation, due to produced geothermal water is injected back to the reservoir, therefore there is an uncertainty regarding the fluid movement in the aquifer which correlates to the heat transfer. Otherwise, lack of published paper related to HSA simulation may cause difficulty in finding reliable data for this study's references.

5. Conclusion

The base model of transient HSA simulation was built by using FEFLOW and generated almost 15,5 GWh with heat production rate at 171 kW, which declined exponentially at 2.303×10^{-5} kW / day. The generation of electrical energy from the thermal energy was not feasible, since the heat and temperature could not fulfil the minimum condition. Therefore, only heat utilization might be applied from this HSA system.

This study has examined the influence of several significant parameters to the HSA system. The production well in this case could produce higher heat production rate by having better aquifer, which has the following criteria: deeper depth, higher thermal conductivity of solid, higher hydraulic conductivity, and lower porosity. These parameters are empirical properties which are possessed by the aquifer formation or lithology layer. When the one of these parameters could not be improved, increasing the flow rate could be an alternative to generate higher heat production rate. However, the increasing flow rate could cause faster decline rate.

The simulation of HSA simulation can be investigated further in term of using more reliable data or performed sensitivity analysis by using quantitative approaches.

6. Appendix

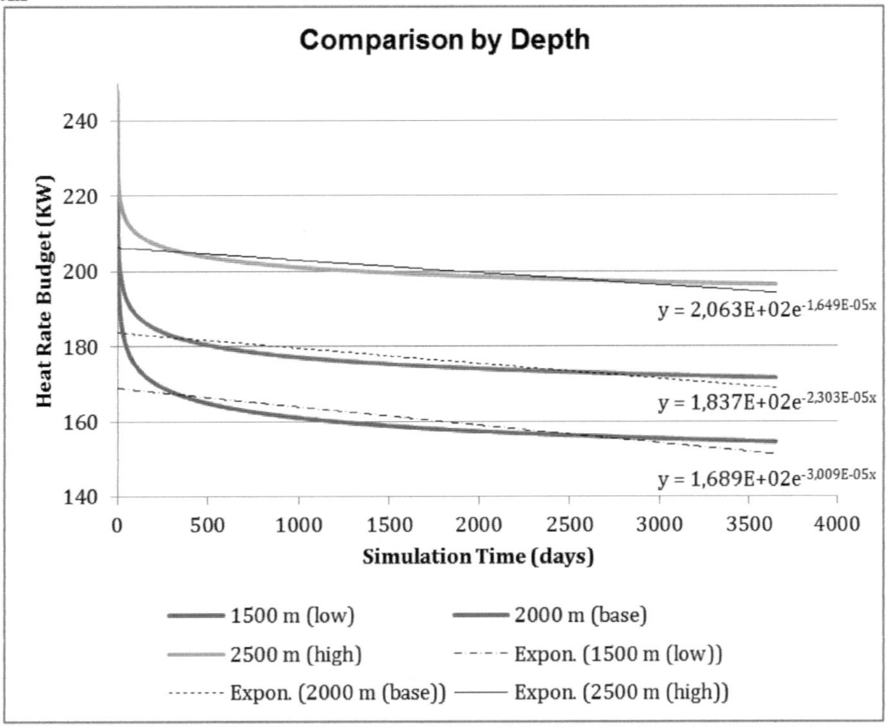

Appendix 1. Heat Rate Budget Comparison by Depth

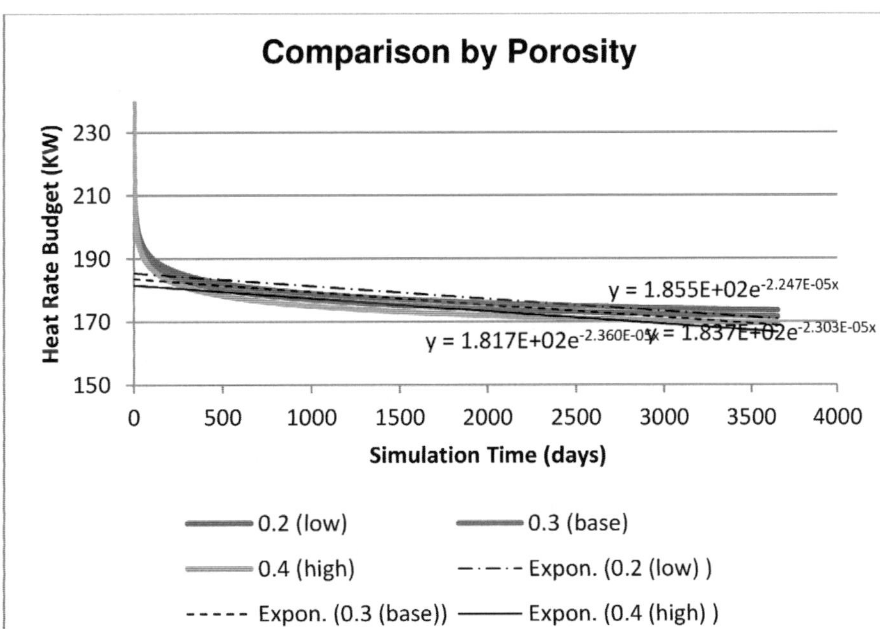

Appendix 2. Heat rate budget comparison by porosity

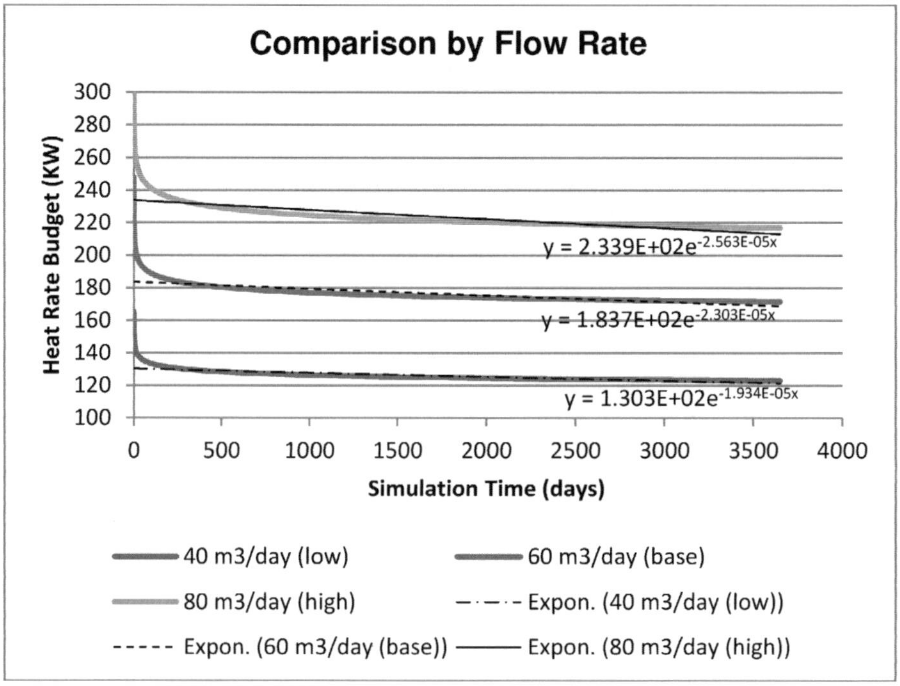

Appendix 3. Heat Rate Budget Comparison by Flow Rate

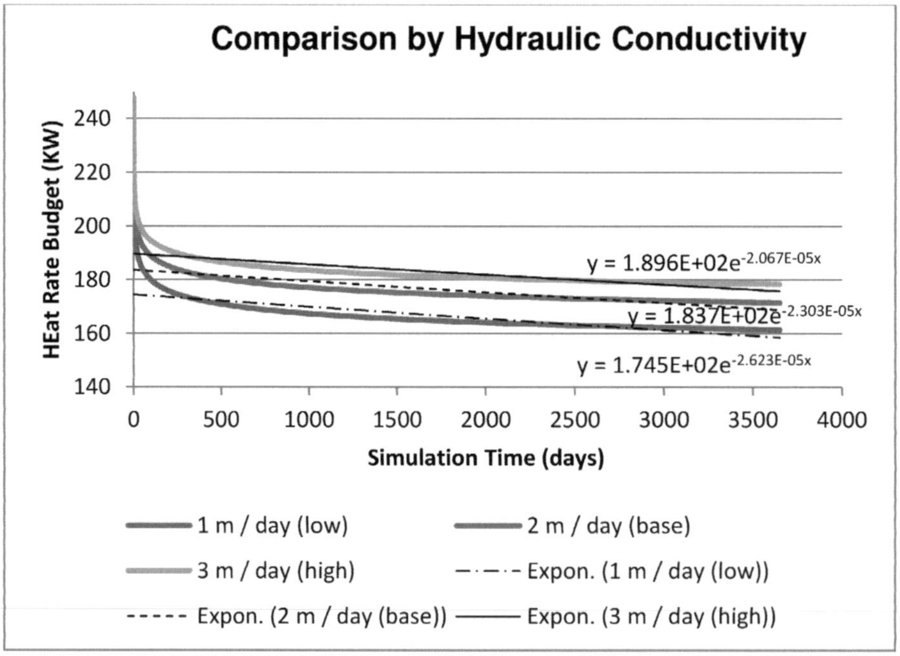

Appendix 4. Heat Rate Budget Comparison by Hydraulic Conductivity

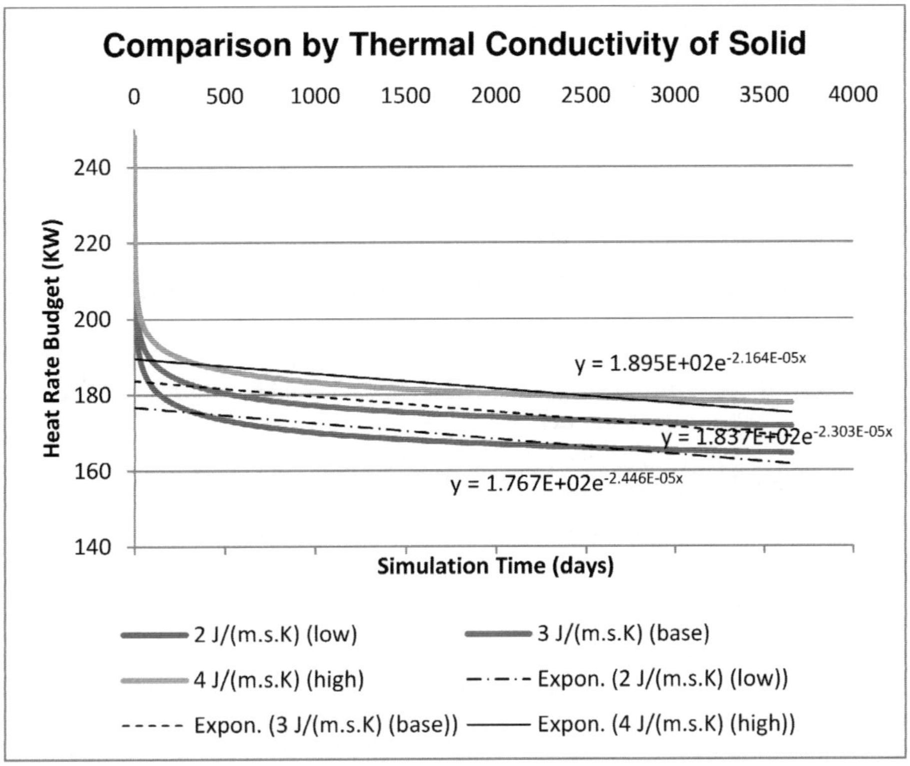

Appendix 5. Heat Rate Budget Comparison by Thermal Conductivity of Solid

References

[1] I. Moeck, "Catalog of Geothermal Play Types Based on Geologic Controls," *Renewable and Sustainable Energy Reviews,* 2014.

[2] V. Claustnitzer, "Introduction to FEFLOW," DHI Groundwater Modelling Centre.

[3] R. DiPippo, "Geothermal Power Plants, Second Edition: Principles, Applications, Case Studies and Environmental Impact," Oxford, Butterworth-Heinemann, 2007, p. 159.

[4] DHI-WASY, "FEFLOW User Manual," Berlin , 2012.

[5] T. Maghiar and C. Antal, "Power Generation from Low Enthalphy," *University of Oradea, Romania,* 2001.

Developing a framework for assessing the impact of geothermal development phases on ecosystem services

Jarot M. Semedi[1,2], Louise Willemen[1], Triarko Nurlambang[2], Freek van der Meer[1] & Raldi H. Koestoer[2]

[1]Faculty of Geo-Information Science and Earth Observation, University of Twente, Enschede, The Netherlands
[2]Sekolah Ilmu Lingkungan, Universitas Indonesia, Jakarta, Indonesia

E-mail: j.m.semedi@utwente.nl; jarot.mulyo@ui.ac.id

Abstract. The 2014 Indonesian National Energy Policy has set a target to provide national primary energy usage reached 2.500 kWh per capita in the year 2025 and reached 7.000 kWh in the year 2050. The National Energy Policy state that the development of energy should consider the balance of energy economic values, energy supply security, and the conservation of the environment. This has led to the prioritization of renewable energy sources. Geothermal energy a renewable energy source that produces low carbon emissions and is widely available in Indonesia due to the country's location in the "volcanic arc". The development of geothermal energy faces several problems related to its potential locations in Indonesia. The potential sites for geothermal energy are mostly located in the volcanic landscapes that have a high hazard risk and are often designated protected areas. Local community low knowledge of geothermal use also a challenge for geothermal development where sometimes strong local culture stand in the way. Each phase of geothermal energy development (exploration, construction, operation and maintenance, and decommissioning) will have an impact on the landscape and everyone living in it. Meanwhile, natural and other human-induced drivers will keep landscapes and environments changing. This conference paper addresses the development of an integrated assessment to spatially measure the impact of geothermal energy development phases on ecosystem services. Listing the effects on the ecosystem services induced by each geothermal development phases and estimating the spatial impact using Geographic Information System (GIS) will result in an overview on where and how much each geothermal development phase affects the ecosystem and how this information could be included to improve national spatial planning.

1. Introduction

Geothermal energy a renewable energy source only can be generated in certain areas. Collisions between tectonic plates form a subduction zone which changes the dense material into buoyant magma which rises through earth surface forming a "volcanic arc" along the subduction zone. Through volcanoes, earth heat source becomes closer to the surface and becomes potential for geothermal energy extraction [1,2]. Geothermal power plants have very low gaseous emissions to the air when compared with all other power generation technologies that emit CO_2 as a normal part of the operation [3]. Geothermal energy potential is widely available in Indonesia due to the country's location in the "volcanic arc".

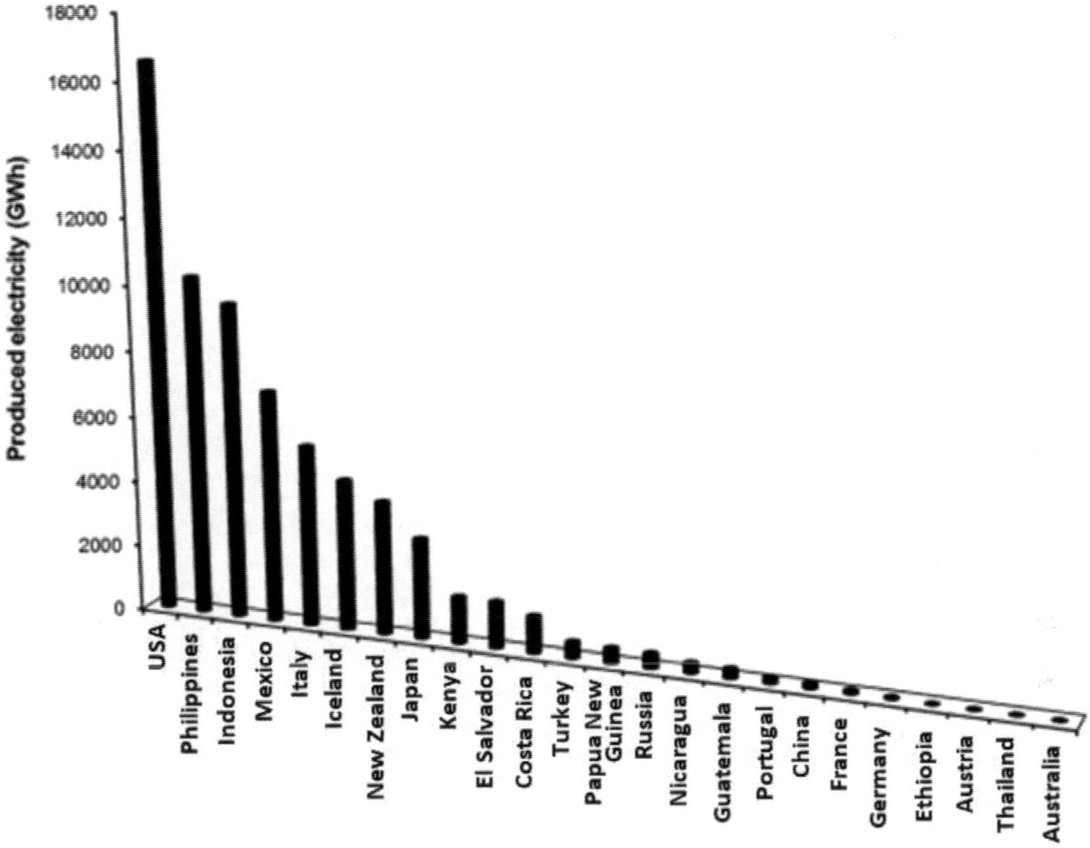

Figure 1. Generated electricity (GWh) from geothermal power plants in 2010 (Bayer et al. 2013)

USA, Philippines, Indonesia, Mexico, Italy, Iceland, New Zealand, and Japan produce more than 90% of the world's geothermal energy [4] (see figure 1.). The Indonesian National Energy Policy has set a target to provide national primary energy usage reached 2.500 kWh per capita in the year 2025 and reached 7.000 kWh in the year 2050 [5]. The growing environmental concerns combined with finite availability call into question the sustainability of an energy strategy based almost exclusively on fossil fuel [6]. 40% of the world's geothermal energy potential is located in Indonesia, and it's a very potential renewable energy in Indonesia to be explored [7,8]. Many technological, economic and environmental aspects of geothermal plants are fundamentally controlled by geological and local factors which make environmental assessment of geothermal development is site specific. These conditions raise the needs to develop geothermal energy as a renewable energy source for electricity in Indonesia.

This conference paper aim is to propose an integrated assessment framework and indicators to measure the impact of geothermal energy development phases on ecosystem services at different national levels.

2. Environmental Impact Assessment of Geothermal Development

Development impacts are often assessed through an environmental impact assessments. Environmental assessment is a process to assess the positive and negative consequences of an activity to the environment. There are two types of environmental assessment, Environmental Impact Assessment (EIA) is applied to project activity level and Strategic Environmental Assessment (SEA) is applied to plan, policy, and program. Environmental Impact Assessment firstly introduced in 1969 and has been tools for decision making until present day, also for geothermal energy project such as in Italy [9,10], Iceland [11], Greece [12], China [13,14], and Indonesia [15].

Geothermal power plants have components, each of which has an influence on the surrounding environment such as production/reinjection boreholes, connecting/delivery pipes, silencers, powerhouse and cooling towers. The effect on the environment varies, some of them only temporarily (during construction or demolition), some of them lasted with power plants operation (e.g. noise nuisance) [4]. Geothermal development phase generally divided into four phases and each phase will give different impact to the environment:

1. Resource exploration and drilling
2. Construction
3. Operation and Maintenance
4. Decommissioning and Site Rehabilitation

The U.S. Department of the Interior through The Tribal Energy and Environmental Information Clearinghouse website (https://teeic.indianaffairs.gov/er/geothermal/impact/index.htm) [16], Bayer et al. [4], and DiPippo [3]. Table 1 shows that the potential environmental impacts from geothermal development phase varies. The environmental impact depends on the type and size of the power plant, the location of facilities with respect to other resources, the number of wells, and drilling technology used.

Table 1. Potential environmental impact of geothermal development phase activity

Potential Impact	Geothermal Development Phase			
	Resource Exploration and Drilling	Construct-ion	Operations and Maintenance	Decommissioning and Site Reclamation
Acoustics/Noise [3,4,16]	Yes	Yes	Yes	Yes
Air Quality [4,16]	Yes	Yes	Yes	Yes
Cultural Resources [4,16]	Yes	Yes	Yes	No
Ecological Resources [3,4,16]	Yes	Yes	Yes	No
Environmental Justice [4,16]	No	Yes	Yes	Yes
Hazardous Materials and Waste Management [3,4,16]	Yes	Yes	Yes	Yes
Health and Safety [3,16]	Yes	Yes	Yes	Yes
Land Use [4,16]	Yes	Yes	Yes	Yes
Paleontological Resources [16]	Yes	Yes	Yes	Yes
Socioeconomics [4,16]	Yes	Yes	Yes	Yes
Soils and Geologic Resources [3,16]	Yes	Yes	No	Yes
Transportation [16]	No	Yes	No	Yes
Visual Resources [3,4,16]	No	Yes	Yes	No

Water Resources (Surface Water and Groundwater) [3,4,16]	Yes	Yes	Yes	Yes
Geological Hazards [3,4]	Yes	Yes	Yes	No

All three reference agreed that geothermal development will give impact related to noise nuisance, ecological resources, waste management, visual resources, and water resources. Although indicators from TEEIC tends to be more detailed compared with Bayer et al. [4] and DiPippo [3], but it does not measure the geological hazards that can be caused by geothermal development activity such as landslide, induced seismicity, and disturbance of hydrothermal manifestations. Meanwhile, DiPippo [3] list of geothermal challenge only focused on ecological point of view and did not consider social-economic-cultural values. Assessment method that combines ecological and socio-economic-cultural concepts are needed to gain a comprehensive overview of the geothermal impact and the impacted.

EIA in Indonesia has become an obligation for every environmental affecting project since the implementation Government Regulation No. 29, 1986. The regulation was replaced by Government Regulation No. 27 Year 1999 and had a second replacement by Government Regulation No. 27 in 2012 with has some addition in environmental permit. EIA guidelines from the Indonesian Ministry of Environment include a comprehensive assessment by combining ecology, socio-economy-culture, and public health aspects [17]. The regulations also put monitoring and evaluation as a reference for project implementations so every environmental impacted activity can be justified. In general, EIA in Indonesia already has a good foundation to assess the impact of the certain project to the environment.

Indonesia EIA practice has not applied optimal public participation as an input in the environmental assessment. This triggers a conflict in society just as happened in Sarulla geothermal field in 2008 [18] and Bali [19] where local communities opposed geothermal development because of lack public participation as screening process of local needs and values related to the environment. Common assessment practice for decision making process often only involving limited actors and mostly using silo based approach and does not integrate between sectors.

3. Ecosystem Services and Participatory Mapping of Geothermal Development

Potential environment impact of geothermal development (see Table 1) are spatial phenomena. Maps in EIA documents are compulsory but are typically only used as a base map showing general spatial information instead of using it as visualization of impact. To identify suitable locations for a geothermal facility is done by identifying locations where the benefits outweigh the conflicts/negative environmental impact.

Baker et al. [20] listed ecosystem services strength that could address the gap in environmental assessment:

1. Ecosystem services is an integrating concept which instead of dealing with environmental properties as an individual but a service that flow from the environment to human well-being and back to the environment.
2. Ecosystem services moved the perspective from environmental properties to environmental benefit.
3. Ecosystem service engage stakeholders as assessment indicators.
4. Ecosystem services may be of particular value where there are clear conflicts between traditional environmental, economic arguments and different spatial level decision making.
5. The ecosystem service framing makes explicit the value of the environment for decision makers.

Ecosystem services approach is an early attempt at outlining how decision makers can take practical steps to restore the health of ecosystem services and make development more sustainable.

Figure 2 shows the main elements of the proposed framework: the driving forces, ecosystem, service provision, human well-being, and societal response. The framework development is based on general Driver, Pressure, State, Impact, Response (DPSIR) framework where geothermal development becomes the main driving force that will affect the ecosystem properties, function, and services.

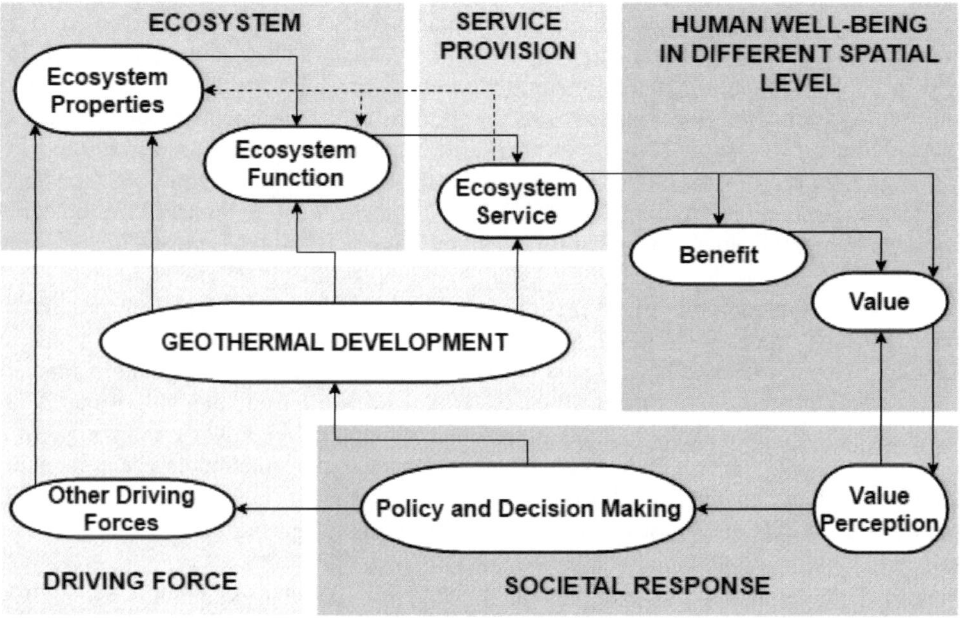

Figure 2. Framework for assessing links between land management, ecosystem services provision, and human well-being (modified from van Oudenhoven et al. [21]). Solid arrows indicate effects; dashed arrows indicate feedbacks

Spatial assessment of geothermal development impacts such as noise, air quality, ecological resource, waste, land use change, geologic resources, and water resources will provide information on how big is the affected areas. As an example, Cai et al. [22] and Farcas & Sivertun [23] able to map the range of noise impact that can be used for mapping the areas impacted from geothermal development phase especially for drilling, construction, and power plant disassembling activity where the noise effect is at the loudest. Spatial analysis can provide information and perspective that cannot be supplied by other analysis [24].

Social-economic-cultural issues of geothermal development are a challenge for EIA of geothermal development in Indonesia because of the different culture and local characteristics throughout the country. Different culture will determine the way of decision making. The public participatory process already is compulsory in Indonesia EIA regulation. Combining public participatory with spatial knowledge will provide richer information about ecosystem services and understand how communities utilize the environment in geothermal proposed area. Brown and Fagerholm [25] reviewed PPGIS/PGIS approaches for mapping ecosystem services. They identify that there are two approaches to measures and quantify ecosystem services in traditional economic values. The first approach is to reveal the importance of natural systems and assets that provide long-term human benefit vis-à-vis. The second approach is to focus on place-based valuation of services in non-monetary terms to provide spatially-explicit guidance for future land use decisions.

Decision making processes do not only need to be integrated between sectors but also need to be integrated between local and national policies. Garcia-Nieto et al. [26] presented that participatory mapping process becomes a powerful method in identifying Service Providing Unit's (SPU's) and Service Benefiting Areas (SBA's) spatially. In some ecosystem services, different stakeholders also show different perspectives related to SBA's spatial scale. Multi stakeholder approach for decision making also was used by Llopis-Albert et al. [27] to facilitate and engage stakeholders participation in the water resources management in Spain.

4. Working Progress

This review identified that spatial analysis is not yet optimal in the EIA process in Indonesia. Visualization of impacts through maps in different scale will help to identify the high and low impacted stakeholders. From a valuation perspective, environmental problems and conflicts originate from trade-offs between values. The urgency and importance to integrate nature's diverse values in decisions and actions stand out more than ever [28]. Ecosystem services mapping will able to identify specific assets of nature that affected by geothermal development.

Ecosystem services approach can explore how current and future trends in the condition of these services will affect geothermal development goals. The findings of this presentation will address the gap between EIA regulation and EIA practice for geothermal development projects in Indonesia and serve as inputs for future research in developing spatial decision support system for geothermal field landscape management. Decision maker can build partnerships across institutional and political boundaries to address the risks and opportunities for ecosystem services that presents caused by environmental affected project.

5. Acknowledgement

This work is part of the initial Ph.D. research joint supervision between the University of Twente – The Netherlands and Universitas Indonesia - Indonesia which is funded by GEOCAP program under Work Package 2.08 to develop a number of cases showing the environmental impact of geothermal exploration and exploitation in a multi-temporal and spatial context.

References

[1] Moeck IS. Catalog of geothermal play types based on geologic controls. Renew. Sustain. Energy Rev. 2014;37:867–82.
[2] Stelling P, Shevenell L, Hinz N, Coolbaugh M, Melosh G, Cumming W. Geothermal systems in volcanic arcs: Volcanic characteristics and surface manifestations as indicators of geothermal potential and favorability worldwide. J. Volcanol. Geotherm. Res. 2016;324:57–72.
[3] DiPippo R. Geothermal Power Plants : Principles, Applications, Case Studies and Environmental Impact. 3rd ed. Waltham, MA. USA: Butterworth-Heinemann, Elsevier Ltd.; 2012.
[4] Bayer P, Rybach L, Blum P, Brauchler R. Review on life cycle environmental effects of geothermal power generation. Renew. Sustain. Energy Rev. 2013;26:446–63.
[5] President of Republic of Indonesia. Peraturan Pemerintah Republik Indonesia Tentang Kebijakan Energi Nasional. Gov. Regul. No. 79, 79 Indonesia; 2014.
[6] Dutu R. Challenges and policies in Indonesia's energy sector. Energy Policy. 2016;98:513–9.
[7] Nasruddin, Idrus Alhamid M, Daud Y, Surachman A, Sugiyono A, Aditya HB, et al. Potential of geothermal energy for electricity generation in Indonesia: A review. Renew. Sustain. Energy Rev. 2016;53:733–40.
[8] Ahsat A, Ardiansyah F. Igniting the Ring of Fire A Vision for Developing Indonesia's Geothermal Power. Jakarta; 2012.
[9] Borzoni M, Rizzi F, Frey M. Geothermal power in Italy: A social multi-criteria evaluation. Renew. Energy. 2014;69:60–73.

[10] Bravi M, Basosi R. Environmental impact of electricity from selected geothermal power plants in Italy. J. Clean. Prod. 2014;66:301–8.

[11] Shortall R, Davidsdottir B, Axelsson G. Development of a sustainability assessment framework for geothermal energy projects. Energy Sustain. Dev. 2015;27:28–45.

[12] Koroneos CJ, Nanaki EA. Environmental impact assessment of a ground source heat pump system in Greece. Geothermics. 2017;65:1–9.

[13] Guo Q, Wang Y, Liu W. Hydrogeochemistry and environmental impact of geothermal waters from Yangyi of Tibet, China. J. Volcanol. Geotherm. Res. 2009;180:9–20.

[14] Sun R, Wang Z. A comprehensive environmental impact assessment method for shale gas development. Nat. Gas Ind. B. 2015;2:203–10.

[15] Ministry of Environment. Peraturan Menteri Negara Lingkungan Hidup Republik Indonesia Jenis Rencana Usaha Dan/Atau Kegiatan Yang Wajib Memiliki Analisis Mengenai Dampak Lingkungan Hidup. Minist. Decree No. 5, 5 Indonesia: Ministry of Environment, Republik Indonesia; 2012.

[16] TEEIC (The Tribal Energy and Environmental Information Clearinghouse). Potential Impacts of Geothermal Energy Development [Internet]. Available from: https://teeic.indianaffairs.gov/er/geothermal/impact/index.htm

[17] Ministry of Environment. Peraturan Menteri Negara Lingkungan Hidup Republik Indonesia Tentang Pedoman Penyusunan Dokumen Lingkungan Hidup. Minist. Decree No. 16, 16 Indonesia: Ministry of Environment, Republik Indonesia; 2012.

[18] Kompas. Amdal Proyek Sarulla Belum Beres. Kompas News Artic. [Internet]. 2008 Nov 11 [cited 2017 Jan 12]; Available from: http://edukasi.kompas.com/read/2008/11/11/21120469/amdal.proyek.sarulla.belum.beres.

[19] Gumelar G. Bali Energy Bermasalah, Proyek PLTP Bedugul Jalan di Tempat [Internet]. CNN Indones. Jakarta; 2016 [cited 2017 Jan 29]. Available from: http://www.cnnindonesia.com/ekonomi/20160830203910-85-154974/bali-energy-bermasalah-proyek-pltp-bedugul-jalan-di-tempat/

[20] Baker J, Sheate WR, Phillips P, Eales R. Ecosystem services in environmental assessment — Help or hindrance? Environ. Impact Assess. Rev. 2013;40:3–13.

[21] van Oudenhoven APE, Petz K, Alkemade R, Hein L, de Groot RS. Framework for systematic indicator selection to assess effects of land management on ecosystem services. Ecol. Indic. 2012;21:110–22.

[22] Cai M, Zou J, Xie J, Ma X. Road traffic noise mapping in Guangzhou using GIS and GPS. Appl. Acoust. [Internet]. Elsevier Ltd; 2015;87:94–102. Available from: http://dx.doi.org/10.1016/j.apacoust.2014.06.005

[23] Farcaş F, Sivertun Å. Road traffic noise: GIS tools for noise mapping and a case study for Skåne region. Int. Arch. Photogramm. Remote Sens. Spat. Inf. Sci. [Internet]. 2005;34:1–10. Available from: http://www.isprs.org/proceedings/XXXVIII/2-W11/Farcas_Sivertun.pdf

[24] Fernandes J, Flynn N, Gibbes S, Griffis M, Isshiki T, Killian S, et al. Renewable Energy in the California Desert: Mechanisms for Evaluating Solar Development on Public Lands. Renew. Energy. 2010.

[25] Brown G, Fagerholm N. Empirical PPGIS/PGIS mapping of ecosystem services: A review and evaluation. Ecosyst. Serv. 2015;13:119–33.

[26] García-Nieto AP, Quintas-Soriano C, García-Llorente M, Montes C, Martín-López B. Collaborative mapping of ecosystem services: The role of stakeholders′ profiles. Ecosyst. Serv. 2015;13:141–52.

[27] Llopis-Albert C, Palacios-Marques D, Soto-Acosta P. Decision-making and stakeholders' constructive participation in environmental projects. J. Bus. Res. 2015;68:1641–4.

[28] Jacobs S, Dendoncker N, Martín-López B, Barton DN, Gomez-Baggethun E, Boeraeve F, et al. A new valuation school: Integrating diverse values of nature in resource and land use decisions. Ecosyst. Serv. 2016;22:213–20.

IOP Conf. Series: Earth and Environmental Science **103** (2017) 012004 doi:10.1088/1755-1315/103/1/012004

Smart geo-energy village development by using cascade direct use of geothermal energy in Bonjol, West Sumatera

Novrisal Prasetya[1], Defry Erwinsyah Umra Lubis[1], Dharmawan Raharjo[1], Nenny Miryani Saptadji[2], Heru Berian Pratama[2]

[1]Study Program of Petroleum Engineering, Faculty of Mining and Petroleum Engineering, Bandung Institute of Technology (ITB)
[2]Study Program of Geothermal Engineering, Faculty of Mining and Petroleum Engineering, Bandung Institute of Technology (ITB)
Jl. Ganesha No.10, Bandung, West Java, Indonesia. 40132

Email: novrisal.prasetya2796@gmail.com

Abstract. West Sumatera is a province which has a huge geothermal potential - approximately 6% of Indonesia's total geothermal potential which equals to 1,656 MWe. One of the significant reserves located in Bonjol subdistrict which accounts for more than 50 MWe. The energy from geothermal manifestation in Bonjol can be utilized prior to indirect development. Manifestation at the rate 3 kg/s and 87 °C will flow to cascading system consisting several applications, arranged in order from high to low temperature to efficiently use the excessive energy. The direct use application selected is based on the best potential commodities as well as temperature constraint of heat source. The objective of this paper is to perform a conceptual design for the first cascade direct use of geothermal energy in Indonesia to establish Bonjol Smart Geo-Energy Village which will be transformed as the center of agricultural, stockbreeding, tourism as well as cultural site. A comprehenssive research was performed through remote survey area, evaluation featured product, analysis of heat loss and heat exchange in cascade system. From potential commodities, the three applications selected are cocoa drying and egg hatching incubation machine as well as new tourism site called *Terapi Panas Bumi*. The optimum temperature for cocoa drying is 62°C with the moisture content 7% which consumes 78 kW for one tones cocoa dried. Whereas, egg incubation system consists of two chamber with the same temperature 40°C for each room and relative humidity 55% and 70%. For the last stage, *Terapi Panas Bumi* works in temperature 40°C. Based on the result technical and economical aspect, it exhibits cascade direct use of geothermal energy is very recommended to develop.

1. Introduction

Indonesia is located along the ring of fire as three tectonic plates collide beneath Sumatera, Java, and Sulawesi. West Sumatera is a province which has a huge geothermal potential - approximately 6% of Indonesia's total geothermal potential which equals to 1,656 MWe. One of the significant reserves located in Bonjol subdistrict which accounts for more than 50 MWe. The downhole temperature of reservoir near the magma intrusion is estimated to be around 180°C while Sungai Takis as the most potential manifestation to be utilized has 87.9 °C of temperature, 6.9 of pH, 3 kg/s of flow rate, and 243 masl [4].

Geothermal energy can be used both directly and indirectly. Indirect use of geothermal energy typically performed to generate electricity, whereas direct uses are addressed to help industrial process such as heating, drying, sterilization and/or pasteurization. Absorbing all the possible of heat from the hot fluid and utilize it as efficient as possible brings out a method termed "Cascade System" used in several stages as for maximizing heat utilization of the working fluid from the previous stage to the next stage. The stages are divided by considering the temperature required which the highest temperature is put at the first stage.

Thus far, many agro industries still use large amount of diesel or even conventional sun drying method as a source to generate thermal energy for conducting those processes. Diesel may be a flexible fuel, nevertheless, high consumption of it can contribute to global warming due to high carbon dioxide emission while sun drying method relies on weather conditions which lead to concern on drying time especially on rainy season which could result in longer time to dry. Even in dry season, it may last seven days to meet specification of water content due to sunlight fluctuation where the longer duration of drying using this method may cause possible damage of cocoa beans due to mold development [5]. The industries usually hatch eggs using bulb as the heat source that can lead to several problems. For instance, unequally distributed egg's temperature requires it to be properly treated inside continously.

Cocoa drying can be a big potential revenue for the region because it will enhance the price of the cocoa commodity. But if using conventional drying method, weather fluctuation can impact the quality of cocoa commodity product which will reduce the farmer's profit. Incubation method for egg's hatching used in Bonjol uses electricity to provide heat. However, every year it is estimated to 115 MWh electricity spent by this method. Due to high price of electricity, high use of it can be a source of economic potential loss for egg hatching business. Tourism and cultural potential in Bonjol has not been fully exploited yet. It only depends on the visitors who visit the Museum.

The objective of this paper is to perform a conceptual design for the first cascade direct use of geothermal energy in Indonesia to establish Bonjol Smart Geo-energy Village. Bonjol Smart Geo-energy Village is a village located in Bonjol which directly utilize a geothermal energy for several needs. Bonjol Smart Geo-energy Village will be transformed as the center of agricultural, stockbreeding, tourism as well as cultural site to increase the welfare of local community.

Direct use of geothermal energy is normally developed in unintegrated manner. It means the utilization is only aimed at a specific process where the waste heat usually unused. But, in this paper, author will develop Bonjol through cascade direct use of geothermal energy to form a "Smart Geo-Energy Village" which means no heat will be wasted.

2. Geothermal Potential in Bonjol

Bonjol geothermal area is located at Bonjol subdistrict, Pasaman regency in the Province of West Sumatera. Geothermal manifestations consist of neutral pH hot springs with temperature ranging from 49.7 °C to 87.9 °C. Geothermal manifestations spread over in center of Bonjol area are predominated by appearance of hot springs in Padang Baru, Takis river, Limau river, and Kambahan river. Hot springs with the highest temperature are located at Limau river, Takis, and Kambahan, which is located in Sumatera depression zone and controlled by Takis fault. Takis hot spring has temperature about ± 87.9 °C and pH 6.9. Limau river hot spring has temperature about ± 73.5 °C, pH 7.3. Kambahan hot spring has temperature about 73.4 °C, pH 7.5 [8].

The geothermal reservoir fluid is chloride type, indicating hot water dominated reservoir. Geothermal system in this area is supposed to be associated to volcanic activities. The youngest volcanic is Binuang dome which is supposed as a heat source for the geothermal system of the area. Gravity data reveal a high anomaly which is interpreted as an intrusive body beneath Takis hot spring. [8]. Heat source potential is estimated ± 3000 m depth as andesitic igneous intrusion that resulted from eruption of Bukit Gajah in quaternary period. The reservoir temperature in the Bonjol area is estimated about 180 °C from water geothermometry and had moderate entalphy. Heat loss factor at

Bonjol area reach up to 1 MWe and the geothermal potential is predicted can produce energy ± 50 MWe [3].

3. Cascade Definition and Characteristics

Cascade utilization of geothermal energy is a reasonable sequential arrangement to use energy at different thermal levels and to obtain several products such as drying fish, balneology, aquaculture, green house, etc. Cascade utilization of geothermal resources allows to obtain multiple products from a single resource or coupling to other renewable sources, thus having a better utilization and energy saving, and avoiding an excess consumption of unrenewable resources.

There are several possibilities to utilize geothermal energy in cascade manner, but in general, it is possible to identify two main groups : cascade systems with electricity production and thermal uses, and cascade systems comprising only thermal uses. Cascade systems with electricity production are suitable when geothermal resource has temperatures greater than 90 °C, with more opportunity of integrate various thermal levels as temperature increases. Cascade systems with only thermal uses are appropriate for geothermal resources with temperatures below 90 °C.

The cascade system for geothermal energy has been developed around the world, such as America, Europe, Africa, Asia, and other regions. In Africa, direct use applications are highlighted for shallow wells, focusing on drying of agricultural products and food. The temperature range suitable for cascading direct uses is 89 °C in wellhead and an outlet temperature for the last level of the cascade of 50 °C [7]. In Indonesia, until now there is no application of cascade system utilization for direct use. If this project can be implemented, it will be the first ever cascade utilization for direct use in Indonesia.

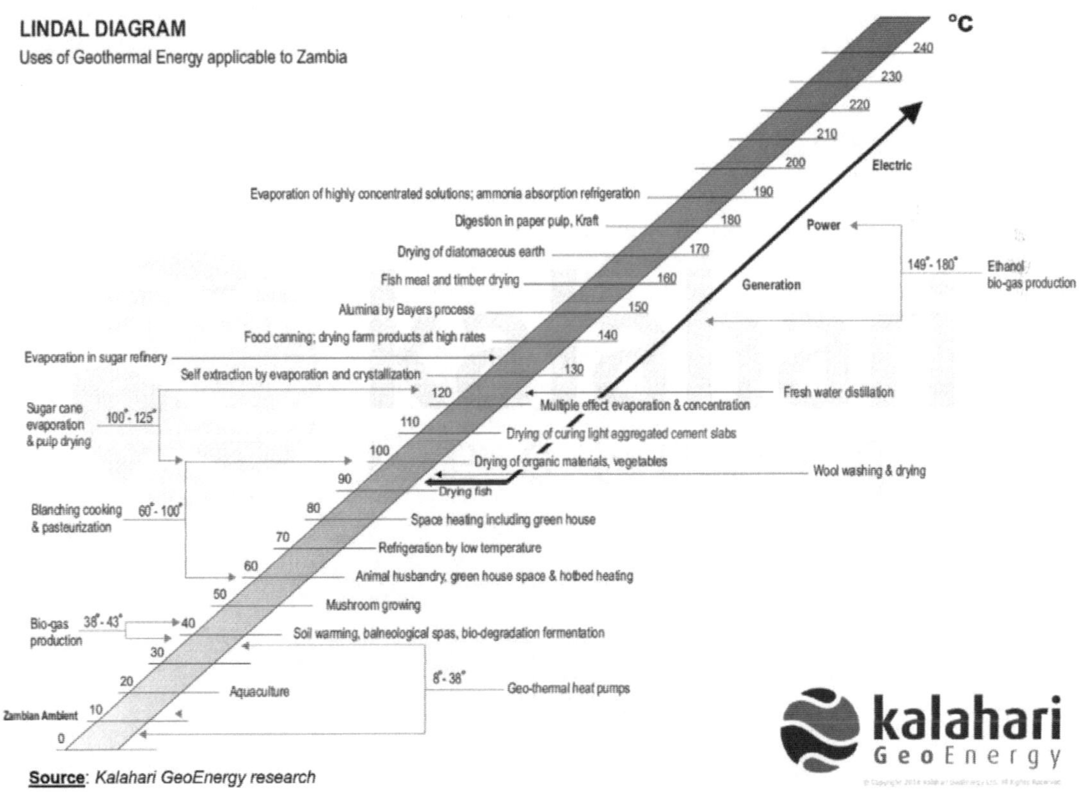

Figure 1. Lindal Diagram

IOP Conf. Series: Earth and Environmental Science **103** (2017) 012004 doi:10.1088/1755-1315/103/1/012004

Cascade direct use is inspired by Lindal diagram (see figure 1) which is a brief of potential direct application of geothermal energy based on the range of working temperature is proposed by an engineer named Baldur Lindal. It delivers insight concerning how to improve direct use geothermal energy efficiency by introducing a wide combination of usage

4. Methodology

4.1. Workflow

The research consists of planning stage, literature study, remote survey area, evaluation and analysis of featured product, analysis of heat loss and heat exchange in cascade system, plan of development and study compilation respectively.

In planning stage, authors determine research limitation, the objective, problem identification and research work flow. After that, literature study is done to address the direct utilization of geothermal energy which have succeeded to be developed in either pilot or commercial scale. It also includes geothermal potential evaluation study in Bonjol, agricultural production data, and other featured product. Information regarding potential is obtained through Pusat Sumber Daya Geologi (PSDG) as the assessor of Bonjol geothermal appointed by Indonesia. Featured regional product data is obtained through government official website and papers.

The next step is remote survey area. The survey is conducted to understand the real condition of the field including resources distance to the nearest social community, accessibility and topography of region. Due to limited time and resource, area survey is conducted remotely through satellite information provided by google earth and street view feature.

After all the required data is collected, authors conduct evaluation and data analysis which integrate the literature and survey data to form an evaluation in the perspective of technical, social and economic aspect. After evaluation and data analysis has been set up, Plan of Development is designed in the use of cascade direct use system. And eventually, all result study is compiled.

4.2. Thermodynamics Principle and Transportation of Heat

4.2.1. Pipeline

The principle of heat transfer in the flowing fluid in pipe is based on conduction and convection process. This phenomenon can be illustrated by figure 2.

In order to evaluate the heat loss along the pipeline, there are three major process should be taken into account : 1). Convection process inside the pipe, 2). Conduction process within the pipe and insulator ,and 3). Convection process from outer wall of pipe to the surrounding area. The heat transfer resistance (R) as the obstraction of driving force are evaluated through these formula:

$$Q = \frac{Twater - Tair}{Ri + Ra + Rb + Ro} \tag{1}$$

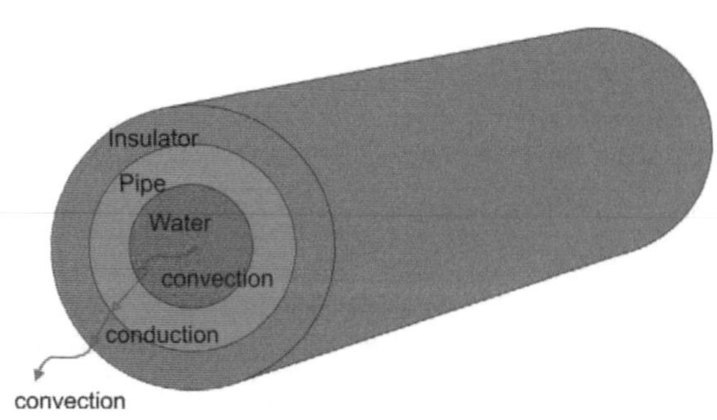

Figure 2. Heat transfer in pipe

4.2.2. Heat Exchanger
By applying the principle of heat transfer in thermodynamics and fluid dynamics in one shell pass and one tube pass in counter-current flow heat exchanger, the temperature outlet of both heat source and cold source in heat exchanger could be determined as below.

$$Q = m\,x\,C\,x\,T \tag{2}$$

$$Q = Uhe\ x\ Ahe\ x\ LMTD \tag{3}$$

Thus, LMTD can be evaluated below.

$$LMTD = \frac{[(Th\ in - Tc\ out) - (Th\ out - Tc\ in)]}{ln\frac{(Th\ in - Tc\ out)}{(Th\ out - Tc\ in)}} \tag{4}$$

Calculating water mass contained in cacao which need to be evaporated

$$M_R = M\ x\ \frac{Q_1 - Q_2}{1 - Q_2} \tag{5}$$

Calculating air rate required to reduce water content from initial humidity to final humidity before entering the drying room.

$$Q_a = \frac{\frac{M_R}{Hr_2 - Hr_1}}{t} \tag{6}$$

The nomenclature for those all symbol is presented in Appendix A.

5. Result

5.1. The Prospect of Direct Utilization of Geothermal Potensial in Bonjol
Based on the literature, the geothermal potential in Bonjol is expected to be over 50 MWe where the manifestations spread over in center of Bonjol. The area is predominated by appearance of hot springs in Padang Baru, Takis river, Limau river, and Kambahan river which the highest temperature are located at Limau river, Takis, and Kambahan in Sumatera depression zone. Heat source potential estimation yields ± 3,000 m for the depth as andesitic igneous intrusion, and the temperature of reservoir is 180 °C which can be used to generate electricity and directly use it. Due to indirect utilization can't be realized in the near future, the manifestation heat can be utilized as the heat source of the cascade direct use geothermal energy system.

However, considering the distance, accessibility, and temperature to the three of these manifestations, the authors decided Takis river hot spring which has temperature about 87.9 °C as the manifestation source. Takis River hot spring is located at a distance of about 1.6 km from the nearest village and has a 243 m elevation. This position is very suitable as a source for the cascade system to be developed because it does not require a pump but it just need pipeline to flow the water from the Takis river to Smart Geo-Energy Village.

Based on literatures, Cascade systems with only thermal uses are appropriate for geothermal resources with temperatures below 90 °C. Due to low manifestation temperature which is 87.9 °C, Takis river as the heat source of cascade system can't be used for electricity but it only can uses for thermal uses specifically in direct use. Based on Lindal diagram, this temperature can be direct use for drying fish, refrigeration, green house, balneology, aquaculture, bi-degradation fermentation, agriculture, etc. In Indonesia, until now there is no application of cascade system utilization for direct use. If this project can be implemented, it will be the first ever cascade utilization for direct use in Indonesia.

The cascade system to be developed consists of three levels, those are cocoa beans drying machine, chicken egg incubation, and *Terapi Panas Bumi*. The selection of this levels is based on the study of agricultural commodity, livestock commodity and tourism sites in Bonjol. In agricultural, Pasaman regency agricultural potential are massively focus on cocoa beans and chicken egg commodity. In 2014, the production of cocoa beans in Pasaman regency is 14,409 ton with selling price Rp 3,000/kg (Bank Data Dinas Pertanian Nasional, 2014). Prospectively, this commodity is probable to be developed. In other hand, Bonjol as a subdistrict in West Sumatera is acknowledged in 2007 to be one of an agricultural region plan. Of course, this project is so profitable to be developed.

In livestock commodity, as shown in Sumatera Barat livestock report by 2014, the production of chicken eggs in Pasaman is 115 ton. The incubation process for chicken eggs goes hand in hand with cocoa drying process which is still using conventional drying method with lamp as the energy source which approximately consumes 9.3 MWh to hatch one tenth of the produced eggs. In this study, the authors' aim is to make a projection that Bonjol as the central drying cocoa beans and chicken egg incubation in Pasaman. By making Bonjol as a center of drying cocoa beans and chicken egg incubation industry as described previously, this project would open up employment opportunities for the community in Bonjol and nearby region.

Astonishingly, tourism potential has not been fully exploited yet showed by the published tourism site merely Museum of Tuanku Imam Bonjol (Web Kabupaten Pasaman, 2014). Pasaman regency known as the area which is passed through by the equatorial line, particularly in Bonjol, may take opportunity for tourism sector to be developed. In this study, the authors wanted to attract attention in the tourism sector by making a *Terapi Panas Bumi* in Bonjol. If this project successful, this will makes a first ever a massage based on geotermal energy in Pasaman regency.

Figure 3. Location of Smart Geo-Energy village (Google Earth Maps)

5.2. Recomendation Location for Direct Use Utilization

In determining the recommended location, the consideration are the manifestation distance, accessibility, tourists travel opportunities, and population residents. The chosen location is as shown in Figure 3. The location is a village so that it can be used as Smart Geo-Energy Village. The distance of Takis River manifestation to the location is approximately 1.6 km to the economic considerations which will be the further project of this paper.

This village is the nearest village to the Takis River hot spring, so economically this village is the most suitable location to be developed. From Figure 3, the length of pipeline needed is about 1.6 km. These locations can be accessed from the Jalan Lintas Sumatera with a distance about 500 m. This location is deemed lucrative to attract travelers because the pathway to this location is in front of the equator monument and Tuanku Imam Bonjol museum which is a crowded tourist spot. Another reason, why tourism will atracted to this village because Bonjol has so many cultural art, such as Randai, Rabab and Lukah gilo which is usually performed in traditional event, culminative point event, and international annual event (Tour de Singkarak).

5.3. Cascade System Design for Direct Use

The geothermal manifestation in Bonjol can't be used to generate electricty. However, based on Lindal diagram and availbility of featured product in Bonjol, the geothermal manifestation will be used for direct utilization by applying cascade system. Cascade system consisting several application, arranged in order from high to low temperature to efficiently use the excessive energy, is designed like in figure 4.

5.3.1. Design for Drying Cocoa Beans

Hot water flows from the takis river hot spring manifestations through the cocoa beans drying system at 71.4 °C and will pass through a heat exchanger. After passing through the heat exchanger, the air will be blown towards the drying chamber with the arrangement tray containing cocoa seeds. The schematic design of cocoa drying machine is shown in figure 5. The performance of cocoa drying system is presented by table 1. Based on the calculation of mass and heat transfer that has been done, obtained the working parameter that can be summarized in table 2.

To avoid case hardening (the outside of the dried seeds while the inside is still moist) , the temperature during the drying process should be maintained at 62 °C. The conclution of the calculated parameters related to cocoa drying is presented in table 2.

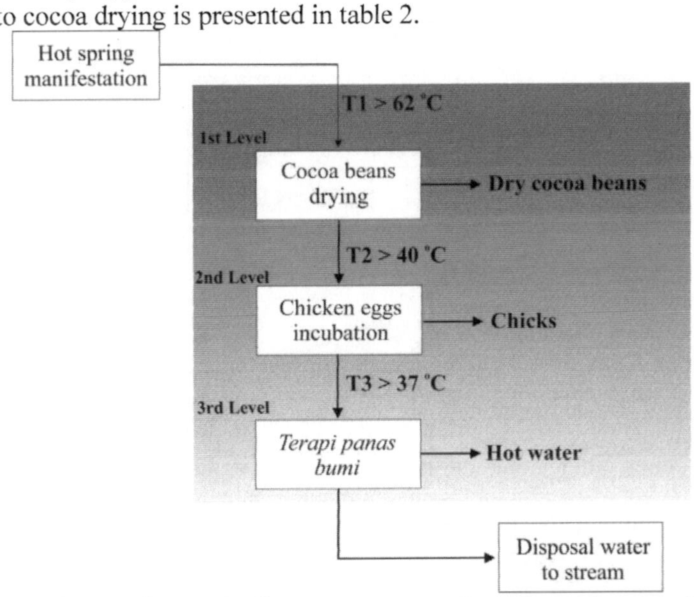

Figure 4. Scheme of cascade direct use system of geothermal energy in Bonjol

IOP Conf. Series: Earth and Environmental Science **103** (2017) 012004 doi:10.1088/1755-1315/103/1/012004

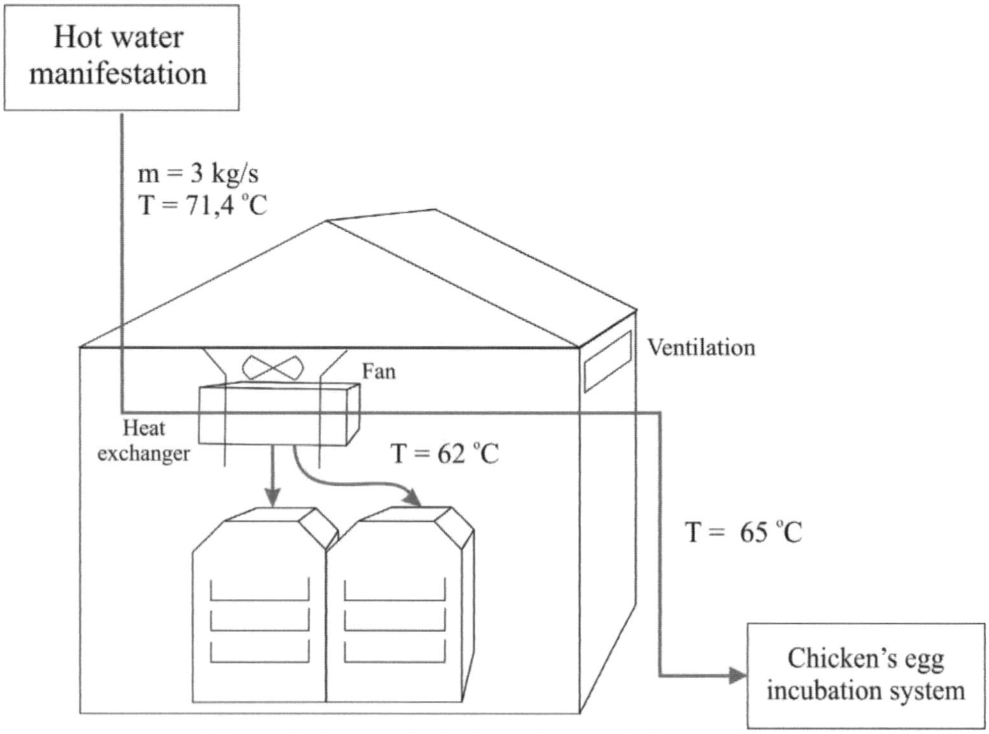

Figure 5. Schematic design of cocoa drying machine

Table 1. Work performance of drying cocoa beans.

Parameters	Value	Units
Capacity of drying cocoa per day	1000,00	Kg
WC initial	0,80	
WC final	0,07	
HR 1	0,75	
HR 2	0,40	
Drying time	7,00	Hours
M water contained	783,92	Kg
M of air needed	0,09	kg / s
Specific Heat Capacity of cocoa	3,10	kJ/kg °C
Tcocoa inlet	28,60	°C
Tcocoa outlet	62,00	°C
Power for heating-up	4,11	kW
Latent heat of evaporation	74,09	kW
Total Power	78,20	kW

Table 2. Work parameters of drying cocoa machine.

Parameters	Air	Water	Units
Tin	28,60	71,40	°C
Tout	62,00	65,16	°C
Pressure	0,98	1,00	Bar
Density	1,09	977,75	kg/m^3
Flow rate	0,09	3,00	m^3/s
Heat capacity		4,18	kJ/kg °C
Energy	78,20	78,20	kW
Velocity	0,74	0,77	m/s
Cross-section area	0,11	0,00	m^2

5.3.2. Design for chicken's egg incubation system

Temperature, humidity, and ventilation are three important parameters to successfully hatch the eggs. The eggs need to be kept at 40°C at all times and incapability to sustain this condition will eventually terminate the embryo. Humidity should be maintained to 40-60% in the first 18 days and 65-75% for the final days before hatching. And ventilation will allow oxygen to enter because the fetuses need fresh air to breathe.

In this scenario, geothermal energy will provide heat energy to maintain the temperature. By using cascade system, the hot water from cocoa drier will be used again to flow in this system.

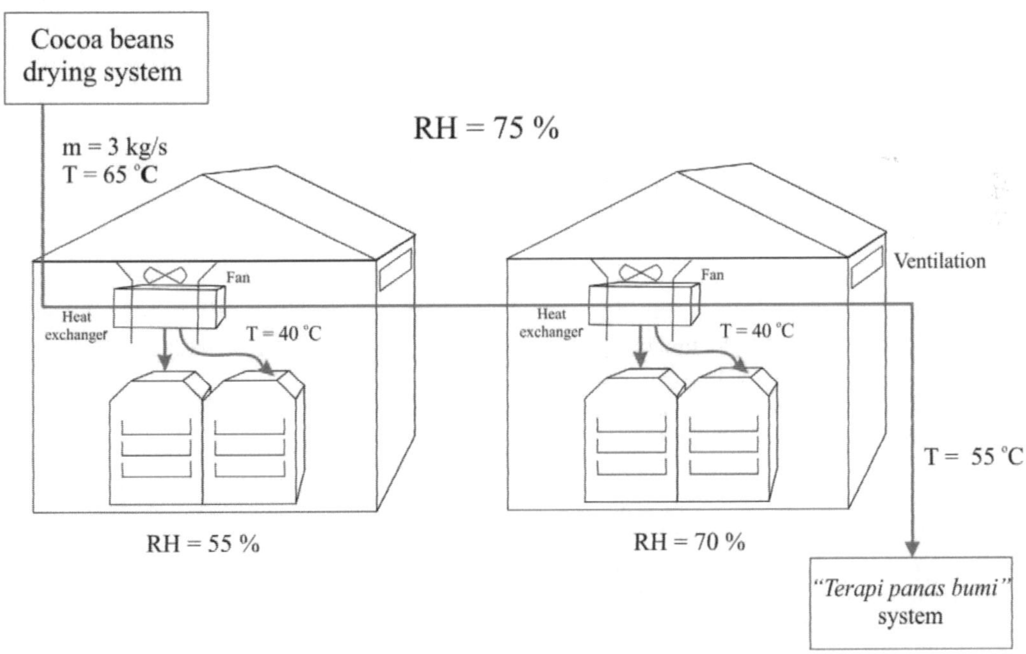

Figure 6. Schematic design of egg incubation machine.

IOP Conf. Series: Earth and Environmental Science **103** (2017) 012004 doi:10.1088/1755-1315/103/1/012004

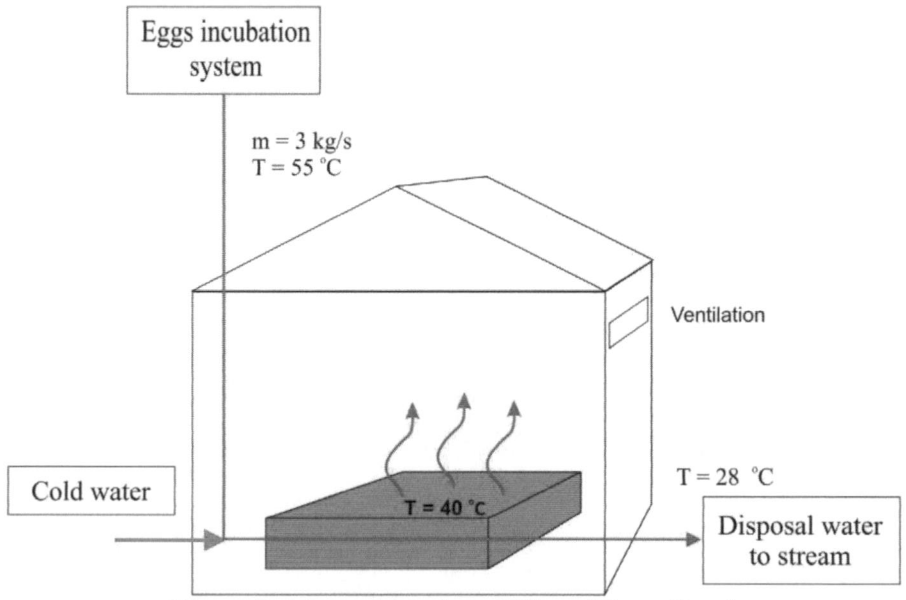

Figure 7. Schematic design of "Terapi Panas Bumi

5.3.3. Design for "Terapi Panas Bumi"

The temperature for *Terapi Panas Bumi* is expected to be 40 °C. The hot water from incubator will flow to pool and accumulate. After that hot water will be flown to river near the location. Due to, no harmful chemical solution, directly flowing to the water is justified. The Schematic design of *Terapi Panas Bumi* is shown in figure 7.

6. Conclusion

The heat transfer calculation shows that it is possible to maintain the required temperature for every stage. Therefore, early development of geothermal energy in Bonjol is encouraged to be established through Cascade system with three stage of uses: Cocoa bean drier, egg hatcher, and tourism potential.

However, further economic studies is required to show whether it is worth it. Hypothetically, the development of Smart Geo-Energy Village should boost the economy of Bonjol and Pasaman District through multiplier effect as well. Firstly, the tourism project will boost the micro businesses in food court, transportation services and others because the expected growth of tourist. Secondly, Bonjol could grow into specialized commodity processing sector because of high quality and low price of drying coupled with the availability of electricity supply in the future. Thirdly, these growths will eventually provide more job opportunity for local people.

7. Further Research Project

The further research project of this paper is :

1. Drier design to calculate volume required to meet the expected drying condition to prevent spoilage of the cocoa.
2. Heat exchanger design.
3. Community acceptance of the project.
4. Cascade system with electricity as well.
5. Project Economic Study.

The objective of economic analysis which we shall explore is to find whether the project is profitable both for investor and community around. We will start by evaluating the cost and revenue stream and the move into net present value as, pay out time and profitability index of the project.

The further project will evaluate cost of development from two source of manifestation and seek for the best scenario and compare it to current method of drying cocoa and hatching egg. After that, author will analyze profitability of *Terapi Panas Bumi* tourism in terms of Net Present Value, Profitability Index, and Internal Rate of Return.

8. Acknowledgment

We express our greatest gratitude to our lecturers Ir. Nenny Miryani Saptadji, Ph.D and Heru Berian Pratama, S.T., M.T for their help on creating this paper.

References

[1] Banjarnahor, Nursanty Elisabeth. 2016. Kajian Pemanfaatan Langsung Fluida Panas Bumi Menggunakan Sistem Kaskade Pada Lapangan Panas Bumi Wayang Windu. Bandung : Institut Teknologi Bandung

[2] Geankoplis, Christie J. 1993. Transport Processes and Unit Operations Third Edition. New Jersey : Prentice-Hall International, Inc

[3] Hutapea, F., Soemintadiredja, P., & Permana, A. 2010. Geothermal Exploration Using Geochemical Method , in Bonjol Area, Pasaman Regency, West Sumatra. Bandung : Institut Teknologi Bandung

[4] Kholid, Muhammad., Iim, Dede., & Widodo, Sri. 2007. Penyelidikan Terpadu (Geologi, Geokimia, dan Geofisika) Daerah Panas Bumi Bonjol , Kabupaten Pasaman Sumatera Barat. Bandung : PSDG

[5] Lasisi, D. 2014. A Comparative Study of Effects of Drying Methods on Quality of Cocoa Beans. Igboora : Oyo State Collage of Agriculture

[6] Oksatriandhi, Benny & Santoso, Eko Budi. 2014. Identifikasi Komoditas Unggulan di Kawasan Agropolitan Kabupaten Pasaman. Surabaya : Institu Teknologi Sepuluh November

[7] Rubio-Maya, C., Diaz, V.M Ambriz., Martinez, E. Pastor., & Belman-Flores, J.M. 2015. Cascade Utilization of Low and Medium Entalphy of Geothermal Resources – A Review. Mexico : Universidad Michoacana de San Nicolas de Hidalgo Group of Energy Efficiency and Renewable Energy & University of Guanajuato Salamanca

[8] Sardjam, B., Munandar, A., & Widodo, S. 2010. Geothermal System in Geothermal Prospect, Pasaman regency, West Sumatra, Indonesia. Bandung : PSDG

[9] Suryanto., Surana, Taufan., Atmojo, Jatmiko Prio., & Prasetyo, Bambang Teguh. 2010. Design of a Geothermal Energy Dryer for Tea Withering and Drying in Wayang Windu Geothermal Field. Jakarta : BPPT

[10] Suryanto., Surana, Taufan., Atmojo, Jatmiko Prio., & Subandriya, Andri. 2010. Development of Geothermal Energy Direct Use in Indonesia. Jakarta :BPPT

APPENDIX A

Properties, symbol and unit

μ	= Dynamic viscosity (kg/m s)
A_{alm}	= Log mean areas for pipe
A_{blm}	= Log mean areas for insulator
A_{he}	= Heat transfer area (m^2)
A_i	= Surface area of inner pipe
A_o	= Surface are of outer insulator
C	= Specific heat capacity of heated fluid (kJ/kg K)
D	= Diameter of pipe contacted by fluid (m)
h_i	= Convective coefficient of water
h_o	= Convective coefficient of air
Hr_1	= Humidity ratio before contact with cocoa
Hr_2	= Humidity ratio after contact with cocoa
k_a	= Mean thermal conductivity of pipe
k_b	= Mean thermal conductivity of insulator
LMTD	= Log mean temperature difference (K)
m	= Flow rate of fluid (kg/s)
m	= Flow rate of heated fluid (kg/s)
M	= Mass of cocoa bean
M_R	= Mass of water content in cocoa to be evaporated
Q	= Heat calculated (kW)
Q_1	= Final water content
Q_2	= Initial water content
Q_a	= Mass of air needed in cocoa drying
r_1	= Radius of outer pipe
R_a	= Heat resistance at pipe due to conduction
R_b	= Heat resistance at insulator due to conduction
r_i	= Radius of inner pipe
R_i	= Heat resistance at inner pipe due to convection
R_o	= Heat resistance at free air due to convection
r_o	= Radius of outer insulator
T	= Temperature of heated fluid (°C)
T_{air}	= Ambient temperature
Tc_{in}	= Cold water inlet temperature (°C)
Tc_{out}	= Cold water outlet temperature (°C)
Th_{in}	= Hot water inlet temperature (°C)
Th_{out}	= Hot water outlet temperature (°C)
T_{water}	= Temperature of water
U_{he}	= Heat transfer coefficient HE (W/m^2 K)

APPENDIX B

Temperature of Fluid Calculation in Each Level of Cascade System

Table 3. Temperature of fluid in each level of cascade system

Level	Fluid	Temperature (°C)
Flowing in Pipe	Water	87.9
		71.4
HE-1 in Drying Cocoa Beans	Water	71.4
		65
HE-2 in Hatching Eggs	Water	65
		55
Mixing Water Before Pool	Water	55
		40
Pool	Water	40
		28

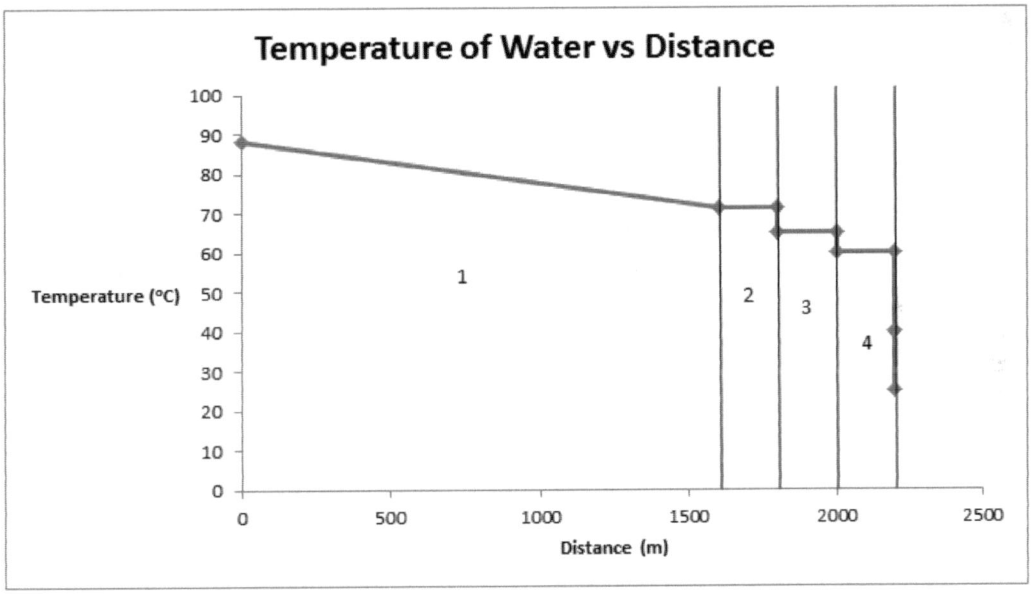

Figure 8. Distribution of temperature of water along the pipeline in cascade system

Where,
1 represents water flow in pipe
2 represents heat exchanger 1 in drying cocoa beans
3 represents heat exchanger 2 in hatching eggs
4 represents hot spring & spa in pool

APPENDIX C

The interface of Macro Excel Program to Calculate Heat Loss in Fluid Flow in Pipe

Heat loss rate of a pipe in an outdoor location

This routine is defined to work for carbon steel pipe dimensions and with water / steam as interior fluid. The pipe is surrounded by an ambient air at a temperature "te" and a clear sky night. This sky condition represents the worst case scenario from heat loss view point.

Input data

1	Pipe nominal diameter	d_n =	4	in	
2	Pipe schedule	Sch =	40	-	
3	Pipe length	L =	1600	m	
4	Pipe absolute rugosity	Rabs =	0.1	mm	
5	Insulation material	Ins :	Mineral Fiber Pipe, C553, Type I & II_23.9 °C to 93.3 °C		
6	Insulation thickness	δ_{ins_mm} =	15	mm	
7	Emissivity of exerior sheet	Emiss =	0.8		
8	Interior fluid	Fluid :	Saturated water		
9	Interior mass flowrate	mKgMin =	120	kg/min	
10	For saturat. steam or water	$p_{satgauge}$ =	1	bar (g)	Note 1
11	For superheated steam or	$p_{UnsatGauge}$ =	1	bar (g)	Note 2
12	subcooled water	t_{Unsat} =	87.9	ºC	
13	Ambient and walls temp.	t_e =	29	°C	
14	Ambient relative humidity	fi =	75	%	
15	Ambient air velocity	vMps =	4.2	m/s	
16	Local height above sea level	Hasl =	243	m.a.s.l.	

Note 1 Saturated steam or Saturated water

If the fluid is "saturated steam" or "saturated water", there is no need to input any value in the cells of "$p_{UnsatGauge}$" and "t_{Unsat}" corresponding to the case of "superheated steam or subcooled water". Any value in those cells will not be considered.

Note 2 Superheated steam or subcooled water

If the fluid is "superheated steam or subcooled water", there is no need to input any value in the cell of "$p_{satgauge}$" corresponding to the case of "saturated steam" or "saturated water". Any value in this cell will not be considered.

Note 3 To run only the array function:

1. Select the range k18:k27
2. Place the cursor at the end of the function

```
=TRANSPOSE(HeatLossOutdoorPipeArray(F11,F12,F13,F14,F15,F16,F17,F18,F19,F20,F21,F22,F23,F24,F25,F26))
```

3. Instead of pressing the Enter-key, press the keys: Ctrl + Shift + Enter

Results using the array function: Module: Outdoor_Pipe_Heat_Loss_Array

Q = HeatLossOutdoorPipeArray(dn_in, Sch, L_m, Rabs_mm, InsMaterial, InsThickness_mm, Emiss, Fluid, m_KgMin, pSatGauge_bar, pUnsatGauge_bar, tUnsat, tAmb, fi_%, v_mps, Ha

hi =	2371.18	W/(m²*K)	(Forced convection inside a pipe with Water/steam, turbulent flow)
he =	23.50	W/(m²*K)	(Combined natural and forced exterior convection coefficient for a horizontal tube)
hrc =	28.30	W/(m²*K)	(Combined radiation convection coefficient)
kpipe =	50.0	W/(m*K)	(Average pipe temperature)
kins =	0.0595	W/(m*K)	(Average insulation temperature)
t1 =	120.25	°C	(Temperature of pipe interior surface)
t2 =	120.20	°C	(Temperature of interface steel pipe - insulation)
t3 =	39.13	°C	(Temperature of insulation exterior surface)
Ue =	3.136	W/(m²*K)	(Global heat transfer coefficient refered to the exterior surface)
Q =	207,925	W	(Heat loss for the pipe of length "L")

Result using the single function: Module: Outdoor_Pipe_Heat_Loss_Q

Q = HeatLossOutdoorPipe(dn_in, Sch, L_m, Rabs_mm, InsMaterial, InsThickness_mm, Emiss, Fluid, m_KgMin, pSatGauge_bar, pUnsatGauge_bar, tUnsat, tAmb, fi_%, v_mps, Hasl_m)

Q = 207,925 W

IOP Conf. Series: Earth and Environmental Science **103** (2017) 012005 doi:10.1088/1755-1315/103/1/012005

Study of sustainable production in two-phase liquid dominated with steam cap underlying brine reservoir by numerical simulation

Heru Berian Pratama[1], **Nenny Miryani Saptadji**[1]

[1]Study Program of Geothermal Engineering, Faculty of Mining and Petroleum Engineering, Bandung Institute of Technology, Indonesia
Jl. Ganesha No.10, Bandung, West Java, Indonesia. 40132

heru.berian@geothermal.itb.ac.id

Abstract. The main issue in the management of the two-phase liquid-dominated geothermal field is rapid decline pressure in the reservoir so that the supply of steam to the power plant cannot be fulfilled. To understanding that, modelling and numerical simulation used reservoir simulators. The model is developed on liquid-dominated geothermal fields are assessed in various scenarios of production strategies (focusing only steam cap, brine reservoir and a combination) and injection strategies (deep and shallow injection, centered and dispersed injection), with the calculation using separated steam cycle method. The simulation results of the model for sustainable production are production 25% from steam cap + 75% from brine reservoir, dispersed and deep reinjection with make-up wells from steam cap results 9 make-up well number. The implementation of production-injection strategy needs to be planned right from the beginning of exploitation so that the strategy can adapt to changes in reservoir characteristics.

1. Introduction

The type of geothermal reservoir that most widely used for power plant is hydrothermal reservoir liquid-dominated. The exploitation of mass and heat from the geothermal fluid will decrease the pressure in the reservoir over time. Thus the pressure drop in the reservoir will have an impact on the formation of boiling zones or boiling will increase. The impacts are an increase in the fraction of steam, dryness, in the reservoir and with good vertical permeability will form a steam cap underlying the brine reservoir.

In general, after steam cap was formed at relatively shallow depths or at the top of reservoir, production will be focused on that area. The drilling point of view, drilling cost, will be lower because the drilling target is at shallow depths. From the aspect number of injection wells, brine from separator decreases as dryness at the wellhead increase thus the number of injection wells will be decrease. Therefore the use of separator becomes less as the increasing of steam flow entering separator. These reasons for economies makes the geothermal developer exploited mass and heat from the steam cap.

Based on the study of two-phase liquid-dominated as described earlier, the steam production from the steam cap can't be offset by the formation of steam cap itself. Therefore, the production of steam will be further decreased and may not sustain to power generation as long as contractual time. Nevertheless the strategy must be balanced with production from liquid reservoir and injection strategies even the steam cap formed. The injection strategies should be optimized so the pressure drop especially in the steam cap or the whole reservoir may be maintained at acceptable point.

The changes of properties in the reservoir can be described by numerical model of geothermal system that has been developed by conceptual model from geosciences data and reservoir engineering data. The first step in numerical simulation is develop a computer model that has a good match with actual data (pressure and temperature), which is a natural state. If the geothermal field hasn't been develop, the model can be running to forecasting with various production-injection strategies to achieve sustainable production.

The numerical simulation researches of two-phase liquid-dominated reservoir are less than both single-phase (liquid) reservoir and vapor-dominated reservoir. These reservoir is interesting and difficult to model because it contains a vapor-dominated underlying liquid-dominated reservoir [1].

Considering the things mentioned above, it is interesting to do a study of model and simulate the two-phase liquid-dominated with steam cap underlying brine reservoir using numerical simulation. These study is based on natural state of reservoir with steam cap underlying brine reservoir [2] and the study of production-injection scenarios at the identical model [3]. By observing changes in pressure, temperature and steam saturation and mass flow in the numerical model as a function of time, therefore these model can forecast the reservoir performance over time with various production-injection strategies.

The objective of comparison the output from the reservoir model for multiple production-injection strategy is to achieve a sustainable production. These strategy is used to maintain installed capacity by adding make-up well. The well to be used as a make-up wells produced mass and heat either from steam cap, brine reservoir or both of them. The objective is to determine a suitable make-up well strategies.

2. Methodology

In general, the methodology in this study is shown in Figure 1. The previous study of synthetic numerical model two-phase liquid-dominated with steam cap underlying brine reservoir for natural state and production-injection strategies have been done [2, 3].

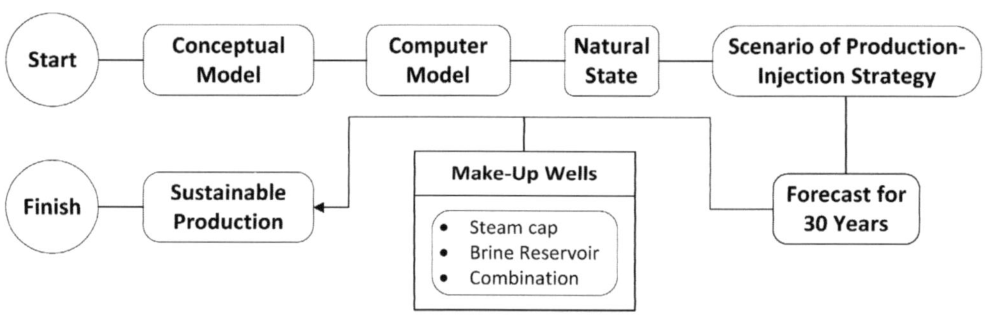

Figure 1. Methodology for sustainable production

The further study is forecast for 30 years of contractual and the next 20 years as prolongation of utilization. The scenario is make-up well from brine reservoir, steam cap or both of them to sustain steam the steam supply for power generation of 220 MW. Furthermore, each strategy will be calculated and analysed to achieve the lowest of decline rate.

3. Conceptual Model

The model of this conceptual is based on liquid dominated geothermal fields that have steam cap at the top of reservoir. The characteristics shown in Table 1. This study used six exploration wells (XXA-1, XXB-1, XXC-1, XXD-1, XXE-1 and XXF-1) and is shown in figure 2. These wells delineate the reservoir. The pressure and temperature profile of XXX-1 and XXB-1 show a high temperature in the center of reservoir and the other wells has lower temperature because at the edge of reservoir.

Table 1. Characteristics of the synthetic model.

Reservoir	Steam cap	Brine Reservoir
Proven Area	13 km^2	23 km^2
Temperature	240°C	240-320°C
Pressure	34 bar	Brine res. = 55 bar
Thickness	500-1000 m	1400-1500 m

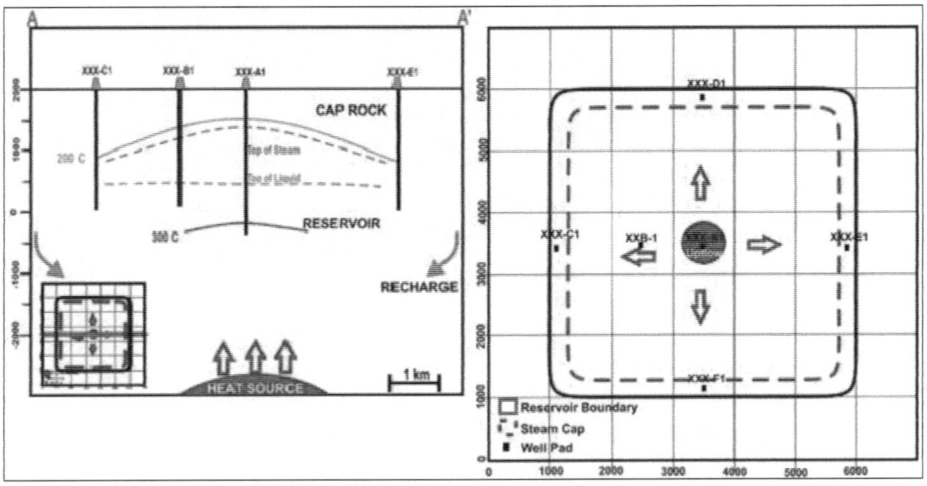

Figure 2. Conceptual model of synthetic reservoir.

4. Natural State

The geothermal system have been assigned in computer model as described in previous study. The properties of the each rocks that represent geological conditions in the computer model used distributed parameter approach (Table 2). The boundary conditions such as outside boundary, heat source and atmosphere was assigned in this model. The objective is give initial conditions to a model therefore the simulator calculate every thermodynamic properties in each blocks.

Table 2. Material data for computer model.

Material Type Legend	Rock Density (kg/m^3)	Porosity	Permeability (m^2)	
			XY	Z
Atmosphere	2600	0.99	1E-10	1E-12
Ground Water	2500	0.02	2E-18	2E-18
Caprock	2600	0.05	1E-18	1E-18
Boundary1	2600	0.001	1E-19	2E-19
Boundary2	2600	0.01	1E-20	1E-20
Heat source	2650	0.07	1E-14	1E-15
Reservoir1	2500	0.25	1E-13	5E-14
Reservoir2	2550	0.2	8E-14	4E-14
Reservoir3	2600	0.15	6E-14	3E-14
Reservoir4	2600	0.15	5E-14	2E-14
Reservoir5	2600	0.1	3E-14	1E-14
Reservoir6	2600	0.1	9E-15	6E-15
Reservoir7	2600	0.1	7E-15	3E-15
Reservoir8	2600	0.09	5E-15	2E-15
Reservoir9	2500	0.05	3E-17	1E-17

The Computer model has 7x7 km2 with reservoir area around 4.5x4.5 km2. Therefore the mesh type applied into a computer model is rectangular and a distributed parameter approach is used. The refine mesh placed in the productive area that have a high temperature and good permeability. The fine mesh has propose of the block is only trough by one well. The fine mesh is 200x200 m and the course is 1000x1000 m and placed on the edge of the computer model as boundary conditions (figure 3).

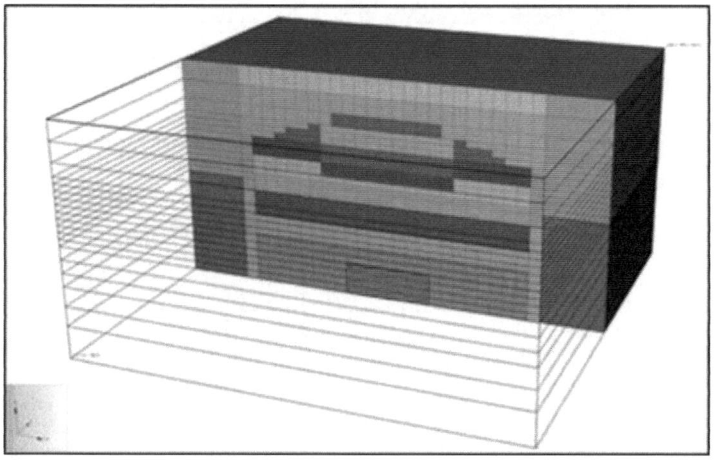

Figure 3. 3D block model at computer model.

The subsurface full of unknown information and it has steam cap underlying brine reservoir therefore the data input is very tricky. The properties in each block should be adjust and these process repeat continues until the reservoir model computer can represent their natural conditions and this process.

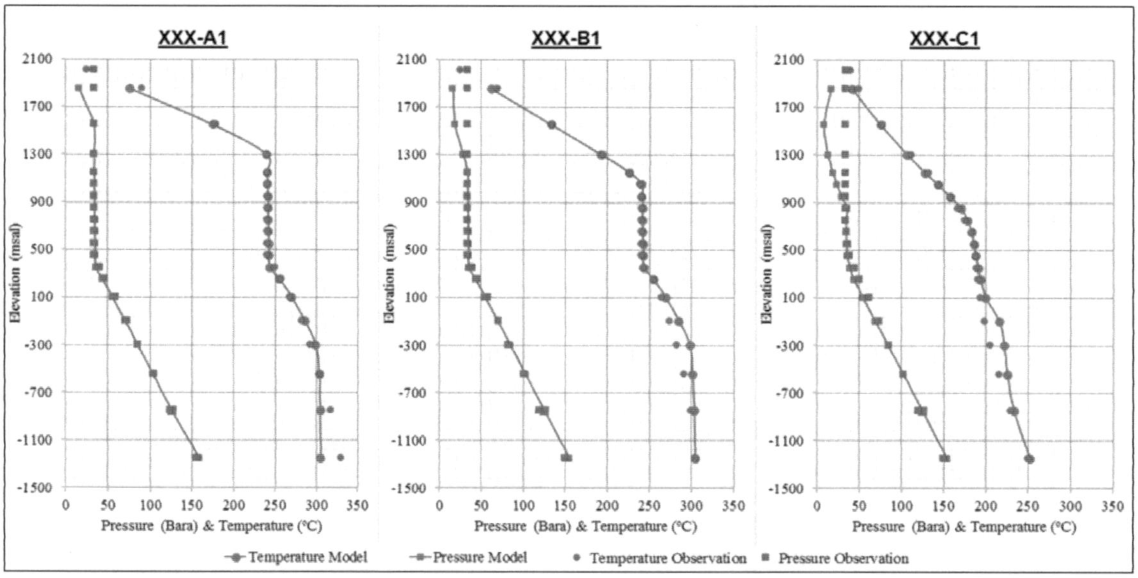

Figure 4. Matching pressure and temperature data between model computer and actual well data.

The output of pressure and temperature from the model shown a perfectly match between actual data (figure 4). The model represent a steam cap underlying brine reservoir. This natural state at steam cap zone has a similarities with conceptual model of vapour dominated proposed by White [4] and enhanced by D'amore and Trusdekk [5] and it is shown in figure 5. The conductive heat transfer occurs from heat source into reservoir and convective entire steam cap reservoir. Steam saturation formed at steam cap zone is 80% and close to value of 85% of vapour dominated geothermal field [6]. These model of have a steam cap covering around 4-23 km2 and thickness around 500-1000 m.

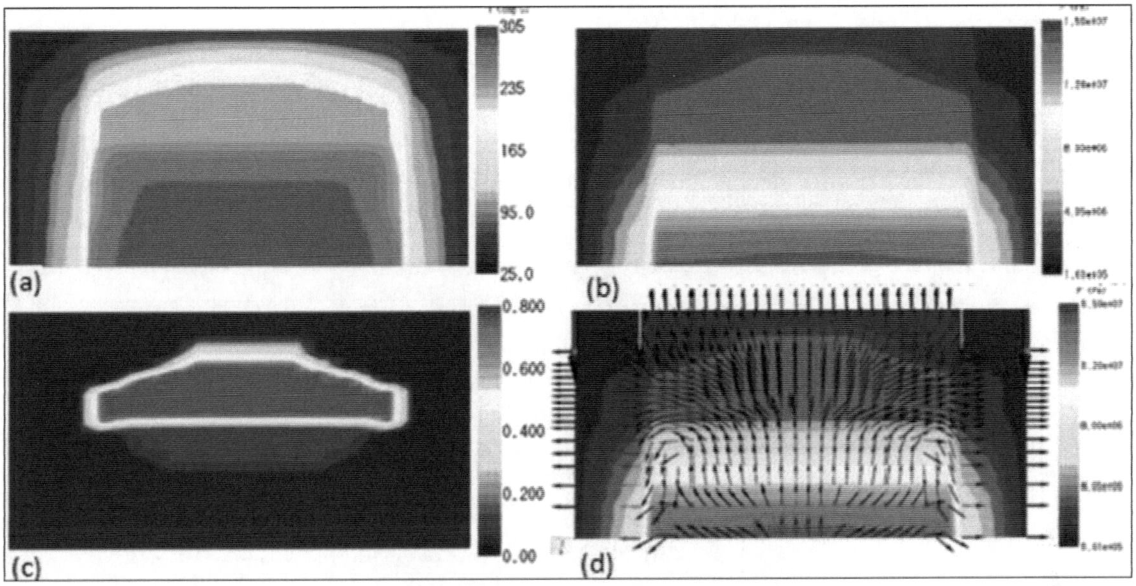

Figure 5. Natural state, (a) temperature, (b) pressure, (c) steam saturation, (d) heat transfer.

5. Forecast

Separated steam cycle has been used in numerical model as previous study (Pratama and Saptadji, 2015) to generated 220 MW. It is described that the study used four production strategies:

1) Production strategy is focused on 100% from steam cap.
2) Production strategy is focused on 100% from brine reservoir.
3) The combination from both 50% of steam cap and 50% of brine reservoir.
4) The combination from both 25% of steam cap and 75% of brine reservoir.

Then the injection used four strategies:

1) Centered injection (single well pad for each brine and condensate)
2) Dispersed injection (multiple well pad for brine injection and single well pad for condensate, both areas are surrounding the reservoir)
3) Shallow injection (both brine separator and condensate injected into the liner of 1100 – 400 msal or 900 – 1600 meter depth).
4) Deep injection (both brine separator and condensate injected into the liner of 300 – (-500) msal or 1700 – 2500 meter depth).

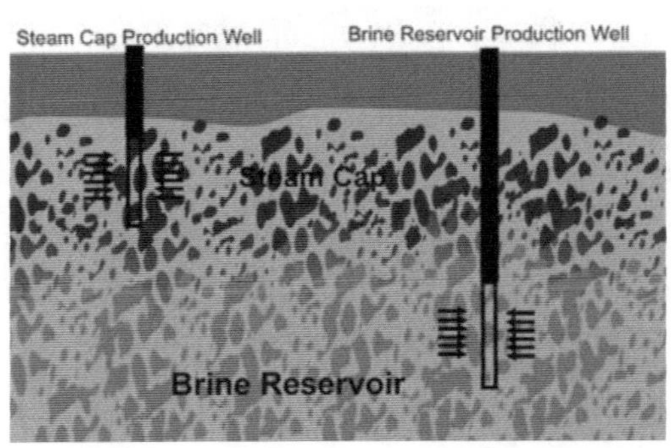

Figure 6. Schematic both of steam cap production well and brine reservoir production well.

The schematic of production heat and mass from steam cap and reservoir shown in figure 6 therefore the placement wells of production and injection shown in figure 7.

Sixteen model and simulation has been carried out and the output compare each other. The numerical model of two-phase liquid-dominated with steam cap underlying brine reservoir has the best production strategy if fluid is produced from 25% of steam cap and 75% of brine reservoir and paired with dispersed and deep injection. Furthermore these scenario of production-injection strategy has the lowest decline in pressure compare to the other scenarios. A higher production rate from brine reservoir will be accelerated the increase of steam saturation that fills into reservoir rocks hence the expansion of steam saturation. Thus it have a rapid decline pressure and increased the boiling, hence the two-phase zone will expand.

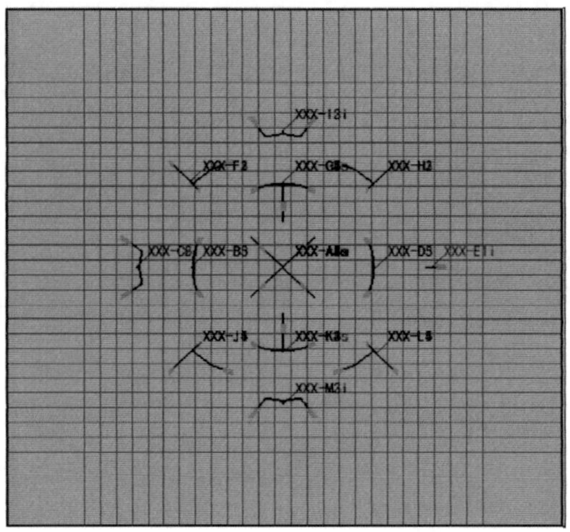

Figure 7. Production and injection placement in the model.

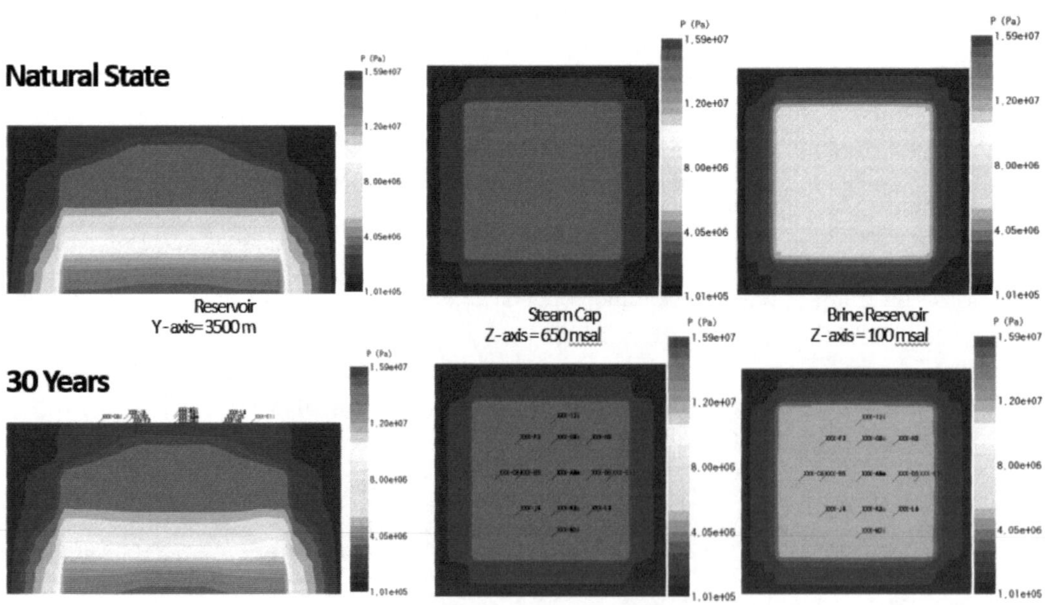

Figure 8. Evolution of reservoir pressure for Production 25% steam cap and 75% brine reservoir with dispersed and deep injection.

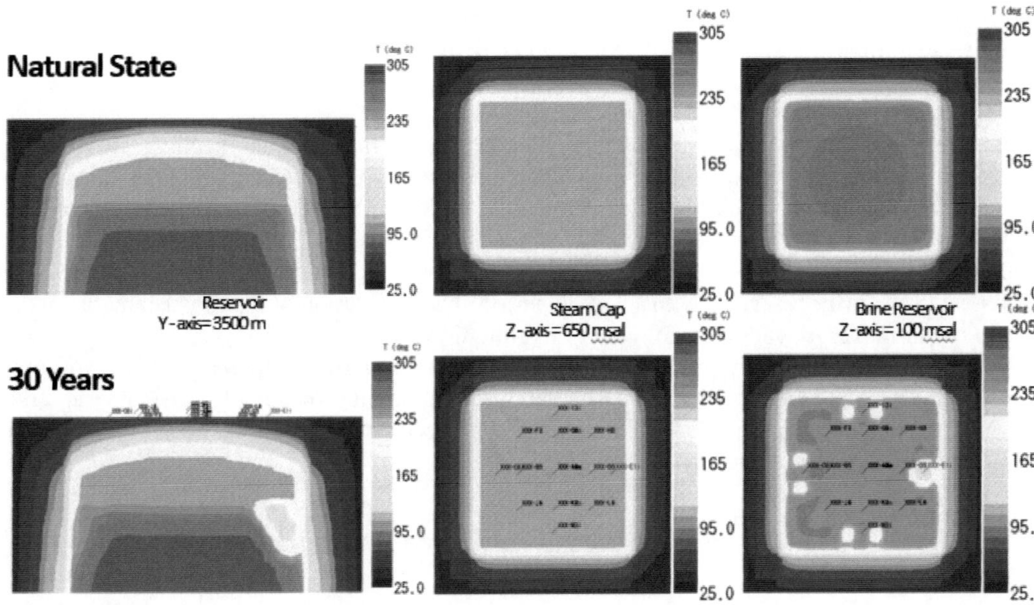

Figure 9. Evolution of reservoir temperature for Production 25% steam cap and 75% brine reservoir with dispersed and deep injection.

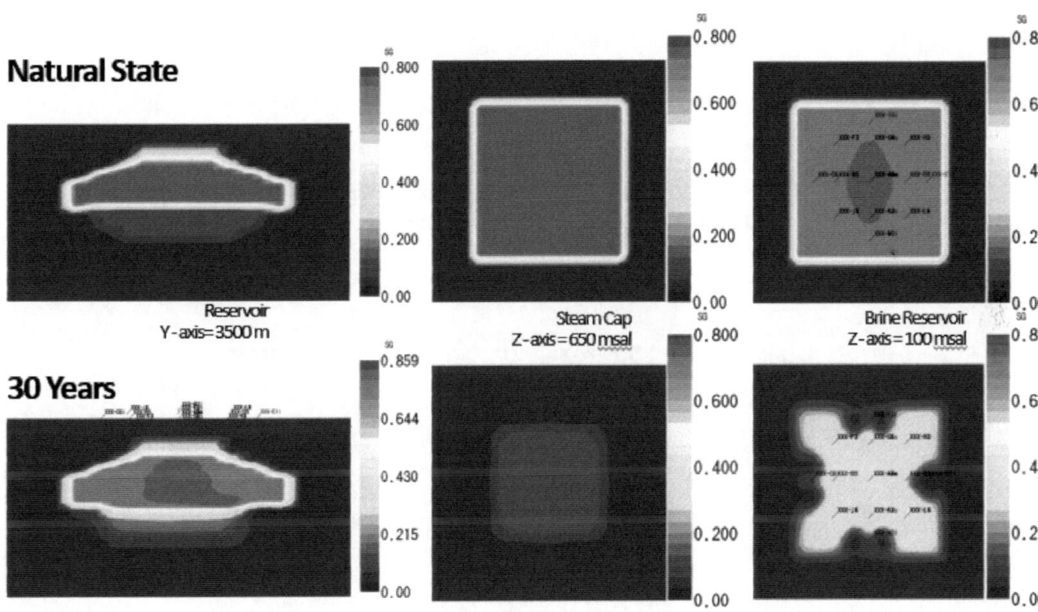

Figure 10. Evolution of reservoir steam saturation for Production 25% steam cap and 75% brine reservoir with dispersed and deep injection.

The evolution of reservoir shown in figure 8, figure 9 and figure 10 for pressure, temperature reservoir and gas saturation respectevely. Dispersed and deep injection generate a relatively lower of pressure drop both in steam cap and brine reservoir. This is because the reinjection of brine separator performed spreading with well pad surrounding the production reservoir. Therefore, this strategy give additional pressure uniformly for balancing the production induced pressure drop and maintain the pressure drop at a lowest level as possibly. While the deep injection strategy improve a thermal recovery because both of brine separator and steam condensate injected at deeper reservoir which have a higher temperature and minimize a cooler reinjected fluid back to production reservoir [3].

6. Sustainability

A decrease in mass flow rate of the geothermal reservoir occurs naturally as injection rate is lower than production rate. Furthermore it caused pressure drop in the reservoir and decline in flow rate. Therefore, in order to maintain the steam supply to power plant is required to add make-up wells. There are three strategy of make-up wells; all make-up wells from steam cap, all make-up wells from brine reservoir and the combination of them. The make-up wells combination is done by opening the make- up wells from steam cap and brine reservoir with a certain period. The simulation output of sustainable prediction is the number of make-up wells over 30 year of exploitation.

The simulation of sustainable production has done to get a production-injection strategy to maintain the steam supply to 220 MW power plant. The production strategy of 25% from steam cap and 75% from brine reservoir compared to 50% steam cap and 75% brine reservoir and both of them paired with dispersed and deep injection strategy.

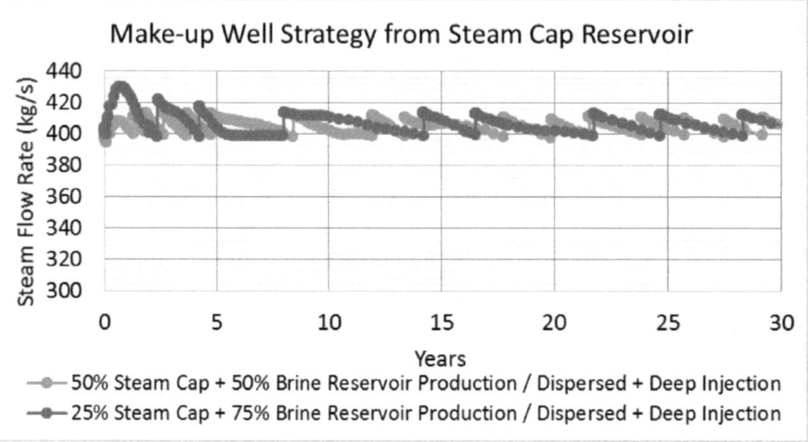

Figure 11. Strategy make-up wells from the steam cap.

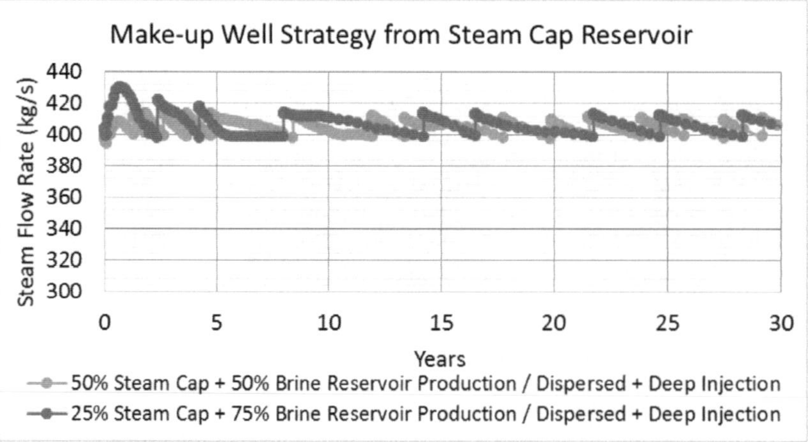

Figure 12. Strategy make-up wells of the brine reservoir.

The graph in figure 11, figure 12 and figure 13 shown the campaign of make-up wells drilling both for early production strategies and make-up wells to sustain 220 MW. The outcome of both scenarios for sustainable production, as mentioned above, production of 25% steam cap has a lowest number of make-up than production of 50% steam cap. Furthermore the best makeup wells is all make-up wells from the steam cap because it has a lowest number of well compare with make-up strategy from all brine reservoir and the combination.

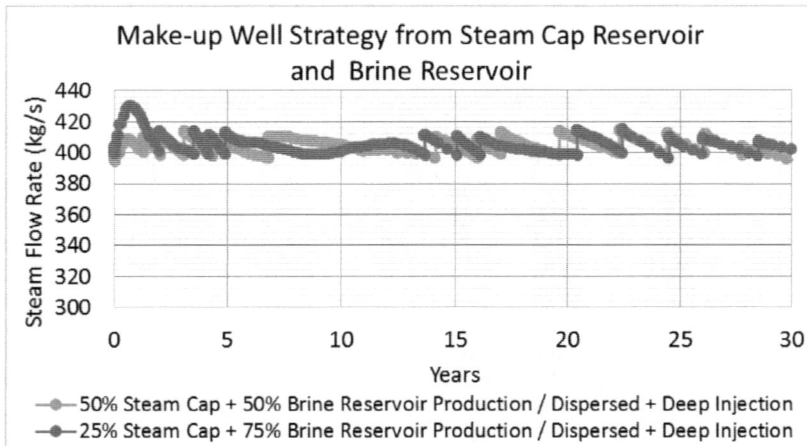

Figure 13. Strategy make-up wells of a combination from steam and brine reservoir cap.

Make-up form steam cap is the best scenario for sustain production over 30 year of exploitation nevertheless it is only visible for production from 25% of steam cap and 75% brine reservoir with dispersed and deep injection. It required 8 make-up wells compare to 15 make-up wells from early production 50% steam cap and 50% brine reservoir with identical injection strategy (figure 14 and figure 15). The higher production from brine reservoir induced pressure drop at deep region therefore increasing boiling process so the dryness increase as a boiling area extended and or thickening. It will supply steam to the steam cap region. These process is the opposite if the main production come from brine reservoir. Make-up well from brine reservoir show a slightly similar results from production 25% steam cap compare to 50% steam cap. It required make-up 12 and 13 wells for 25% steam cap and 50% steam cap respectively. The combination make-up well strategy shown a similar result with make-up wells from brine reservoir.

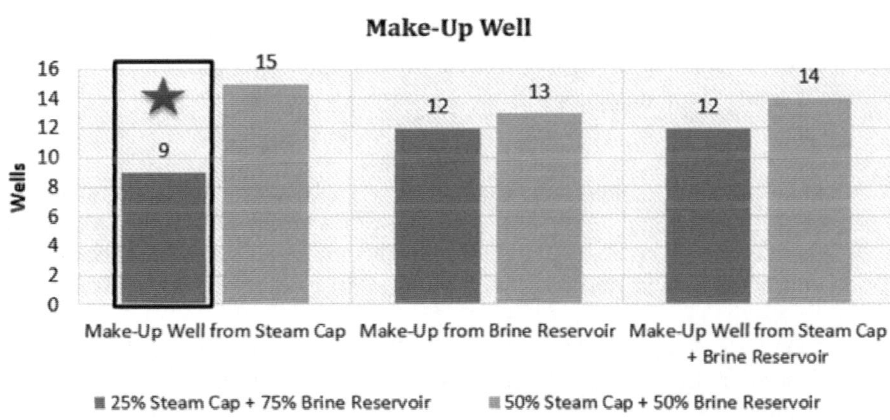

Figure 14. The requirement number of make-up well for each production and injection strategies.

In early production, it is recommended that produced heat and mass from deep reservoir or brine reservoir thus the boiling increase as the pressure in the brine reservoir decreasing. It caused the increasing dryness at boiling area as boiling process become greater. Therefore the boiling area will be expand and thickening. Nevertheless the production from steam cap becoming more economical as the steam from these area is supported by the greater boiling zone.

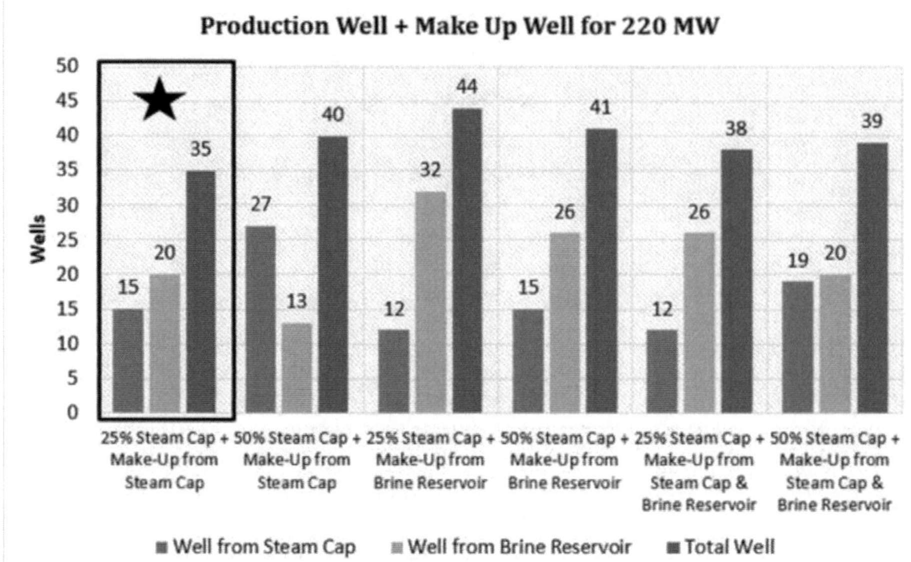

Figure 15. The total production wells and injection wells for each production and injection strategies.

7. Conclusion

1. The sustainable production is the production strategy from 25% steam cap and 75% brine reservoir combined with dispersed and deep injection.
2. The optimized strategy of make-up well is make-up all from steam cap with 9 wells for 30 years of exploration.
3. Total wells for 30 years of sustainable production are 35 wells with 15 wells from steam cap and 20 years from brine reservoir.

Acknowledgment

This research was supported by Beasiswa Unggulan Program, KEMENDIKBUD (Biro Kerjasama Luar Negeri). We thank our colleagues from Institut Teknologi Bandung (ITB) who provided insight and expertise that greatly assisted the research.

References

[1] O'Sullivan, M.J., Karsten, P., Lippmann, M.J., 2000, Geothermal Reservoir Simulation: The State-of-Practice and Emerging Trends, World Geothermal Congress 2000, Kyusu-Tohoku, Japan.

[2] Pratama, H.B. and Saptadji, N.M., (2016), Numerical Simulation for Natural State of Two-Phase Liquid Dominated Geothermal Reservoir with Steam Cap Underlying Brine Reservoir, 5th IIGW, Bandung, Indoensia.

[3] Pratama, H.B. and Saptadji, (2015), Study of Production-Injection Strategies of Synthetic Geothermal Reservoir Liquid-Dominated Model with Numerical Simulation, 37th NZGW, Rotorua, New Zeland.

[4] White, D.E., Muffler, L.J.P., Truesdell, A.H., 1971. Vapor-dominated hydrothermal systems compared with hot-water systems. Econ. Geol.

[5] D'Amore, F., Truesdell, A.H., 1979, Models for steam chemistry at Larderello and The Gesyers. Proceedings, 4th Workshop on Geothermal Reservoir Engineering, Stanford University.

[6] Grant, M.A., Donaldson I.G., Bixley P.F (1st edition 1982, 2nd edition 2011): Geothermal Reservoir Engineering, Academic Press, Oxford.

Basic overview towards the assessment of landslide and subsidence risks along a geothermal pipeline network

Astisiasari[1], **Cees Van Westen**[1], **Victor Jetten**[1], **Freek van der Meer**[1], **Dyah Rahmawati Hizbaron**[2]

[1]Faculty of Geo-Information Science and Earth Observation, University of Twente, Enschede, the Netherlands

[2]Faculty of Geography and Environmental Science, University of Gadjah Mada (UGM), Yogyakarta, Indonesia

a.astisiasari@utwente.nl

Abstract. An operating geothermal power plant consists of installation units that work systematically in a network. The pipeline network connects various engineering structures, e.g. well pads, separator, scrubber, and power station, in the process of transferring geothermal fluids to generate electricity. Besides, a pipeline infrastructure also delivers the brine back to earth, through the injection well-pads. Despite of its important functions, a geothermal pipeline may bear a threat to its vicinity through a pipeline failure. The pipeline can be impacted by perilous events like landslides, earthquakes, and subsidence. The pipeline failure itself may relate to physical deterioration over time, e.g. due to corrosion and fatigue. The geothermal reservoirs are usually located in mountainous areas that are associated with steep slopes, complex geology, and weathered soil. Geothermal areas record a noteworthy number of disasters, especially due to landslide and subsidence. Therefore, a proper multi-risk assessment along the geothermal pipeline is required, particularly for these two types of hazard. This is also to mention that the impact on human fatality and injury is not presently discussed here. This paper aims to give a basic overview on the existing approaches for the assessment of multi-risk assessment along geothermal pipelines. It delivers basic principles on the analysis of risks and its contributing variables, in order to model the loss consequences. By considering the loss consequences, as well as the alternatives for mitigation measures, the environmental safety in geothermal working area could be enforced.

Keywords: landslide, subsidence, geothermal, pipelines, disaster, risk, assessment

1. Introduction

Geothermal energy has long been known as a renewable energy resource with the environmental friendly benefits. However, exploiting geothermal energy could also bring a considerable consequence to the environment without implementing a proper risk management plan. Some environmental challenges that are common are: pollution (thermal, noise), wildlife disturbance, and geological hazards (seismic, landslide, subsidence) (DiPippo, 2012a). Moreover, some geothermal fields show a noteworthy number of disasters, especially landslides and subsidence that may bring disastrous consequences (see Table 1).

In a geothermal working area, the pipeline transfers geothermal fluids (steam, liquid, or mixture) from well-pads to the power plant, in the process of generating electricity. However, this essential function may pose a threat to its vicinity, manifested in a pipeline failure. Or else, the infrastructure may be impacted from disasters that may cause pipeline collateral damage. As a primary hazard, pipelines may deteriorate over time and cause failure, manifested in leakage, fracture, burst, etc. While as a secondary hazard, the disaster-induced threat (e.g. landslides, earthquakes, flash-floods, and subsidence) could also generate a pipeline failure.

As a prime infrastructure, geothermal pipelines therefore require an assessment of multi-risk that specifically addresses the risk analysis and its contributing variables, to model the consequences. This multi-risk assessment conveys the possible impact that presents not only the probability of spatial-temporal occurrences, but also the consequence to monetary loss. Besides, risk assessment also encompasses the alternative decisions or strategies for handling the risks.

The importance of multi-risk assessment is within the scope of risk management framework, as proposed by UNISDR (2009); that is to purposively consider the perceived risk, and therefore could enforce the environmental safety. It provides risk information that would be required for selecting mitigation measures, e.g.: site development planning (pipe route, structure construction), operational measures (monitoring, preparedness, insurance, safety standard, regulation), and structural mitigation (slope reinforcement, drainage) (DNV & Muhlbauer, 2012; ENV-1, C-CORE, D.G. Honegger Consulting, & SSD, 2009).

Table 1. Landslide and Subsidence Incidents in Geothermal Working Areas

Landslide	Subsidence
Kamchatka, Russia June, 2007 16.3 million m^3 volume (Gvozdeva, Frolova, & Zerkal, 2015)	Wairakei, New Zealand (1956 – 1971) Subsidence rate: 10 m/yr over 30 km^2 area (Bromley, Currie, Jolly, & Mannington, 2015)
Sungai Penuh – Province of Jambi (January, 2013) Exploration drilling site 4 died, 5 injured, 1 missing http://www.thejakartapost.com/news/2013/01/28/west-sumatra-landslide-leaves-7-dead.html	California, US (1991 – 1994) Subsidence rate: 16 – 18 mm/yr, (Massonet, Holzer, & Vadon, 1997)
Leyte, Philippines (March 1st, 2013) Landslide and sulfuric fumes 5 dead, 9 missing, 31 injured http://www.rappler.com/nation/22868-5-dead-in-lopez-s-leyte-power-plant-incident	Salak – Province of Jawa Barat (1994 – 2008) 16 cm subsidence in 15 yrs (Nordquist, Acuña, & Stimac, 2010)
Wayang Windu – Province of Jawa Barat (May, 2015) 4 died, 14 buried, 55 evacuated http://jurnaljakarta.com/berita-4460-pipa-geothermal-di-pangalengan-meledak-akibat-tanah-longsor.html	Wayang Windu – Province of Jawa Barat (1995 – 2012) 8 cm subsidence in 17 yrs (Masri, Barton, Hartley, & Ramadhan, 2015)
Hululais – Province of Bengkulu (April 2016) Exploration drilling site 1 died, 4 buried, 4 injured http://en.tempo.co/read/news/2016/04/28/307766683/Landslide-in-Bengkulu-Geothermal-Site-Kills-1-Buries-4	Kamojang – Province of Jawa Barat (2006 – 2007) Subsidence at 6cm (Zacnudin, Kadir, Santoso, Abdassah, & Kamah, 2010)

2. Components of the Disaster Risk Assessment

The risk-related terminology is described by UNISDR (2009). Disaster is the adverse disruption to the set of elements (physical, social, and environmental), once exposed by the hazard occurrence. While risk quantifies the probability of adverse consequences due to the exposure to hazardous events. UNISDR describes disaster risk analysis as a method to estimate the probability and consequence of risk, by analysing the component variables of hazard, vulnerability, and coping capacity. Moreover, risk assessment also evaluates the alternative decisions or strategies, related to the estimated risks (UNISDR, 2009; van Westen et al., 2011).

In multi-risk assessment, the loss consequence of several hazard types are estimated through scenarios of probable occurrences from each single hazard over periods of time. To this account, the

IOP Conf. Series: Earth and Environmental Science **103** (2017) 012006 doi:10.1088/1755-1315/103/1/012006

risk could be presented into three basic aspects, that are (See Figure 1) (Dezfuli et al., 2011; DNV & Muhlbauer, 2012):

- Scenario → a set of adverse situations (damage, failure, destruction, fatality, injury, pollution), resulting in the exposure of assets. These scenarios are used to quantify the consequences for different events.
- Likelihood → the probability appraisal on how likely each scenario may occur. It expresses the frequency in terms of probability or returning period, using historical occurrence data
- Consequence → the quantified impact from each scenario that is implied in loss. The cost of the loss could be monetarized into direct and indirect cost (Doro-on, 2014; Muhlbauer, 2004):
  - Direct cost, e.g.: damages to property, fatality and injury, environmental remediation, profit loss, repair cost, market share disruption, fine and penalty
  - Indirect cost, e.g.: customer discontent, litigation ramification, political reaction, downstream business disruption (agriculture production)

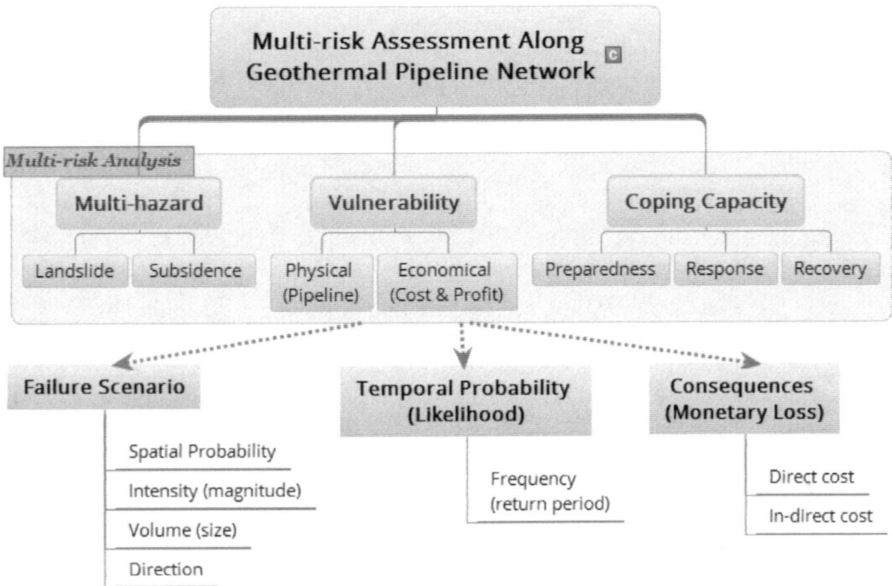

Figure 1. Components of Disaster Risk Assessment

To depict the monetary value of risk, risk curve is used to plot the probability of the scenario against its estimated loss. Figure 2 illustrates the loss values that are represented by y-axis as the consequence value (monetary loss), and x-axis as the temporal-occurrence probability. The area under the curve presents the estimated loss value of total risk (van Westen et al., 2011).

Disaster reduction measures are needed to minimize and reduce the monetary loss. To evaluate the effectiveness of mitigation measures, Cost-Benefit Analysis (CBA) is applied. CBA conveys the beneficial appraisal of financial investment on risk reduction measures, e.g.: structural mitigation, monitoring instrument, and warning system. These risk reduction measures could also apply as a variable of coping capacity.

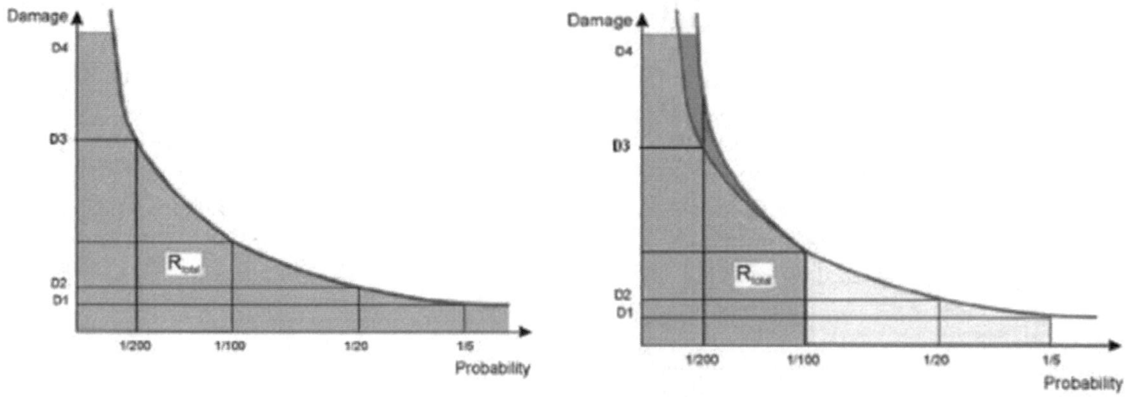

Figure 2. Risk Curve of Monetary Loss (Blue Area), as the Function of Damage Loss over Probability Occurrence: without Risk Mitigation (Left), with Risk Reduction Investment (Right). Source (van Westen et al., 2011)

3. Approaches for the Assessment of Landslide and Subsidence Hazards

3.1. Landslide

Landslide is the phenomenon of soil or rock movement downslope, generated by the force of gravity, which may occur in areas with susceptible slopes, geology, and other precursor events (e.g. earthquake, extreme precipitation) (DiPippo, 2012b; Highland & Bobrowsky, 2008).

Landslide in the geothermal fields occur more frequently due to the presence of hydrothermal alteration zones, and highly weathered materials (Huang & Tian, 2006). This hydrothermal alteration may produce thick soil; and with weathering effects could contribute to slope instability (DiPippo, 2012b).

Many approaches are available to assess landslide susceptibilities in qualitative and quantitative ways, that can be seen from Table 2 (ENV-1 et al., 2009; Jaiswal et al., 2010; Pioquinto & Caranto, 2005; Rex, Devin, & Edwin, 2008):

Table 2. Approaches for Landslide Susceptibility

Methods	Approaches
Qualitative	Geomorphological Analysis • Landscape observation (field or remote sensing technique) • Relatively implicit and rapid • Expert determination
	Variable Weighting & Parameter Scoring • Indexing the contributing factors through criteria matrix • Tools, e.g. SMCE • Expert determination
Quantitative	Statistical Analysis • Parameters comparison for hazard distribution or density, • Univariate or Bivariate (comparing each parameter influence separately), • Multivariate (comparing parameters concurrently. Correlation coefficient and degree of significance are analysed)
	Geotechnical Modelling (Deterministic / Probabilistic) • E.g. slope stability, hydrological analysis • Scenarios of probable events • Mathematically model the driving force (e.g. saturation level, weathered ground, pore pressure, and aquifer level, rainfall, etc.)

Source (ENV-1 et al., 2009; Jaiswal et al., 2010; Pioquinto & Caranto, 2005; Rex et al., 2008)

3.2. Subsidence

Land subsidence is the deformation phenomenon of surface sinking, that may happen due to natural causes (e.g. subsurface erosion, magma process) or by human activity (mining, groundwater or gas extraction) (ENV-1 et al., 2009; Highland & Bobrowsky, 2008; Werner & Friedman, 2010)

Fluid pressure supports the overburden cap soil. Subsidence is likely to happen in the formations with lithostatic fluid pressure, than with hydrostatic fluid pressure. The fluid withdrawal greater than recharge can decrease the pressure. Subsidence can take place in poorly compacted soils like thermal clays; and less frequent in competent formations (where permeability of fracture dominates) (DiPippo, 2012b).

Some approaches to assess subsidence susceptibility are by field reconnaissance of ground displacement, e.g. through:

- Feature and geological observation (sinkhole, fault, wells, submergence),
- Ground-based levelling (geodetic measurement),
- Geophysical or geotechnical analysis for gravity measurement;
- Remote sensing technique by radar or Lidar interpretation

4. Geothermal Pipeline Impact Analysis (Pipeline Failure Probability)

Pipelines may experience disruption or failure due to extreme stress or impact, e.g. mass movement or subsidence. Subsequently, the pipeline failure may also cause environmental consequences due to hazardous material release that may contain flammable, toxic, corrosive, or carcinogenic substances (see Figure 3).

This pipeline damage or failure can be categorized as: leakage, spill, release, crack, breakage, fracture, blockage, rupture, collapse, or burst. Based on the analysis of historical failures, there are four types of pipeline failure mechanisms, that are caused by: (1) third party or external force, like ground movement; (2) corrosivity from material product, atmosphere, and subsurface; (3) pipeline structural design, related to construction, fatigue, integrity; (4) incorrect operation or human error (Doro-on, 2014; EGIG, 2015; Muhlbauer, 2004)

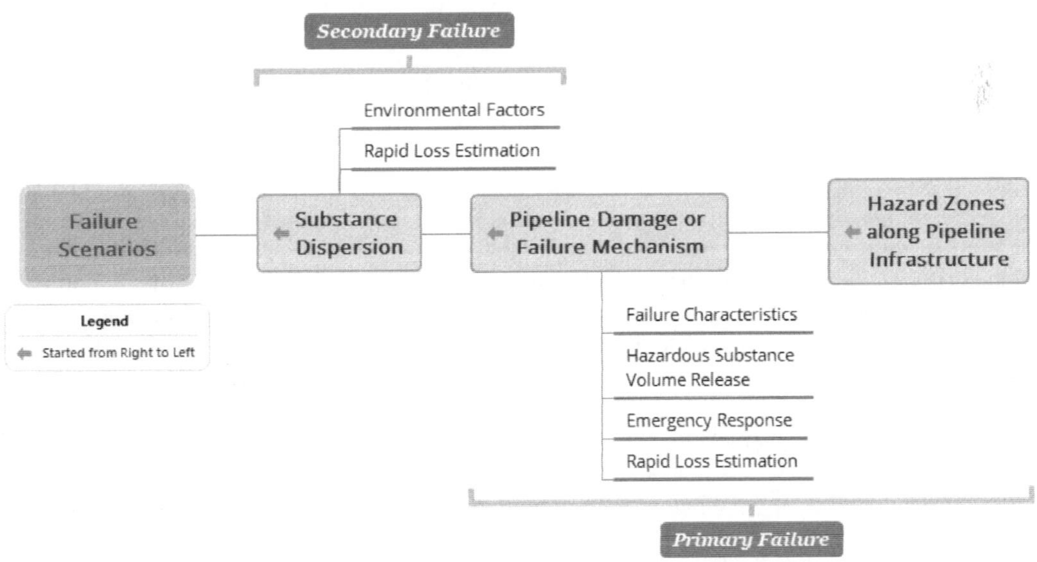

Figure 3. Event Tree for Disaster Induced Pipeline Failure

Therefore, to develop the scenarios for pipeline failure due to the disaster impact, one needs to take into consideration the variables of: (1) pipeline physical characteristic, that is for the resistance and

IOP Conf. Series: Earth and Environmental Science **103** (2017) 012006 doi:10.1088/1755-1315/103/1/012006

survivability of pipeline from the exposure event; (2) exposure, as in the severity and frequency of events over times; (3) mitigation, that is for the effectiveness percentage of defence measure (DNV & Muhlbauer, 2012; ENV-1 et al., 2009).

4.1. Primary Failure

The physical pipeline characteristics can be used as a base for the analysis of element at risk. Pipeline engineering principle on its structural design is associated with the pipeline strength, toughness, safety, integrity, and value; in response to the abnormal stress from disaster exposure. Therefore, pipeline physical characteristics can account as a basic consideration for failure event scenarios. Table 3 shows the main physical characteristics of pipeline and its parameters (Liqiong Chen et al., 2014; DNV, 2010; EGIG, 2015; ENV-1 et al., 2009; Miranda & Lopéz, 2011; Muhlbauer, 2004).

Table 3. Physical Characteristics of Pipeline for the Element at Risk Analysis

Physical Characteristic	Parameters
Construction Design	• Material and coating specifications; • Pipe support: surface, buried, hung, elevated, berm, culvert, etc; • Pipe-bends or flexibility: umbilical, curving angle; • Seamless, joints, submerged, electric resistance; • Utility or component support: valves, fittings, anchor, tank, vessel, pump • Design code for safety standard and tolerance
Dimension size	• Diameter; • Wall thickness; • Length
Load content	• Temperatures & Pressure; • Product type (gas, liquid, combination); • Flow rate and volume (flow rate by time) • Density and viscosity • Solubility • Hazardous type (ignitability, reactivity, toxicity, corrosivity, carcinogenicity)
Route and layout	• Location of separators, power plant, pipes; • Slope, elevation, land-cover, landowner, soil type
Construction, Support load, and Technical structure	• Construction Year • Joints or connectors, swan neck, • Expansion loops, anchors, stopper, elbow section, valves; • Weight support, shock absorber; • Ground allowance level: stress & seismic, leakage & break; • Corrosion allowance level • Minimum yield strength (MPa)

There is always a certain level of uncertainty on how pipelines behave or respond, once subjected to extreme conditions or abnormal stress. To this account, interaction between soil and pipelines is considered in the soil spring stiffness analysis. It presents soil loads direction acts towards the pipeline resistance (ENV-1 et al., 2009). Therefore, besides the pipeline physical characteristic, soil properties should also be analysed. This geotechnical analysis should model soil spring stiffness through soil variables i.e.: friction, cohesion, saturation, and permeability.

4.2. Secondary Failure

The dispersion of hazardous material release may contaminate the environment. The spatial impact to the environment may differ, based on: (1) failure characteristics (type, size, volume released); (2) environmental conditions (weather, morphology, geology, soil); (3) preparedness and response (detection and warning system).

Generally, the environmental effects from pipeline failure can be characterized from several proxies, e.g.: infiltration, drainage, groundwater, weather, landcover, and slope. These can be used to determine hazard zones that are in approximate distance from the possible occurrence of pipeline failure.

Hazard zones along the pipeline network may have different characteristics in terms of pipeline structure and environmental characteristics. Therefore, grouping the pipeline into segments is a practical way to assess the risk. Basically, the segmentation could be done by: (1) fixed-length, or (2) dynamic segmentation on certain criteria or parameters. The hazard susceptibility analysis can then identify the hazard density for each pipeline segments (See Figure 4).

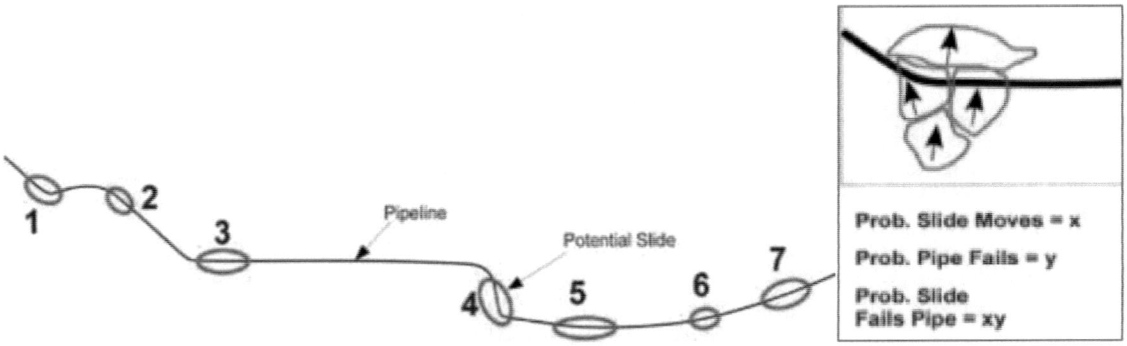

Figure 4. Pipeline Dynamic Segmentation Based on Hazard Susceptibility Zone
Source (ENV-1 et al., 2009)

4.3. Likelihood (Temporal Probability) of Pipeline Failure

After the susceptible areas are identified, pipeline failure frequency is analysed. It is the temporal probability appraisal on how each failure scenario is likely to occur. The likelihood requires a set of parameters of historical occurrence data to be associated with pipeline physical characteristics and mitigation measures. The linkage correlations between these likelihood parameters are analysed statistically (See Figure 5) (ENV-1 et al., 2009)

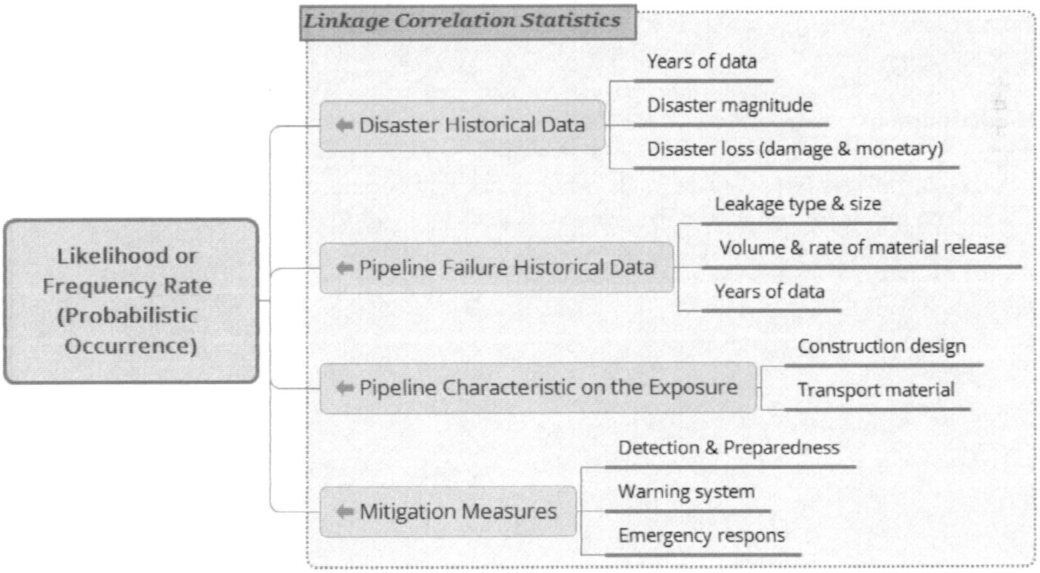

Figure 5. Determining the Likelihood for the Probable Occurrence Scenario through Parameters

5. An Approach for the Assessment of Landslide and Subsidence Risks Along the Geothermal Pipeline

The basic overview of an approach towards the multi-risk assessment along the geothermal pipelines could be seen through the general conceptual framework from Figure 1. It basically combines the predictive risk scenarios from different types of disaster (See Figure 6) (van Westen et al., 2010; W. Marzocchi et al., 2009):

A. Data preparation for each variable (hazards, vulnerabilities and elements at risk, coping capacity).
B. Hazard assessment to identify the predictive scenario of:
   a. Susceptible hazards location
   b. Temporal probability, return period (frequency of failure occurrence),
   c. Hazard characteristics: magnitude and intensity, volume (size), change and direction
C. Vulnerability assessment of pipeline element at risk
   a. Pipeline physical characteristics
   b. Pipeline economical value
   c. Pipeline failure characteristics: hazardous substance release, failure type and size
D. Coping capacity, as measures for the disaster risk reduction (preparedness, response, and recovery)
E. Risk assessment, by integrating variables spatially to assess the probabilistic consequence at respective return periods (See Figure 6)
   a. Loss scenario (monetary value) to pipeline and possibly to environment
   b. Acceptance level of acceptable risk

6. Discussion

The assessment of multi-risk along the geothermal pipelines is still not sufficiently studied and understood. There is a need to develop new methods to address the disaster risk assessment in geothermal fields; that is taking into account the hazard, vulnerability, and capacity analysis for enforcing the environmental safety. Therefore, it is important to find the best practical method to model the probable scenarios in terms of spatial-temporal probability of pipeline failure and possible monetary loss.

Acknowledgement

This paper is a part of a joint Ph.D. research program between University of Twente – The Netherlands and University of Gadjah Mada – Indonesia, that is funded by the GEOCAP program and LPDP Scholarship, for the research under the Work Package 2.08 that is related to the environmental impact in a spatial-temporal context.

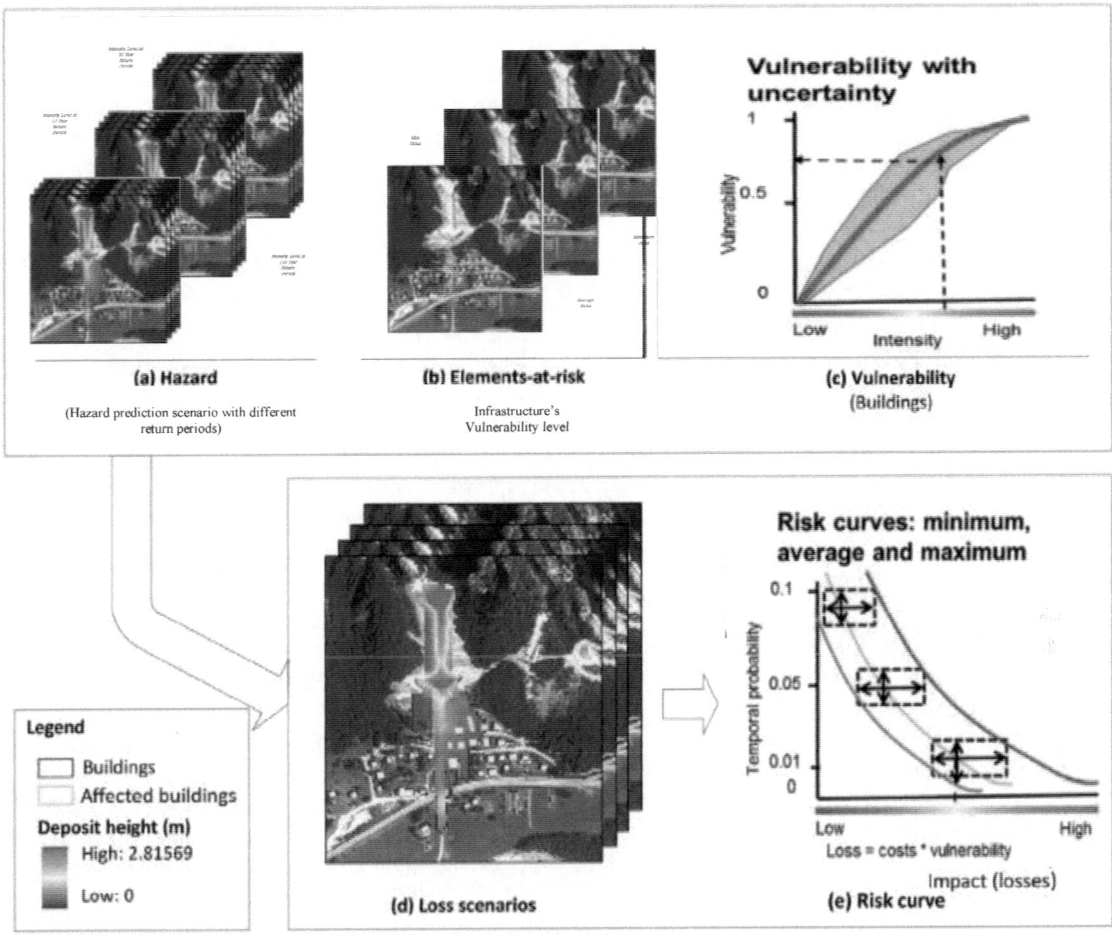

Figure 6. Approach for the Assessment of Multi-risk, Presented as scenario and its loss value by Respective Returning Periods. Source (Aye et al., 2016; Chen et al., 2016)

References

[1] Aye, Z. C., Jaboyedoff, M., Derron, M. H., Van Westen, C. J., Hussin, H. Y., Ciurean, R. L., … Pasuto, A. (2016). An interactive web-GIS tool for risk analysis: A case study in the Fella River basin, Italy. *Natural Hazards and Earth System Sciences*, *16*(1), 85–101. https://doi.org/10.5194/nhess-16-85-2016

[2] Bromley, C. J., Currie, S., Jolly, S., & Mannington, W. (2015). Subsidence: an Update on New Zealand Geothermal Deformation Observations and Mechanisms. In *World Geothermal Congress*.

[3] Chen, L., van Westen, C. J., Hussin, H., Ciurean, R. L., Turkington, T., Chavarro-Rincon, D., & Shrestha, D. P. (2016). Integrating expert opinion with modelling for quantitative multi-hazard risk assessment in the Eastern Italian Alps. *Geomorphology*, *273*, 150–167. https://doi.org/10.1016/j.geomorph.2016.07.041

[4] Chen, L., Wu, S., Lu, H., Huang, K., Lv, Y., & Wu, J. (2014). Stress Analysis of Buried Gas Pipeline Traversing Sliding Mass, (1), 257–261. https://doi.org/10.2174/1874149501408010257

[5] Dezfuli, H., Benjamin, A., Everett, C., Maggio, G., Stamatelatos, M. G., & Youngblood, R. (2011). *NASA Risk Management Handbook* (1.0). Washington, D.C.

[6] DiPippo, R. (2012a). *Geothermal Power Plants : Principles , Applications , Case Studies and Environmental Impact Third Edition*. https://doi.org/10.1016/B978-0-08-098206-9.00024-5

[7] DiPippo, R. (2012b). *Geothermal Power Plants: Principles, Applications, Case Studies and Environmental Impact. Geothermal Power Plants* (3rd Editio). Elsevier. Retrieved from http://www.sciencedirect.com/science/article/pii/B9780750686204500186%5Cnhttp://www.sciencedirect.com/science/article/pii/B9780750686204500149%5Cnhttp://www.sciencedirect.com/science/article/pii/B9781856174749500388%5Cnhttp://www.sciencedirect.com/science

[8] DNV. (2010). Risk Assessment of Pipeline Protection. *Recommended Practice DNV-RP-F107.*

[9] DNV, & Muhlbauer, W. K. (2012). Pipeline Risk Assessment: the Essential Elements. *Pipeline and Gaz Journal*, 7.

[10] Doro-on, A. M. (2014). *Risk assessment and Security for Pipelines, Tunnels, and Underground Rail and Transit Operations*. Taylor and Francis Group.

[11] EGIG. (2015). *Gas Pipeline Incidents (9th Report of the European Gas Pipeline Incident Data Group, Period 1970 - 2013)*. Groningen. Retrieved from http://www.egig.eu

[12] ENV-1, C-CORE, D.G. Honegger Consulting, & SSD, I. (2009). *Guidelines for Constructing Natural Gas and Liquid Hydrocarbon Pipelines Through Areas Prone to Landslide and Subsidence Hazards. Design, Materials, and Construction Committee of Pipeline Research Council International, Inc.* Retrieved from http://citeseerx.ist.psu.edu/viewdoc/download?doi=10.1.1.250.5780&rep=rep1&type=pdf

[13] Gvozdeva, I. P., Frolova, J. V., & Zerkal, O. V. (2015). Slope Processes Hazards in Geothermal Areas: A Case Study of the Geysers Valley, Kamchatka. *World Geothermal Congress 2015*, (April), 5.

[14] Highland, L. M., & Bobrowsky, P. (2008). The Landslide Handbook — A Guide to Understanding Landslides. *Landslides*, 129. https://doi.org/Circular 1325

[15] Huang, S., & Tian, T. (2006). STUDY OF ENVIRONMENTAL IMPACT IN GEOTHERMAL DEVELOPMENT AND UTILIZATION. In *the 7th Asian Geothermal Symposium* (pp. 35–44).

[16] Jaiswal, P., Van Westen, C. J., & Jetten, V. (2010). Quantitative assessment of direct and indirect landslide risk along transportation lines in southern India. *Natural Hazards and Earth System Science*, *10*(6), 1253–1267. https://doi.org/10.5194/nhess-10-1253-2010

[17] Masri, A., Barton, C., Hartley, L., & Ramadhan, Y. (2015). Structural Permeability Assessment Using Geological Structural Model Integrated with 3D Geomechanical Study and Discrete Facture Network Model in Wayang Windu Geothermal Field, West Java, Indonesia. In *Geothermal Reservoir Engineering*.

[18] Massonet, D., Holzer, T., & Vadon, H. (1997). Land Subsidence Caused by the East Mesa Geothermal Field, California, Observed Using SAR Interferometry.

[19] Miranda, J. L. H., & Lopéz, L. A. A. (2011). Piping Design : the Fundamentals.

[20] Muhlbauer, W. K. (2004). *Pipeline Risk Management Manual: Ideas, Tecniques, and Resources* (3rd ed.). Gulf Professional Publishing.

[21] Nordquist, G. A., Acuña, J., & Stimac, J. (2010). Precision Gravity Modeling and Interpretation at the Salak Geothermal Field , Indonesia. In *World Geothermal Congress 2010* (pp. 25–29).

[22] Pioquinto, W. P. C., & Caranto, J. A. (2005). Mitigating the Impact of Landslide Hazards in PNOC-EDC Geothermal Fields. *World Geothermal Congress 2005*, (April).

[23] Rex, L. B., Devin, L. G., & Edwin, L. H. (2008). *Landslide and Land Subsidence Hazards to Pipelines. Landslide and Land Subsidence Hazards to Pipelines*. Retrieved from http://pubs.usgs.gov/of/2008/1164/pdf/OF08-1164_508.pdf

[24] UNISDR. (2009). 2009 UNISDR Terminology on Disaster Risk Reduction. Geneva: UNISDR. Retrieved from http://www.unisdr.org/files/7817_UNISDRTerminologyEnglish.pdf

[25] van Westen, C. J., Alkema, D., Damen, M. C. J., Kerle, N., & Kingma, N. C. (2011). *Multi-Hazard Risk Assessment (Distance Education Course Guide Book)*. ITC University of Twente.

[26] van Westen, C., Quan Luna, B., Vargas Franco, R., Malet, J., Jaboyedoff, M., Horton, P., & Kappes, M. (2010). Development of training materials on the use of Geo-information for Multi-Hazard Risk Assessment in a Mountainous Environment. *Proceedings of the Mountain Risks International Conference*, (June 2016), 469–475. Retrieved from http://www.researchgate.net/publication/234139496_Development_of_training_materials_on_the_use_of_geo_-_information_for_multi_-_hazard_risk_assessment_in_a_mountainous_environment/file/d912f50f86e5713aeb.pdf

[27] W. Marzocchi, Mastellone, M. L., Ruocco, A. Di, Novelli, P., Romeo, E., & Gasparini, P. (2009). *Principles of Multi-risk Assessment: interaction amongst natural and man-induced risks.* https://doi.org/10.2777/30886

[28] Werner, E. D. ., & Friedman, H. P. (Eds.). (2010). *Landslides : Causes , Types, and Effects*. New York: Nova Science Publishers, Inc.

[29] Zaenudin, A., Kadir, W. G. A., Santoso, D., Abdassah, D., & Kamah, Y. (2010). Determination of Negative Density Changes in the Kamojang Geothermal Field Using Time- Lapse Microgravity Analysis. In *World Geothermal Congress*.

[30] http://en.tempo.co/read/news/2016/04/28/307766683/Landslide-in-Bengkulu-Geothermal-Site-Kills-1-Buries-4

[31] http://jurnaljakarta.com/berita-4460-pipa-geothermal-di-pangalengan-meledak-akibat-tanah-longsor.html

[32] http://www.fema.gov/hazus

[33] http://www.rappler.com/nation/22868-5-dead-in-lopez-s-leyte-power-plant-incident

[34] http://www.thejakartapost.com/news/2013/01/28/west-sumatra-landslide-leaves-7-dead.html

Probabilistic approach of resource assessment in Kerinci geothermal field using numerical simulation coupling with monte carlo simulation

Iki Hidayat[1], Sutopo[1,2], Heru Berian Pratama[2]

[1] Study Program of Petroleum Engineering, Faculty of Mining and Petroleum Engineering, Bandung Institute of Technology
[2]Study Program of Geothermal Engineering, Faculty of Mining and Petroleum Engineering, Bandung Institute of Technology
Jl. Ganeca No. 10, Bandung, West java, Indonesia. 40132

Email : ikihidayah@gmail.com

Abstract. The Kerinci geothermal field is one phase liquid reservoir system in the Kerinci District, western part of Jambi Province. In this field, there are geothermal prospects that identified by the heat source up flow inside a National Park area. Kerinci field was planned to develop 1×55 MWe by Pertamina Geothermal Energy. To define reservoir characterization, the numerical simulation of Kerinci field is developed by using TOUGH2 software with information from conceptual model. The pressure and temperature profile well data of KRC-B1 are validated with simulation data to reach natural state condition. The result of the validation is suitable matching. Based on natural state simulation, the resource assessment of Kerinci geothermal field is estimated by using Monte Carlo simulation with the result P10-P50-P90 are 49.4 MW, 64.3 MW and 82.4 MW respectively. This paper is the first study of resource assessment that has been estimated successfully in Kerinci Geothermal Field using numerical simulation coupling with Monte carlo simulation.
Keywords: Kerinci, numerical simulation, resource assessment, monte carlo

1. Introduction

Kerinci geothermal field is located in the Kerinci district, western part of Jambi province (5 km to the NE of Sungai Penuhtown) and approximately 525 km from Jambi capital city. This field has been developing by PGE since 2008 due to the energy shortage and to meet power demands that increase 9-10% per year in South Sumatra. Jambi province needs 275 MW of electrical power in the next five years, almost double increase from the current peak load of 150 MW. The electricity power in Kerinci area generated by the old diesel fuel engine, black out almost every night and no industrial growth, so that the local government support PT. PGE to explore the geothermal resources of Kerinci (Silitonga, 2015). Figure 1 presents the location of the Kerinci Geothermal Field.

IOP Conf. Series: Earth and Environmental Science **103** (2017) 012007 doi:10.1088/1755-1315/103/1/012007

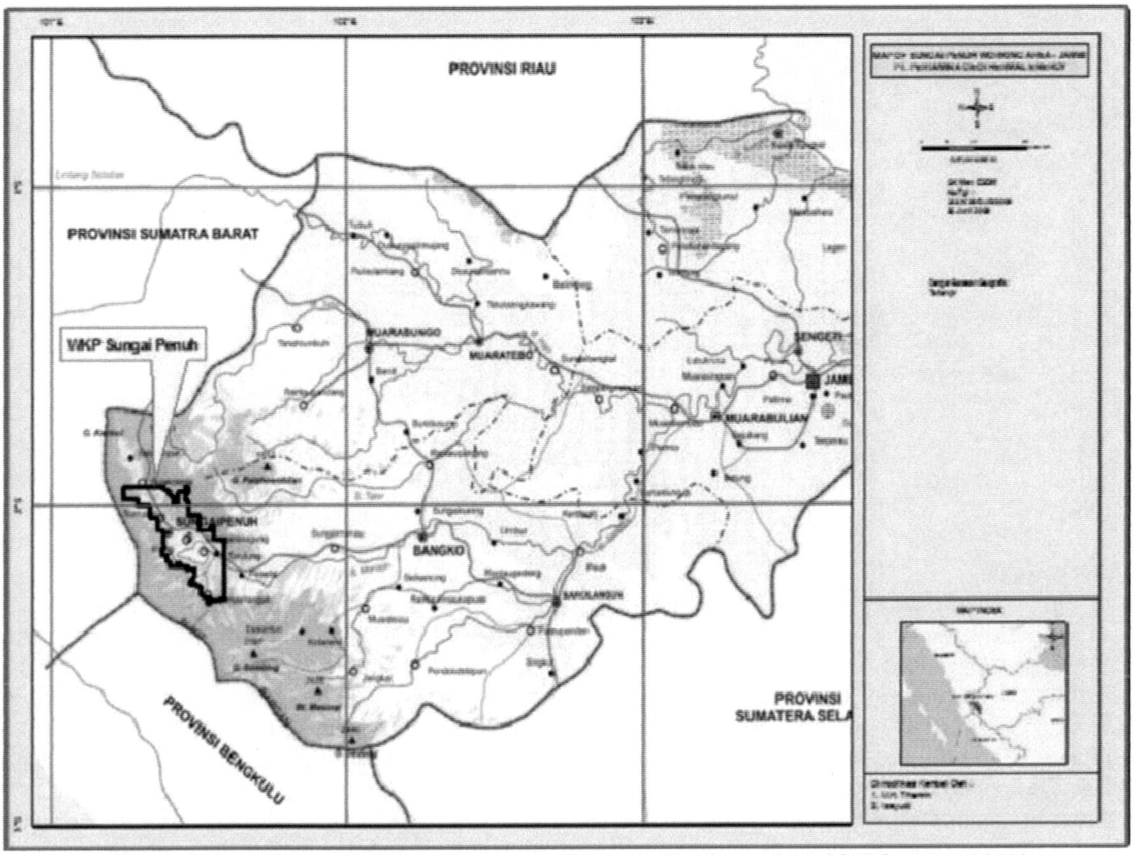

Figure 1. *Location and surface features of Kerinci Geothermal Field (Silitonga, 2015).*

2. Objectives

The objectives of this study are following :

1. To develop numerical model of Kerinci Geothermal Field which achieved natural state condition by using TOUGH2 software.
2. To validate pressure and temperature profile of the model and measured data from KRC B-1 wells
3. To characterize reservoir that including area, thickness, pressure, temperature, and permeability distribution.
4. To determine geothermal energy reserve of numerical model of Kerinci Geothermal Field by heat stored method with deterministic and probabilistic approach.

3. Conceptual Model

Conceptual model is the key to create numerical modelling. It is generated based on geological, geophysical, geochemistry and well survey data. The conceptual model of Kerinci geothermal field contain heat source location beneath M. Kunyit, iso temperature distribution, recharge location, distribution of surface manifestation, and well location (figure 2). The heat tends flowing to the NE. Permeable zone is about lower than 700 m. The conceptual model is used for numerical model as main based.

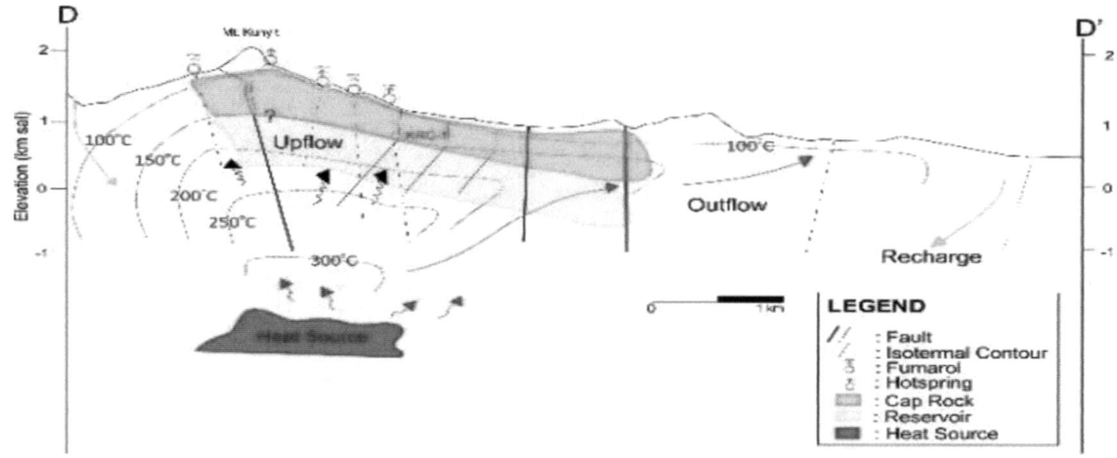

Figure 2. Conceptual Model of Kerinci Geothermal Field in vertical cross section (Modified JICA, 1989)

4. Methodology

In development of numerical model of natural state, there are several steps that must be undertaken. These steps are designed based on conceptual model to change parameters in TOUGH2 software and presented by flow chart at figure 3.

The first step is to create computer model based on conceptual model. The next step, input data including material properties and boundary condition. Then, Run the model and result of numerical model will be calibrated. If the calibration is reached, natural state model is obtained. But if not, back to input data to change the material until natural state.

From the natural state model, the author characterize the reservoir to obtain parameters value as data input for resource estimation. The resource estimations are used heat stored as a deterministic approach and heat stored with Monte Carlo simulation as probabilistic approach.

5. Numerical Model

5.1. Grid Model

The model of grid is developed in three dimensional. The model is aligned at 50° NE with a covering total area of 4.75 km × 3.5 km or equal to 16.25 km^2 Elevation of this model is started from 1410 masl to -2000 masl or total thickness is 3410 m. The model consists of 7820 (23 × 17 × 20) blocks. The grid blocks is presented detail in table 1.

The numerical model used EOS1 (water, water with tracer) type and used single porosity system to simplify the model. The 3D numerical model is presented in figure 4.

The model of grid system is used to follow parameter approaching. The main reservoir used smallest size and the boundary condition used biggest grid block size. The parameters that used through the model are heterogeneity approaching.

This numerical model is used to simulate reservoir performance by properties input including rock and fluid. The best model is obtained by good match pressure and temperature profile of simulation and measured well data. In Kerinci Field, there is one exploration well (KRC-B1) that has pressure and temperature profile data to be matched with simulation data.

5.2. Material Properties

Materials are used to define the permeability and other properties in an analysis. After the model of grid was built, this numerical model creates materials to approach the simulation with actual condition with rock properties based on conceptual model. Materials can be assigned by layer, by region, or to individual block.

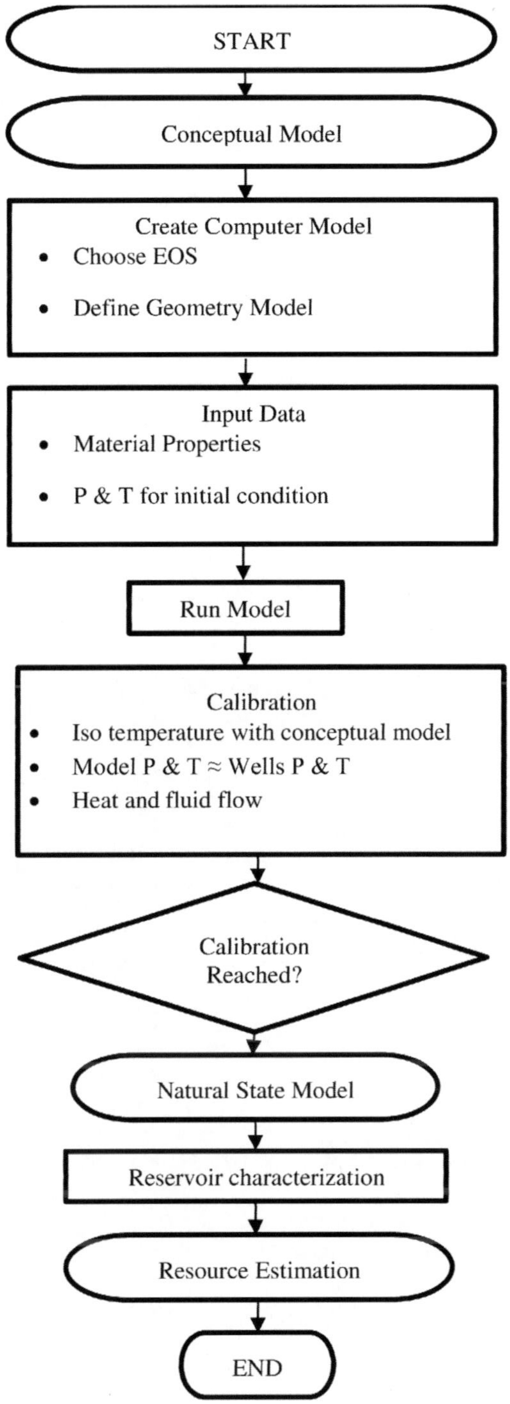

Figure 3. Flow chart of numerical model of natural state.

Table 1. Grid block system

Direction (*x,y,z*)	Grid blocks	Grid block size (m)
x	1	250
x	1	225
x	19	200
x	1	225
x	1	250
y	1	250
y	15	200
y	1	250
z	1	10
z	4	150
z	8	100
z	4	200
z	3	400

The properties that must be inputted are specific heat, wet heat conductivity, rock density, porosity, permeability (*x, y, z* direction). The main parameter in material properties to develop numerical model of natural state is permeability structure.

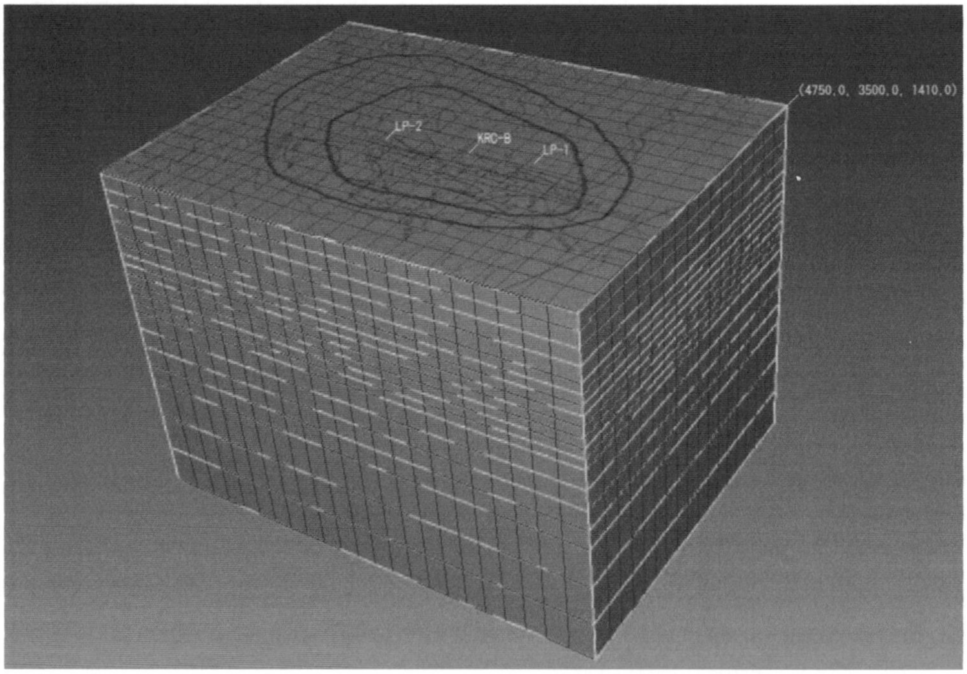

Figure 4. The 3D Numerical Model of Kerinci Field

5.3. Boundary and Initial Condition

5.3.1. Initial Condition
Initial condition are used to define the initial state of each block of numerical model. In the condition, it is depend of EOS and phase type. In this numerical model is used EOS1 and single-phase based on actual

condition. The next step, define initial pressure and temperature condition in all grid blocks. The pressure and temperature is stated by equation with depending some variables. The equation is presented by equation (1) and (2). The equation (1) and (2) is used by gradient of well data to approach in numerical model. The equation of gradient pressure :

$$P = 1.14 \times 10^7 - 8000z \quad (1)$$

And the equation of gradient temperature :

$$T = 81.4 - 0.04z \quad (2)$$

Where : P is pressure (Pascal), T is temperature (°C) and z is depth (m).

5.3.2. Top Boundary

Top of numerical model of natural state is located at the water table and there is atmospheric condition assumed by pressure at 1 bar ($1.01 \times 10^5\ Pa$) and temperature at 25°C. In this boundary, the materials used big volume factor, a multiplier on the volume that used to obtain the final volume. It means, the value of pressure and temperature in this materials along simulation tends to be constant from initial condition. In below of atmosphere layer is groundwater layer that used another big volume factor. But the volume factor is smaller than atmosphere volume factor to make simulation process to be continuously convergent.

5.3.3. Side Boundary

This numerical model is assumed all side of boundaries in closed condition (no mass and heat flow). This boundary is used low permeability (10^{-18} m^2 or 0.001 mD) and big volume factor (10^{20}).

5.3.4. Bottom Boundary

Bottom boundary was set heat sources and basement rock materials. Heat sources material was set by pressure and temperature based on JICA report (1989). From JICA (1989), temperature heat sources is around 300 – 320°C. The pressure and temperature of heat source is greater than surrounding materials and the value pressure at 252 bar and temperature at 300°C tend to be constant.

Table 2. Layer Properties for the initial numerical model.

Layer	Top, Bottom (masl)	Layer	Top, Bottom (masl)
Atmosphere	1410, 1200	Reservoir6	300, 200
Ground water	1400, 1250	Reservoir7	200, 100
Cap Rock1	1250, 1100	Reservoir8	100, 0
Cap Rock2	1100, 950	Reservoir9	0, -200
Cap Rock3	950, 800	Reservoir10	-200, -400
Reservoir1	800, 700	Reservoir11	-400, -600
Reservoir2	700, 600	Reservoir12	-600, -800
Reservoir3	600, 500	Basement Rock1	-800, -1200
Reservoir4	500, 400	Basement Rock2	-1200, -1600
Reservoir5	400, 300	Basement Rock3	-1600, -2000

Table 3. Layer Properties for the initial numerical model.

Material	Color	Permeability-*x* (mD)	Permeability-*y* (mD)	Permeability-*z* (mD)	Porosity
Atmosphere		1.0×10^9	1.0×10^9	1.0×10^9	0.99
Ground Water		0.025	0.025	0.025	0.07
Cap Rock		0.05	0.05	0.05	0.05
Reservoir 1		10	10	20	0.1
Reservoir 2		10	10	10	0.1
Basement Rock		8	8	4	0.1
Side Boundary		0.001	0.001	0.001	0.05
Heat Source		10	10	10	0.1

6. Natural State

6.1. Numerical Model Calibration

The main purpose in this numerical model is to achieve the natural state. So that, this paper in model calibration is focusing to match the natural state. This natural state has been described completely in previous paper at IIGCE event 2016 (Iki, 2016). There are two general steps to achieve natural state model, those are first estimates of model parameters from the conceptual model and calibration of the permeability structure by natural state modelling. In the first estimates, next step the numerical model is ran by TOUGH2 software. The first important in running of simulation is ensuring the numerical model to achieve steady state condition in the end of simulation running. It means, the value of DT (solution time increment) is increasing to converge.

After steady state condition in simulation result, compared with actual pressure and temperature profile. If the result of matching between simulation data and actual data is not good, the parameters in TOUGH2 were adjusted in and iterative process.

The following are several ways to adjust this numerical model in calibration of natural state :

1. The pressure and temperature of heat sources are adjusted.
2. The size grid blocks of heat sources
3. Permeability structure in *x, y, z* direction is adjusted in accordance of layer location.
4. Construction of suitable material in each layer based on conceptual model.
5. The amount of materials are made in accordance based on conceptual model
6. The amount of layers in the reservoir

After the result of pressure and temperature profile are good fit, the next step to ensure direction of fluid and heat transfer movement is right. The model must be adjusted recharge location that affect direction of fluid movement. In caprock and boundary zone, there are no fluid movement although opposite direction because the zones are impermeable rock. In direction of heat transfer, the movement will pass through permeable zone by convection but heat transfer will pass by conduction through impermeable zone.

6.2. Comparison of Pressure and Temperature Profiles

Kerinci geothermal field has three exploration wells but two of them are slim hole well. To compare pressure and temperature profiles between simulation and actual data, the author used KRC-B1 well for calibration process because available actual data considering. The result in comparison of pressure and temperature numerical model Kerinci geothermal field from KRC-B1 well is presented in previous paper respectively.

Based on figure 5, the result of pressure profile between actual data and simulation data is very good matching. In this figure, there is one simulation data in -500 m elevation but there is no actual data. It means that the simulation data is drilling well to -500 m. The gradient pressure in KRC-B1 well is about 0.08 bar/meter. But this numerical model need more well data to validate gradient temperature each location. If gradient pressure in two or more wells data tends to be similar, it means numerical model has one heat source and interconnected. But for this case, the author assumes that Kerinci geothermal field has one heat source only.

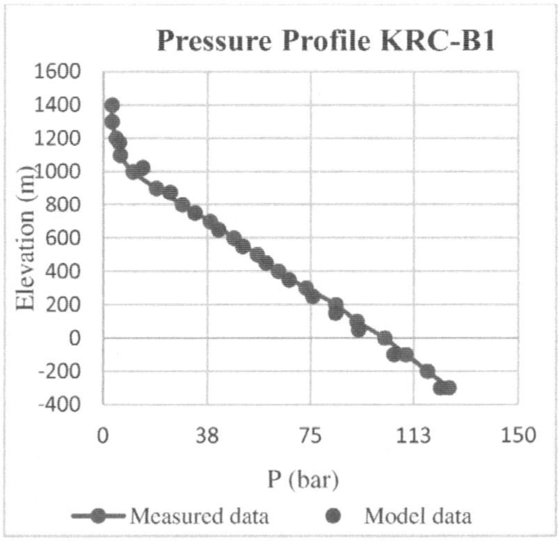

Figure 5. Comparison pressure profile of KRC-B1.

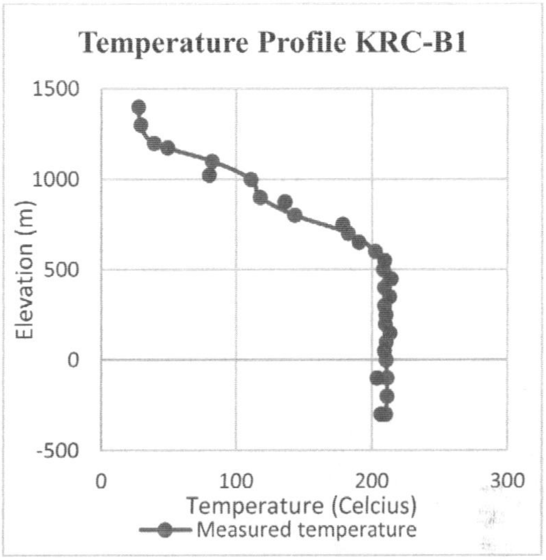

Figure 6. Comparison temperature profile of KRC-B1.

In figure 6, the result of temperature profile between actual data and model data is suitable matching. But there are a little bit different in temperature reservoir. In the reservoir elevation, the temperature actual tends to constant (209 - 211°C) but in the temperature model is varying (207 - 217°C). It is caused probably the materials of this numerical model is less of many kinds in construction. But the more well data are used, the better numerical model is validated. The bellow, comparison of conceptual model and natural state result are shown in figure 7. Size of numerical model area is shown in E' to E vertical line.

6.3. Heat and Fluid Flow

Heat and fluid flow is one of calibration processes in the numerical model. Based on conceptual model, there are two recharges location in Kerinci geothermal field. Those recharges is located in southwest and northeast. It will affects direction of heat and fluid flow. In result of numerical model of Kerinci geothermal field, heat from heat sources flows to upward through permeable zone. But heat flow was halted in impermeable zone including side boundary and caprock. At near both materials, the heat flow is reversed flowing to bottom and establish a circle. In another point, heat tends to move from heat sources location towards the east. It is caused of permeable zone in that area. In reservoir zone, the heat transfer was occurred by convection. But in caprock and side boundary, heat transfer was occurred by conduction.

The result of fluid flow in numerical model of Kerinci geothermal field, the path flow is similar with the heat flow. Basically, the fluid flow will be followed by heat flow, it is called convection. Heat is still flowing in impermeable zone by conduction but fluid could not flow in impermeable zone. As a result, there are other fluid from recharge that cause fluid flow downward.

In numerical model is a little different about topography with conceptual model. The fact of surface in Kerinci field is not flat, mainly in the top of Kunyit Mountain. Assumption in top elevation of numerical model is flat at highest elevation of well (1410 masl). The purpose of the assumption is simplification to construct the numerical model and the top of numerical model is used to accommodate wells. One of types of top boundary conditions is used along a boundary about fixed value of temperature (Franco, 2013).

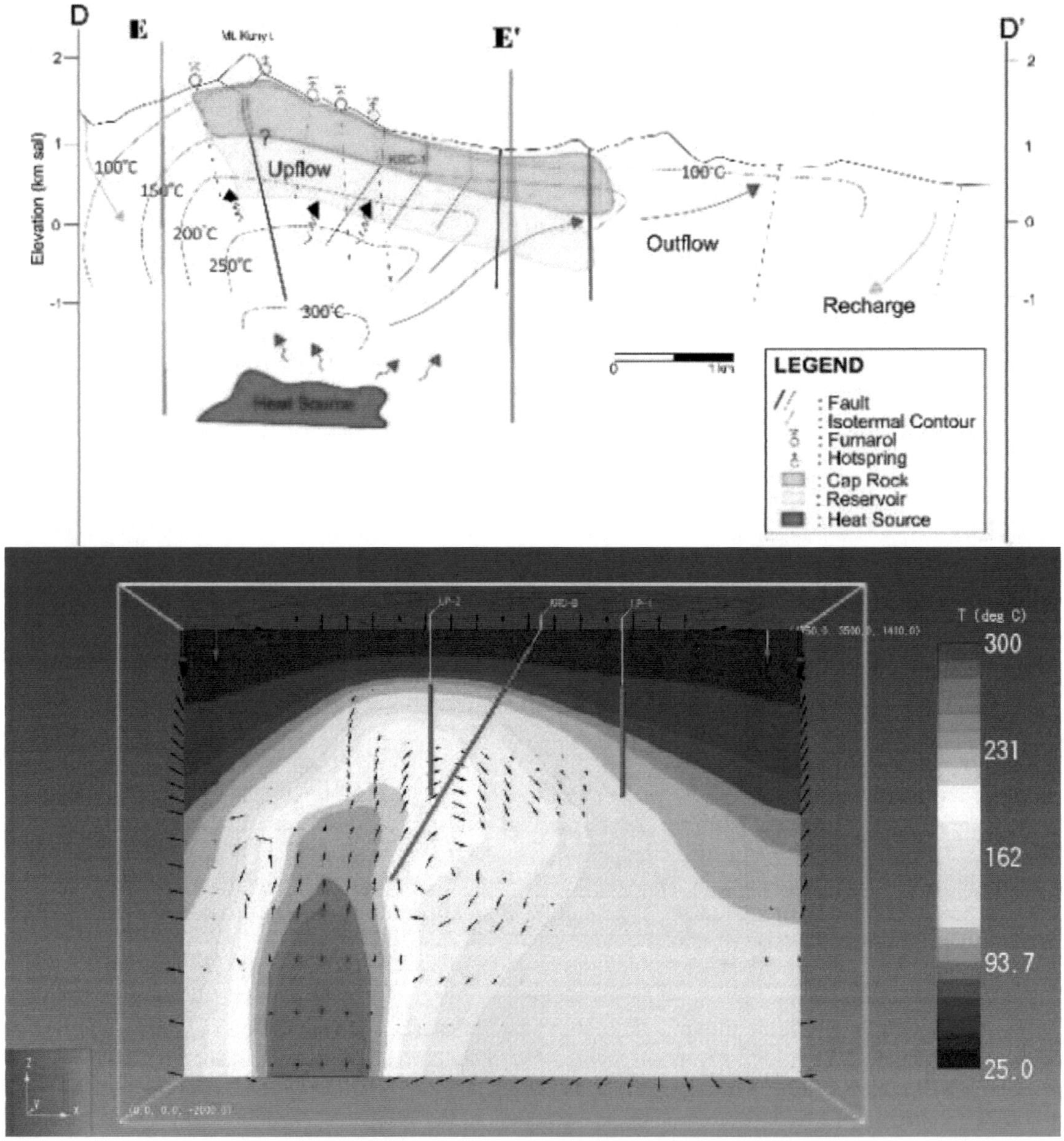

Figure 7. Comparison conceptual and natural state result

Before the boundary condition is defined, first the definition of geological element or topography (Franco, 2013). The consequently, fluid will flow from top of Kunyit Mountain to underneath and the pressure in around the top of mountain will be higher than surrounding.

From conceptual model of Kerinci geothermal field, it was existed Duabelas fault. But in the numerical model is not exist because the author assumed that the fault area is neglected because it is very small and the material is homogeneity with reservoir material. In fact in fault area, the permeability is higher than surrounding.

6.4. Permeability Distribution

Permeability structure is most influence in the numerical model. It will affect heat and fluid flow, pressure distribution and temperature distribution. In numerical model of Kerinci geothermal field, permeability is important parameter to achieve natural state. This numerical model needs heterogeneity in *x, y, z* -direction permeability. It is caused temperature profile in Kerinci geothermal field is not symmetry as well as also heat and fluid flow.

The reservoir permeability is the main permeability in numerical model of Kerinci geothermal field. The result of this numerical model, highest permeability value reservoir is in Reservoir-1 material. The value of *x, y* -direction is a half of value of *z* -direction in reservoir-1 material. In Reservoir-2, the value of *x, y, z* direction is the same. In basement rock permeability is lower than Reservoir-2. The value of permeability *x, y* -direction in basement rock is lower than z-direction. It is caused there are big effect of gravity in bottom of reservoir. So that the value of *z* - direction permeability in basement rock is small.

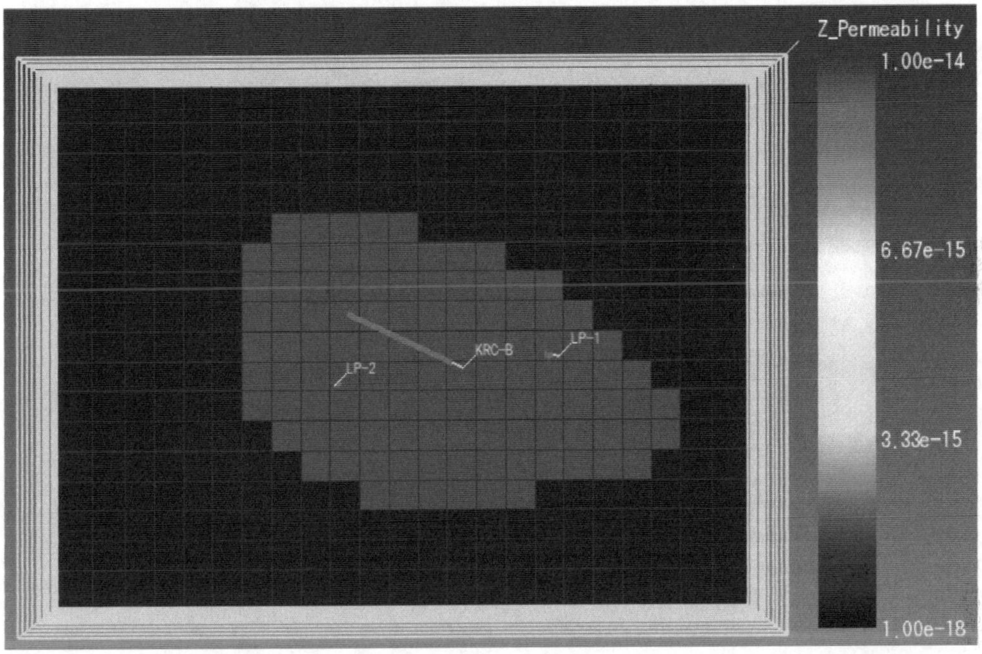

Figure 8. Permeability distributions at 50 masl (reservoir-1 and side boundary)

6.5. Reservoir Characterization

Based on result of numerical modelling of natural state, the model could determine reservoir characterization. The area reservoir in this model covered about proven and probable area. The proven area is bounded by three exploration wells. The proven area is covered about 5.5 km^2. Probable area is bounded by surface manifestation and the area is covered about 8.5 km^2. Reservoir has thickness about 1500 m. Permeability range in the reservoir is varying between 5×10^{-15} m^2 (5 md) and 1×10^{-14} m^2 (10 md).

This model has single phase liquid, so that the system reservoir in the Kerinci geothermal field is one phase liquid reservoir where temperature reservoir is lower than the saturation temperature, presented in figure 11.

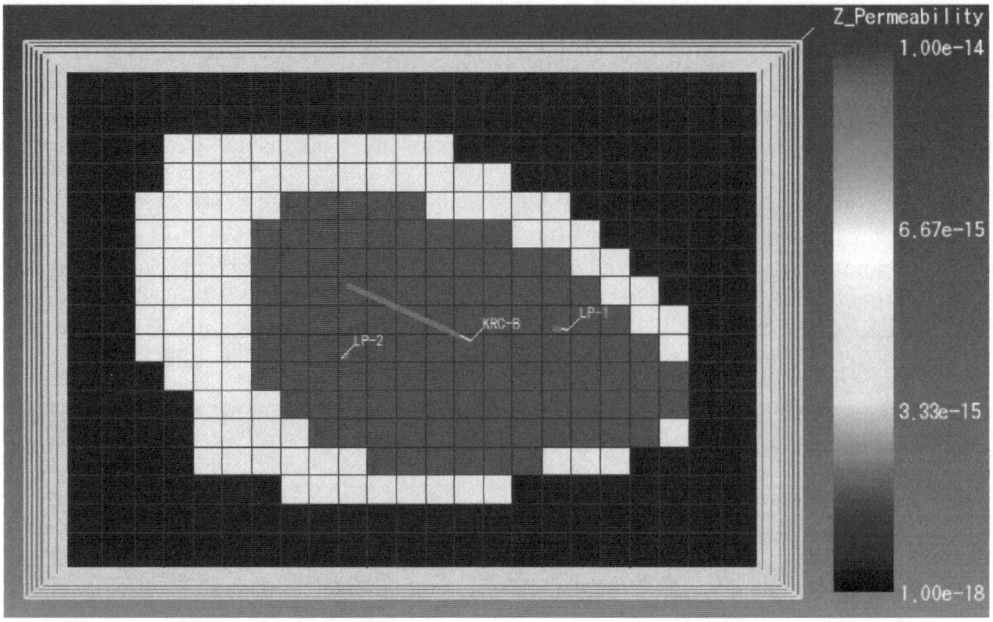

Figure 9. Permeability distributions at -50 masl (reservoir-2 and side boundary).

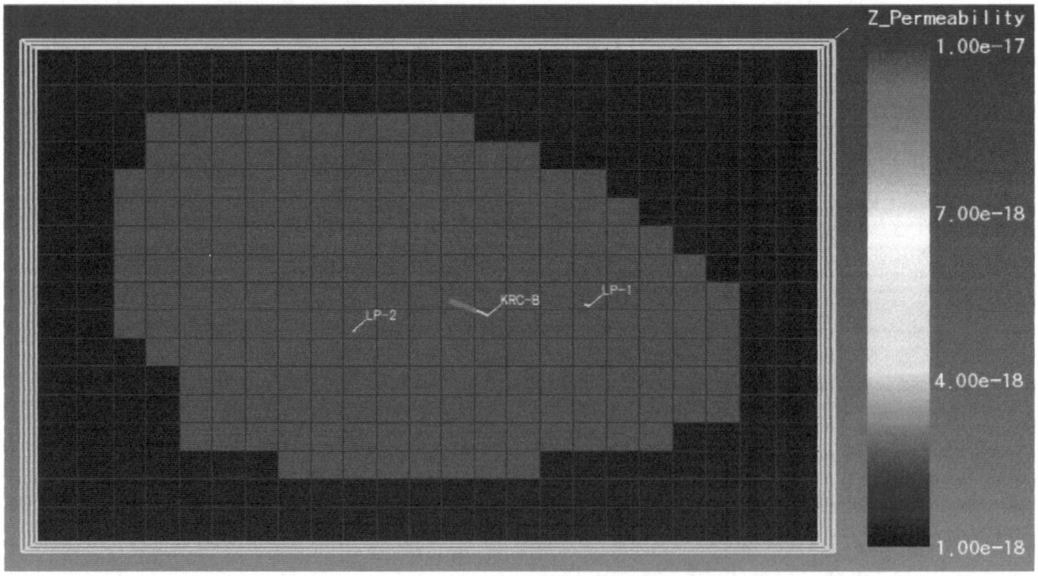

Figure 10. Permeability distributions at -875 masl (reservoir-2 and side boundary).

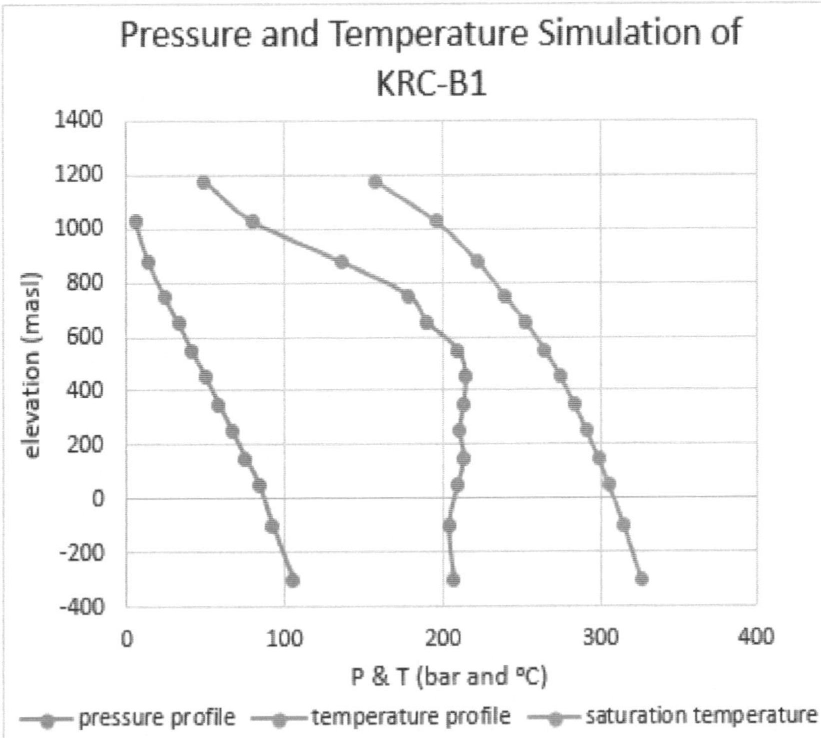

Figure 11. Pressure, temperature, and saturation temperature KRC-B1

One phase liquid could be shown with saturation from result of numerical model. It is clearly that the gas saturation of natural state Kerinci is 0. The following is saturation gas in result of numerical model in figure 12.

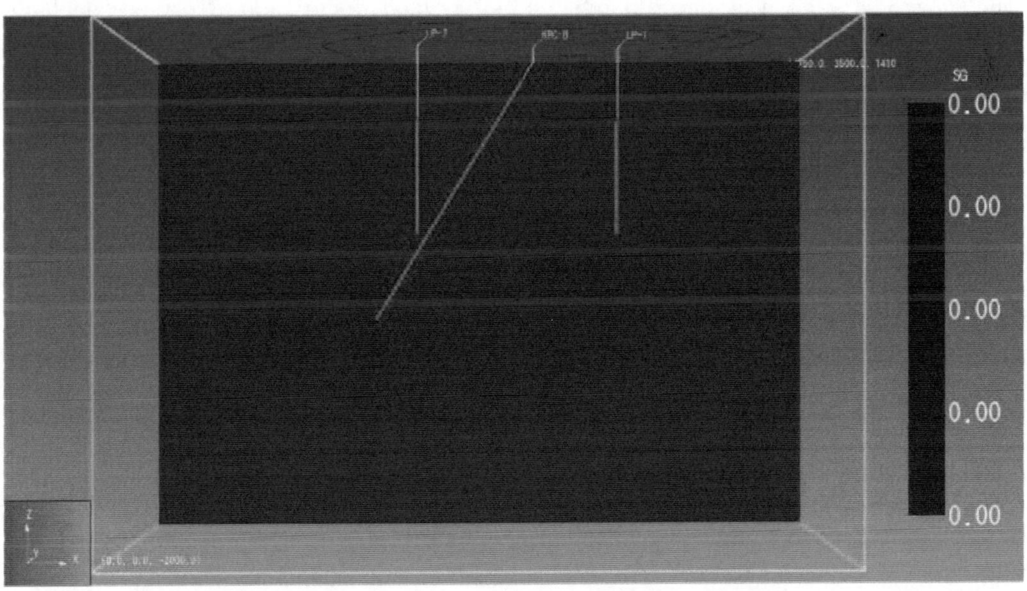

Figure 12. Gas saturation of natural state Kerinci

6.6. Resource Estimations

The estimation of resource in Kerinci geothermal field used heat stored principal. There are two methods including deterministic and probabilistic approach. Deterministic approach used volumetric method and probabilistic approach used Monte Carlo simulation.

The volumetric method needs some parameters including area, depth, porosity, rock density, heat rock capacity, initial and final water saturation, initial and final temperature, recovery factor, conversion factor, and project time. The parameters is obtained from conceptual model, reservoir characterization, general assumptions. The result of resource by volumetric method is 57 MWe. The detail of result is presented in figure 13. From the result of volumetric method as deterministic approach, the author could analyze the sensitivity of parameters. The most influencing parameters are area and recovery factor.

Monte Carlo simulation is calculation of iterations from probabilistic model that simulate properties physic in distribution value for estimating geothermal resource by heat stored principal. In this simulation, the author estimate minimal, maximal, and mostly value in each parameter of geothermal resouce.

The proven area in Kerinci geothermal field was bounded by three exploration wells. It means the area has proven in geothermal resource by drilling well. So that, the proven area is obtained. The probable area is obtained from reservoir characterization based on surface manifestation distribution.

The thickness of reservoir is obtained from high temperature profile. From high temperature profile, the thickness is from 500 masl down to -500 masl. But the thickness is added by reservoir-2 material based on temperature distribution in result of numerical model. The thickness of the reservoir is approximately about 1500 m. Rock porosity is obtained from relation of rock type and porosity value (B. Layman and Soemarinda, 2003). The rock porosity is obtained about 0.15. Recovery factor is obtained from correlation between recovery factor and porosity (After Muller, 1978), the value is about 0.375. Conversion factor is obtained from correlation between conversion factor and reservoir temperature (Nathenson, 1975 and Bodvarson, 1974), the value is about 10%. The project time of Kerinci geothermal field is assumed about 30 years.

In this Monte Carlo simulation, the numerical model of Kerinci geothermal field was resulted the value geothermal resource by the type of degree confidence rank (P10, P50, and P90). The highest degree confidence respectively from P10, P50, and P90. Input parameters value for Monte Carlo simulation is shown in figure 18.

The result of Monte Carlo simulation, the value of Kerinci geothermal P10 is 49.4 MWe, P50 is 64.3 MWe, and P90 is 82.4 MWe, presented also in figure 14. Based on Monte Carlo simulation calculation, the geothermal potential of kerinci geothermal field is lower than 55 MWe for highest degree confidence.

Parameter	MIN (MWe)	MAX (MWe)	MOST (MWe)
Area (km2)	53	81	57
Thickness (m)	54	61	57
Rock Density (kg/m3)	57	59	57
Porosity (fraction)	49	66	57
Rock Heat Capacity (kJ/kg.C)	55	60	57
Temperature Initial (C)	53	62	57
Temperature Final (C)	57	57	57
Water Saturation Inititial (fraction)	57	57	57
Water Saturation Final (fraction)	61	57	57
Recovery Factor (fraction)	38	77	57
Electric Efficiency (fraction)	57	59	57
CLEAR	RUN SENSITIVITY		

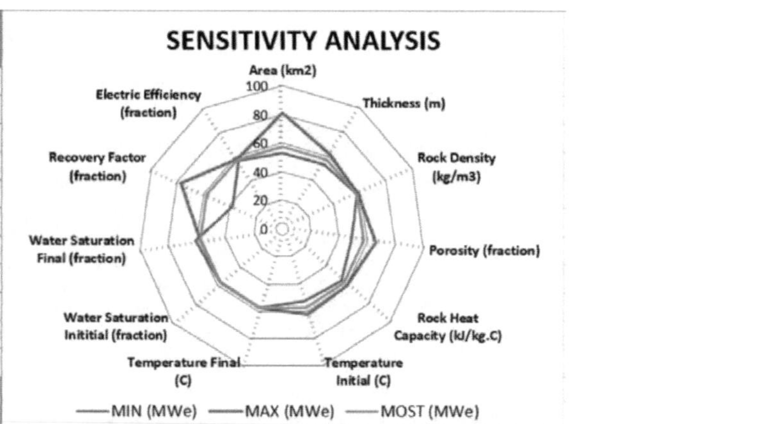

Figure 13. Resource calculation results based on deterministic heat stored method with sensitivity analysis

Parameter	Min	Max	Most	Remark
Area (km2) =	5.5	8.5	6.0	From natural state model
Thickness (m) =	1500	1600	1550	From material reservoir model
Rock Density (kg/m3) =	2500	2600	2500	Based on JICA 1989
Porosity (fraction) =	0.10	0.2	0.15	Based on Layman, 2003
Rock Heat Capacity (kJ/kg.C) =	0.9	1.1	1	Relation with reservoir temperature
Temperature Initial (C) =	210	220	215	From natural state model
Temperature Final (C) =	180	180	180	General assumption
Water Saturation Initial (fraction) =	1	1	1	General assumption
Water Saturation Final (fraction) =	0.20	0.30	0.30	General assumption
Recovery Factor (fraction) =	0.25	0.50	0.38	Based on After Muller, 1978
Electric Efficiency (fraction) =	0.1	0.1	0.1	Nathenson, 1975
Life Time (years) =	30			Government rule

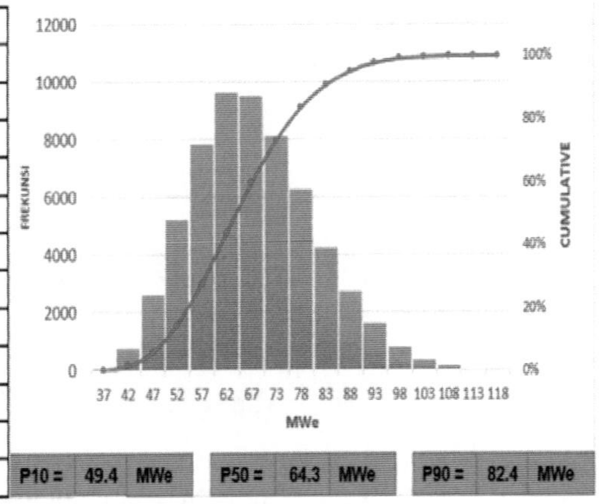

Figure 14. Resource calculation results based on Monte Carlo simulation.

7. Conclusion

From this numerical modelling the following conclusions can be summarized :

1. The first numerical model of natural state of Kerinci geothermal field has been successfully developed.
2. Pressure and temperature profile between model data and KRC-B1 well data is very good matching.
3. From reservoir characterization is obtained proven and probable area, thickness reservoir, reservoir permeability range, temperature reservoir and system reservoir fluid.
4. This numerical model of natural state Kerinci geothermal field could generate electrical around 57 MW with heat stored method as deterministic approach. But the probabilistic method using heat stored with Monte Carlo simulation for P10-P50-P90 are 49.4 MW, 64.3 MW and 82.4 MW respectively.

8. Recommendation

1. To ensure geothermal potential, this numerical model should be more validated of using other well data.
2. This numerical model should be improved by validating with existence of fault in around of reservoir.
3. This numerical model should be considered about topography and surface manifestation distribution.
4. This numerical model should be applied by using experimental design to generate probabilistic resource assessments result.
5. For further study, this numerical model could be forecasted in the future scenario.

9. References

[1] JICA, 1989. *"The Feasibility Study on Kerinci Geothermal Development Project in The Republic of Indonesia Final Report"*. JICA library. March , 1989

[2] Silitonga, 2015. "*Exploration Strategy in Kerinci Geothermal Prospect located inside the National Park of Indonesia"*. Proceeding World Geothermal Congress. Melbourne, Australia, 19-25 April 2015

[3] Franco, A and Vaccaro M. 2013.*"Numerical simulation of geothermal reservoirs for the sustainable design of energy plants : A review"*. Elsevier Ltd : 993

[4] Iki, Sutopo, and Heru. "*Numerical Modelling of Kerinci Geothermal Field, Jambi, Indonesia*". Proceeding the 4th Indonesia International Geothermal Convention & Exhibition 2016, Jakarta Convention Center, Indonesia

IOP Conf. Series: Earth and Environmental Science **103** (2017) 012008 doi:10.1088/1755-1315/103/1/012008

Aerated drilling cutting transport analysis in geothermal well

Aris Wakhyudin[1], **Deni Setiawan**[2], **Oscar Dwi Marjuan**[1]

[1]Bandung Institute of Technology, Bandung, Indonesia
[2]PT. Mitra Konsultansi Indonesia, Tangerang Selatan, Indonesia

Email: wakhyudin.aris@gmail.com

Abstract. Aeratad drilling widely used for geothermal drilling especially when drilled into predicted production zone. Aerated drilling give better performance on preventing lost circulation problem, improving rate of penetration, and avoiding drilling fluid invasion to productive zone. While well is drilled, cutting is produced and should be carried to surface by drilling fluid. Hole problem, especially pipe sticking will occur while the cutting is not lifted properly to surface. The problem will effect on drilling schedule; non-productive time finally result more cost to be spent. Geothermal formation has different characteristic comparing oil and gas formation. Geothermal mainly has igneous rock while oil and gas mostly sedimentary rock. In same depth, formation pressure in geothermal well commonly lower than oil and gas well while formation temperature geothermal well is higher. While aerated drilling is applied in geothermal well, Igneous rock density has higher density than sedimentary rock and aerated drilling fluid is lighter than water based mud hence minimum velocity requirement to transport cutting is larger than in oil/gas well drilling. Temperature and pressure also has impact on drilling fluid (aerated) density. High temperature in geothermal well decrease drilling fluid density hence the effect of pressure and temperature also considered. In this paper, Aerated drilling cutting transport performance on geothermal well will be analysed due to different rock and drilling fluid density. Additionally, temperature and pressure effect on drilling fluid density also presented to merge.

1. Introduction

In geothermal drilling, many challenges will be faced. One of them is the reservoir pressure. Formation pressure found in geothermal drilling is smaller than the hydrostatic pressure of the mud used when drilling. Therefore, when drilling mud is applied, its pressure will exceed the formation pressure. Then drilling will be lost and enter to the formation itself. This is certainly not desirable. Aerated drilling method is not a new method used in drilling applications. However, this method can be used to address some of the problems that occurred in the geothermal field.

In this paper, it will discuss about the use of method of aerated drilling in the geothermal field. The density of the rock in geothermal field that will face is relatively bigger than the rock in the field of oil and gas. The greater the density of the rock that will face, the harder it will drill bit to penetrate the formation. Also, the drill cutting will be increasingly difficult to be transported to the surface.

In the other hand, the use of aerated drilling method requires a limited amount of gas that can be used to lower the density of the mud that pump into the annulus. Therefore, this paper will be modelled on a technical restriction that can be applied to drill using aerated drilling techniques.

2. Geothermal Well Drilling

The Geothermal well drilling process is basically similar to the process of drilling the wells of oil/gas, both in terms of the stages of the process, technology, tools, as well as its drilling expert. There are two major challenges in drilling geothermal wells at once that distinguishes it from drilling in oil and gas wells, especially, in terms of temperature and lost circulation problem.

Lost circulation occurs because the target in a geothermal well are fractures. The fracture is connected to a heat source. When the fracture is appeared through in the drilling process, most likely drilling mud will enter the fracture instead of returning to the surface. This lost circulation condition is technically giving some negative impact on the drilling process and need to be tackled.

In terms of temperature, the target of geothermal wells has a high temperature, because the temperature is what the energy that wants to be extracted. The higher the temperature is, the more economical the project of geothermal well it will be. However, in terms of the drilling process, It will become more challenging because the technology adopted from the drilling of oil and gas industry is actually designed for a relatively lower temperature, and it is often a barrier.

The conditions met in geothermal operation are [5]. Some of these conditions are as follows:

a. Lost Zones: Partial and lost circulation zones range from 7 to as much as 42 requiring cement plugging.
b. Hole Temperature: Bottom hole temperature range from 300°C and 425°C (compared to low or ambient temperature in oil-gas wells) at depth of 7000 to 11000 feet.
c. Lithology: Soil formation are hard silicified volcanic rocks as compared with generally sedimentary rock formation in oil and gas lithology.
d. Mud Density: Mud weight requirements are already on reduced density level such as 8.8 to 10.5 ppg compared with more than 13 ppg for oil and gas.
e. Hole Acidity: Hole condition is highly corrosive with pH range of 3 to 5 pH which can be aggravated by introduction of oxygen from air injection during mud circulation.
f. Blind Drilling: When zones of total lost circulation are encountered on the 8 ½" hole interval, the use of mud is stopped and blind drilling with water is conducted.

3. Managed Pressure Drilling (MPD)

MPD is drilling technique that the mud pressure in bottom hole is kept constant using certain technique. It is important to anticipate several problems that can be encountered such as drilling in narrow pressure window and loss circulation zone. Based on International Association of Drilling Contractors (IADC), MPD is an adaptive drilling process used to more precisely control the annular pressure profile throughout the wellbore to obtain several objectives:

1. To ascertain the downhole pressure environment limits
2. To manage the annular hydraulic pressure profile

4. Underbalance Drilling

Underbalance technique is using light drilling fluid where drilling fluid pressure is low than the pore pressure of formation rock. The fluids that commonly used such as air, gas, aerated water, and foam. Underbalance drilling technique can increase penetration rate, minimized differential sticking, improve formation evaluation, reduce formation damage, and prolonged bit life. The disadvantages of underbalance drilling are corrosion problem, high cost and difficult hydraulic calculation.

5. Aerated Drilling

5.1. Volume Requirement

Methods to reduce the hydrostatic pressure of the wellbore is to add gas into the system. Extra gas will result in a pressure drop due on replacing fluid pushed by the gas so that the density of the system will decrease by itself. The most important parameter in making the design of aerated drilling cuttings is to determine the carry capacity on a mixed system of drilling fluid. For a specific gas injection rate and injection rate of the fluid must be able to lift the whole cutting perfectly formed at the bottom of the well to the surface.

Produced cutting in geothermal drilling is relatively large and has a density that is larger than oil and gas field. Therefore, it will take the value of the minimum velocity rate of the fluid mixture and gas to lift the cutting to surface. In the application of aerated drilling is almost certain that the regime of the flow is turbulent. Then, the primary consideration in the designing flow rate is to include turbulent factor in the calculation. The basic step in planning for aerated drilling is shown in figure 1.

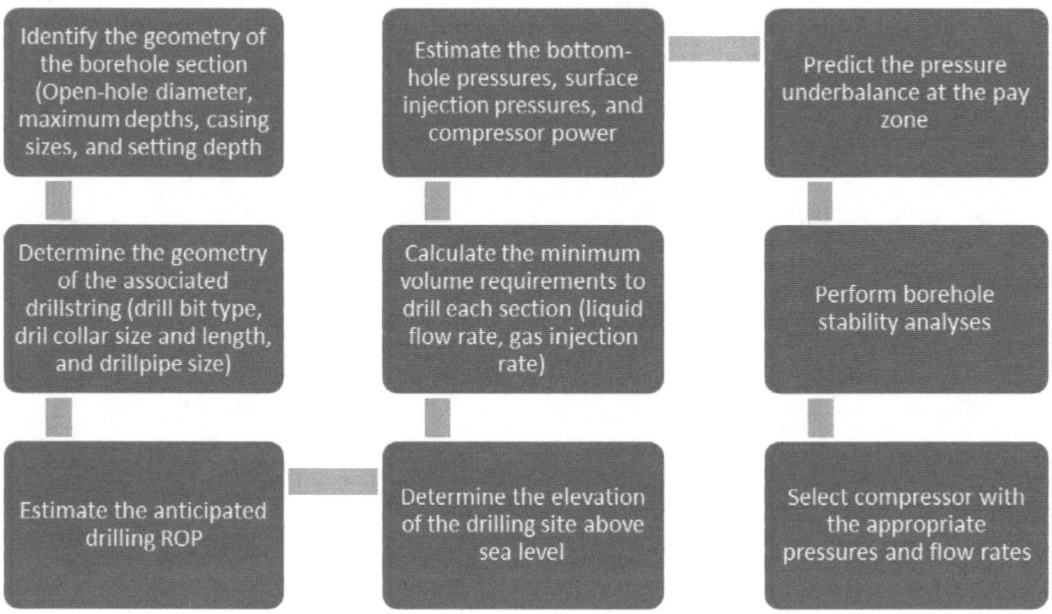

Figure 1. Workflow for calculating volume requirement in aerated drilling techniques [4]

5.2. Cutting Transport

5.2.1. Cuttings-Carrying Capacity. The cutting-carrying capacity of the aerated mud is defined as the maximum cutting size that can be lifted by the aerated drilling mud. Rittenger's formula [2] can be used to calculate the slip velocity, as follows:

$$V_{sl} = 7.3x\left(\frac{D_c x\,(\rho_c - \rho_f)}{\rho_f}\right)^{0.5} \quad (1)$$

The required cutting transport velocity, V_{tr} depends on the rate of penetration, R, and the allowable cutting concentration, C_c in the annulus.

$$V_{tr} = \frac{R}{C_c} \quad (2)$$

$$V_{tr} = \frac{\pi \, x \, d_b^2}{4xC_c xA} \tag{3}$$

The volumetric flow rate of liquid Q_l is defined as follows:

$$Q_l = \frac{2.2283}{10^3} \cdot Q_m + \frac{1.5597}{10^3} \cdot Q_i \tag{4}$$

The volumetric flow rate of gas Q_g is defined as follows:

$$Q_g = 6.7846 \, x \, 10^{-2} \cdot T \cdot \frac{Q_{air\ inj}}{P} \tag{5}$$

The average velocity can be obtained using the following equation,

$$V_f = \frac{c''}{P} + d'' \tag{6}$$

Where,

$$c'' = \frac{9.77TQ_{air\ inj}}{A} \tag{7}$$

$$d'' = \frac{0.33Q_m + 0.22Q_i}{A} \tag{8}$$

Maximum allowable cuttings settling velocity can be determined using the following equation,

$$V_{sl} = V_f - V_t \tag{9}$$

The density of gas can be also estimated using the ideal gas law,

$$\rho_g = S_g \cdot \frac{P}{53.3 \cdot T} \tag{10}$$

The volume fraction of liquid at certain point for every depth of interest is defined by,

$$f_l = \frac{Q_l}{Q_l + Q_g} \tag{11}$$

5.2.2. Cuttings Transport Efficiency. Test Cuttings transport efficiency is a function of the drilling mud and depends on several factors:

a. The density of the drilling mud

The addition of drilling mud will increase the buoyance force. Every particle of cutting is acting to opposite direction against gravity forces. However, the lifting capacity will be effective to drive out the mud and lift the cutting to the surface.

b. Torque effect
Rotation of drillpipe will affect on sludge removal capacity in laminar or turbulent flow regime. Effects of torque (power play) will cause the particles tend to spin in terms of turning drops up due to variations in velocity of the mud.

c. Viscosity and gel strength
Some mud which have low viscosity and gel strength will give great impact to the particles in the same annular velocity and circulation time. Based on the experiments conducted by Bruce and William, the mud with low viscosity and gel strength only has a small capacity to transport particles to the surface.

d. Distribution of velocity in the annulus
Large cutting lifting capacity can be achieved with turbulent flow than the laminar flow which has a low viscosity. This is because the effects of turbulence sludge tend to minimize settled cutting in the annulus near the pipe or the wall of the wellbore.

e. Dimensional of cutting
Bit design will determine the size and shape of the cutting. Physical magnitude of cutting will greatly affect the capacity of cutting transport. The thickness of cuttings having a large diameter tend to be difficult removed from the wellbore, because the cuttings are going back down to the bottom of a well

6. Case Study

Aerated drilling modeling is based on synthetic data and combined with field data. The well data was released by [1]. The well X-1 was drilled vertically in 2002. At the time, field Y was an untapped resource and an exploration well was needed for more understanding and knowledge of the deep geothermal system. The well location was chosen based on results from a number of surface exploration surveys, which indicated that the reservoir temperature was above 280°C at depth. In the drilling report, increased fluid circulation loss was detectable down to 1600-1700 m depth, which indicates a feed zone below that, [1].

The study was focused to the section 9 5/8 in. The rate of penetration is in the range of 20 to 30 ft/hr for every depth of interest. The summary data that is used can be seen in table 1. The OD and ID of drill collar were 2 13/16 in and 4 ½ in, respectively. Modeling process in aerated drilling will be using a variation of gas injection rate ranged between 250 to 600 scf/min.

Table 1. Well X-1 Data, [1]

Depth	Pressure (bar)	Pressure (psia)	T (C°)	T (Rankine)
0	0	14.7	25	536.67
250	31	449.5	198	848.07
500	52	754	200	851.67
750	71	1029.5	185	824.67
1000	87	1261.5	210	869.67
1250	100	1450	205	860.67
1500	130	1885	170	797.67
1750	146	2117	220	887.67

From Well X-1, nine warm-up profiles and three injection profiles were available figure 2. Only six

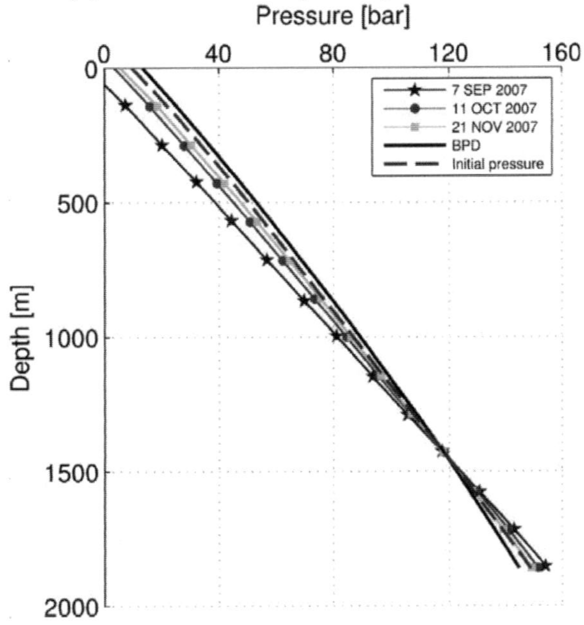

Figure 2. Pressure Profiles and Initial pressure estimation for Well X-1 [1]

warm-up profiles can be used for the interpretation of formation temperature, since the three others either do not have enough data or there has not passed enough time from when the well was closed to the measurement (shut-in time). Those six profiles are the latter measurement from 16 September 2002, from 10 October 2002, 18 August 2003, 22 August 2003 and the two from 7 September 2004. A feed zone is visible at 1900 m depth in the injection profile from 4 September, [1].

7. Result and Discussion

The back pressure is the most important key in maintaining pressure regime. At some point, gas and liquid tend to split up the quality of 0.8 (percentage of gas in the fluid). If the quality is above 0.8, then the worst will happen as the separation of gas and liquid, emerging slugs of water and can cause pressure surges in the hole, as well as gas or air can be a continuous phase so that the cutting cannot be lifted to the surface. figure 3 shows the operating zone that is the zone that can be safe in the process of aerated drilling.

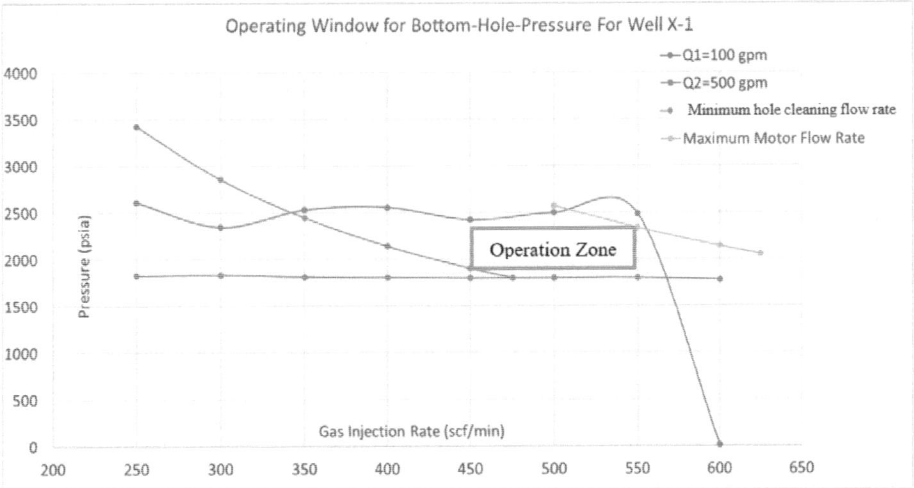

Figure 3. Operating Window for Bottom-Hole-Pressure for Well X-1

The modeling was done using two different pump flow rate. The first uses a flow rate of 100 gpm with gas injection rate variation ranges from 250-600 scf/min. Secondly, by using a flow rate of 500 gpm pump with gas injection rate variation of the same. Should be noted that the flow rate of the pump is at the minimum or maximum interval desired pressure and is under formation pressure (underbalanced drilling). figure 4 shows the gas injection rate required for each increase in temperature that will be encountered. the higher the temperature, the less for gas injection is needed.

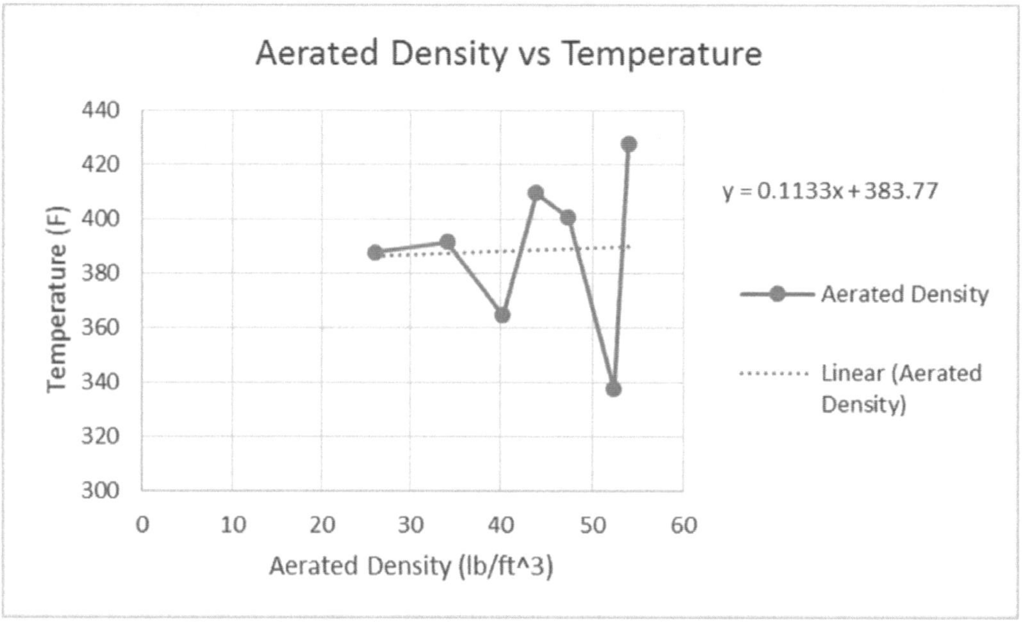

Figure 4. Aerated Density vs Temperature for Well X-1

figure 5 is a curve between the cutting size of the rock with the circulating pressure. In this study used the assumption that the density of the rock is ranging for 2.80 - 2.83 (SG). The size of cutting which can be lifted to the surface in the operating zone is 2.5 to 4-in.

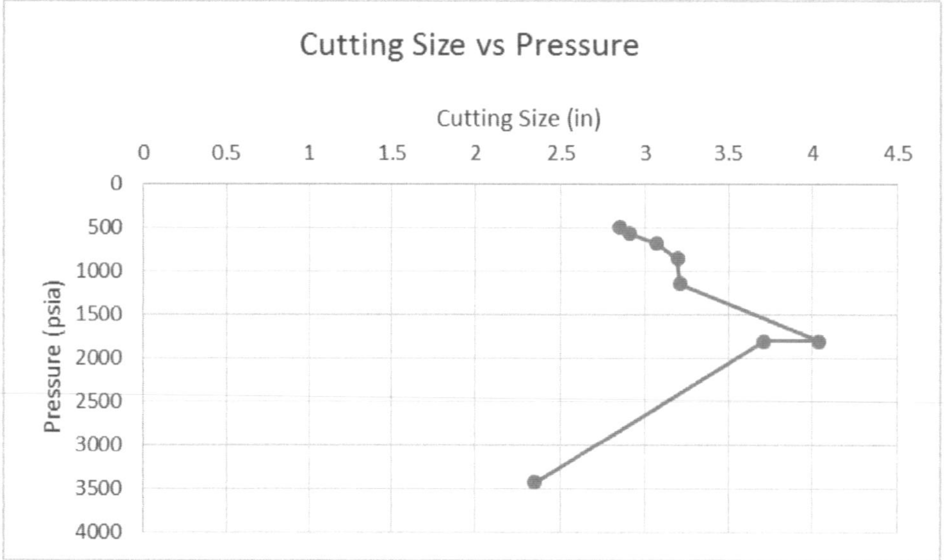

Figure 5. Cutting Size vs Pressure for Well X-1

Gas injection rate is very important on several factors such as pressure drop in the well, changing in the flow pattern, and the viscosity of the fluid phase and the charges. There are always limitation on how much gas can be added to the system to reduce the pressure as well. In figure 3 we can see that there are two limitations that cuts the respective curve given flow rate is currently 100 gpm and 500 gpm. The first limitation is the minimum flow rate to clear the hole of the cutting. The second limitation is the maximum flow rate that can be generated by the provided pump. Also in figure 6 can be seen that the higher the flow rate of gas that is injected, the maximum cutting required cutting settling velocity and transport velocity will increase. There are anomalies which can be seen on the profile of the maximum cutting settling velocity that is after the injection rate of 500 gpm maximum cutting settling velocity will continue to decline.

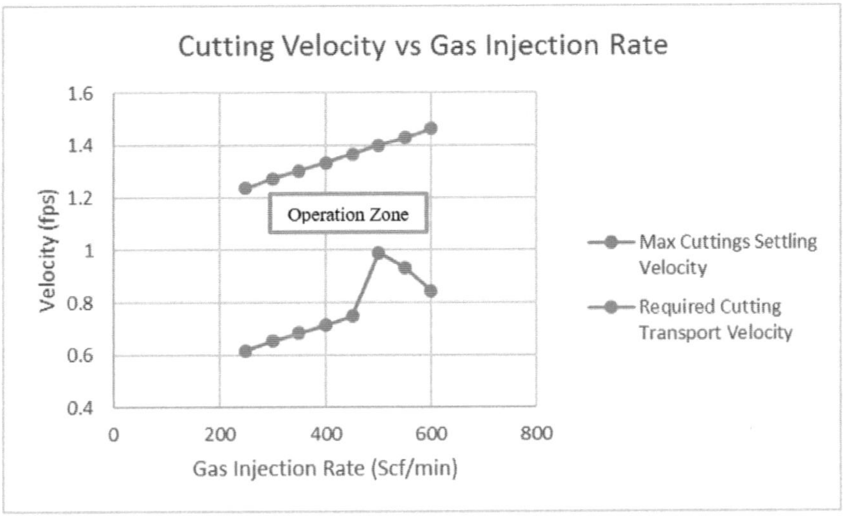

Figure 6. Cutting Velocity vs Gas Injection Rate for Well X-1

There is a different operation zone if the temperature is stated as constant. As shown in figure 7, the operation zone is limited to the small gas injection rate but the higher pressure required in circulating drilling mud.

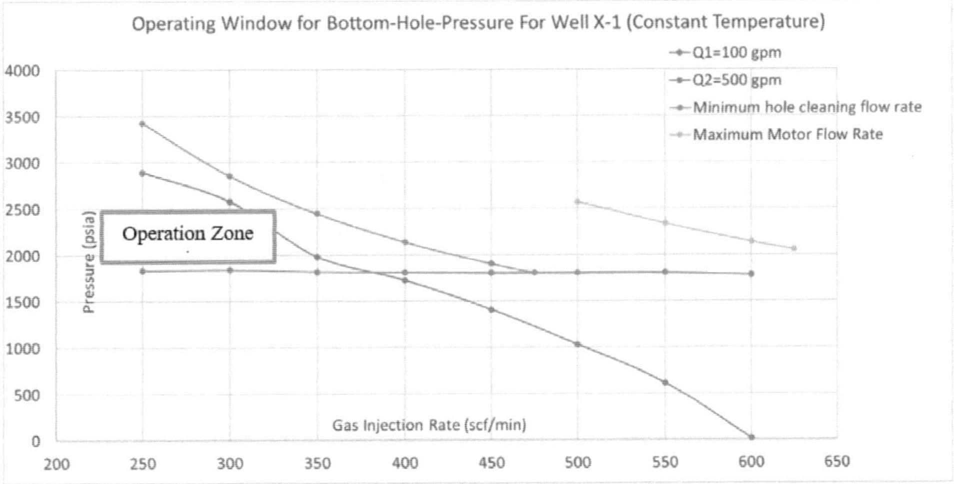

Figure 7. Operating Window for Bottom-Hole-Pressure for Well X-1 (Constant Temperature)

8. Conclusion

From this study, there are several lessons learned. Cutting-carrying capacity is depending on how much the gas is injected to the system of the drilling mud. Also, temperature and pressure play the role to create the large operation zone which is the safe zone for applying aerated drilling method. The operation zone is the range of required gas to be injected as well as the cutting will be effectively lifted to surface with the desired cutting size and the pump power which is available.

Nomenclature

$V_{sl} = Cutting\ Terminal\ Settling\ Velocity\ (fps)$
$D_c = Cutting\ Diameter\ (in)$
$\rho_c = Cutting\ Density\ \left(\frac{lb}{ft^3}\right)$

$\rho_f = Fluid\ Density\ \left(\frac{lb}{ft^3}\right)$
$R = Rate\ of\ Penetration\ \left(\frac{ft}{h}\right)$
$V_t = Cutting\ Transport\ Velocity$ (fps)
$C_c = Cutting\ Concentration$ (%)
$A = Surface\ Area\ (in^2)$
$d_b = Wellbore\ Diameter\ (in)$
$Q_l = Volumetric\ Flow\ Rate\ of\ Liquid\ (gpm)$
$Q_i = Volumetric\ Flow\ Rate\ of\ influx\ (gpm)$
$Q_m = Mud\ Flow\ Rate\ of\ Liquid\ (gpm)$
$Q_{air\ inj} = Air\ Injection\ Rate\ (gpm)$
$P = Pressure\ (psia)$
$T = Temperature\ (F^o + 460)$
$V_f = Average\ Velocity\ in\ the\ top\ of\ the\ drill\ collar$

References

[1] Gislason, G. (2014). Interpretation of Formation Temperature and Pressure of Wells in The Peistareykir Geothermal Field. Faculty of Earth Sciences, University of Iceland.
[2] Gray, K. E. (1958, January 1). The Cutting Carrying Capacity of Air at Pressures Above Atmospheric. Society of Petroleum Engineers.
[3] Guo, B., & Rajtar, J. M. (1995, September 1). Volume Requirements for Aerated Mud Drilling. Society of Petroleum Engineers. doi:10.2118/26956-PA
[4] Guo, B., & Ghalambor, A (2002). Gas Volume Requirements for Underbalanced Drilling. PennWell Corporation. Tulsa. Oklahoma.
[5] Rizo, T. M., & Cuenca, A. P. (1984, January 1). Aerated Fluid Drilling Observations In Geothermal Operation In Luzon, Philippines. Society of Petroleum Engineers. doi:10.2118/12455-MS
[6] Sharma, Y., Kamp, A., Yonezawa, T., Rivero, L. M., Kobayashi, A., & Gonzalez, J. (2000, January 1). Simulating Aerated Drilling. Society of Petroleum Engineers. doi:10.2118/59424-MS

Application of experimental design in geothermal resources assessment of Ciwidey-Patuha, West Java, Indonesia

Ali Ashat[1], **Heru Berian Pratama**[1]

[1]Geothermal Engineering Study Program, Institut Teknologi Bandung, Indonesia
Jl. Ganesha No.10, Bandung, West Java, Indonesia. 40132

heru.berian@geothermal.itb.ac.id

Abstract. The successful Ciwidey-Patuha geothermal field size assessment required integration data analysis of all aspects to determined optimum capacity to be installed. Resources assessment involve significant uncertainty of subsurface information and multiple development scenarios from these field. Therefore, this paper applied the application of experimental design approach to the geothermal numerical simulation of Ciwidey-Patuha to generate probabilistic resource assessment result. This process assesses the impact of evaluated parameters affecting resources and interacting between these parameters. This methodology have been successfully estimated the maximum resources with polynomial function covering the entire range of possible values of important reservoir parameters.

1. Introduction

Patuha geothermal field is located in west Java, Indonesia, about 50 km southwest of the city of Bandung. Patuha field is one of vapor-dominated geothermal system in Indonesia with liquid deep zone included and temperature about 210 – 230°C covering 20 km2. The steam cap about 700 m is underlying on huge deep liquid water zone that makes this system unique to be built in numerical simulation. Recently, Patuha field is operated by PT. Geodipa Energi which has produced 55 MWe to PLN in 2015.

The uncertainties in both subsurface parameters and the development to implement in Ciwidey geothermal field resources can be accommodate used probabilistic assessment. The geothermal developer has done this through using heat stored method with Monte Carlo simulation. The range of parameter values in these method covers the range of uncertainty for that particular variable. The proposed method for this study is experimental design with reservoir numerical simulation. Experimental design is a schematic was of simultaneously testing multiple variable that effect a response. A full factorial was used in this study. The objective of this study are build numerical simulation of Ciwidey geothermal field based probabilistic resource assessment.

2. Methodology

Experimental design (ED) is the process of planning a study to meet specified objectives. ED was used to reservoir simulation with probabilistic approach in order to study the reservoir model from uncertainty parameters. Therefore the field capacity as the amount of electricity generation (MW) from heat and mass production can be determined.

These study used a full factorial of 3 factor. The relationship of 2 level between all the 3 parameter may be estimated by a response polynomial or proxy equation (Equation 1).

$$Y = \beta_0 + \beta_1 A + \beta_2 B + \beta_3 C + \beta_4 AB + \beta_5 AC + \beta_6 BC + \beta_7 ABC \quad (1)$$

Where $\beta 0$ is the average value of the response and β_1, β_2,..., β_7 is regression linear. The polynomial describes the behavior of the responds between parameter including the interaction effects between parameters.

The parameters that have high uncertainty are liquid saturation (SL), porosity (ϕ), and permeability (k_{xyz}). These are the reference model parameters and it chosen for ED parameters.

The methodology shown in **Figure 1**. The conceptual model gives the general description to develop numerical model using TOUGH2 with interface. While the ED and polynomial equation generated from commercial experimental design.

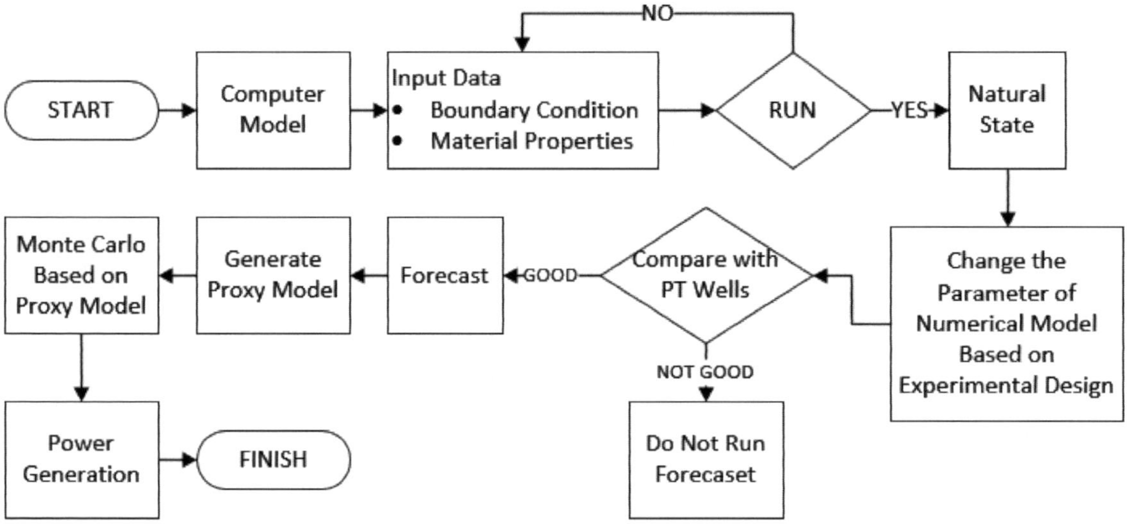

Figure 1. Methodology.

3. Conceptual Model

Patuha Geothermal System is consist of three reservoirs, there are associated with area of G. Patuha, G. S. Patuha and extend toward W-NW part and G. Urug (**Figure 2**), according to the resistivity survey. It is also confirmed by thermal gradient measurement at 150 m depth shows three anomaly area of high temperature [1] and ALOS PALSAR satellite imagery analysis [2] shows three lineament trending feature are probably identified as reservoir zone. The surface manifestation in Kawah Putih such as fumarole and crater lake is indicated as upflow zone of this system. The presence of fumaroles and boiling steam-heated waters in Kawah Cibuni and Kawah Ciwidey would be expected to upflow of vapor dominated reservoir, that allow the steam to rise toward the surface through permeable zone. The northeastern part of G. S. Patuha or definitely in area between wellpad 4 and wellpad 2 is probably interpreted as upflow zone due to increased pressure and temperature within well of PPL-02, liquid zone temperature of well is statically compared to others and also confirmed by chemical isotope and gas characteristic [3]. Eventhough, no surface feature as consequent the direct discharge of deep reservoir around this area.

The magmatic vapor plume in Kawah Putih contributed to supply the deep steam reservoir of this system [4] that pressure and temperature within steam reservoir increasing respect to magmatic plume. It is probably seperate with the vapor reservoir zone on the west and eastern part, due to geological structure controlled by Cimanggu Fault in the western and Cileueur Fault in the eastern part (figure 2). Ciwidey Fault and trending lineament of NE-SW and NW-SE around Kawah Ciwidey region are expected as permeable pathways of production well in this area, correspond to the surface geological structure mapping [5] and gravity interpreation from West JEC [3] as seen in figure 2.

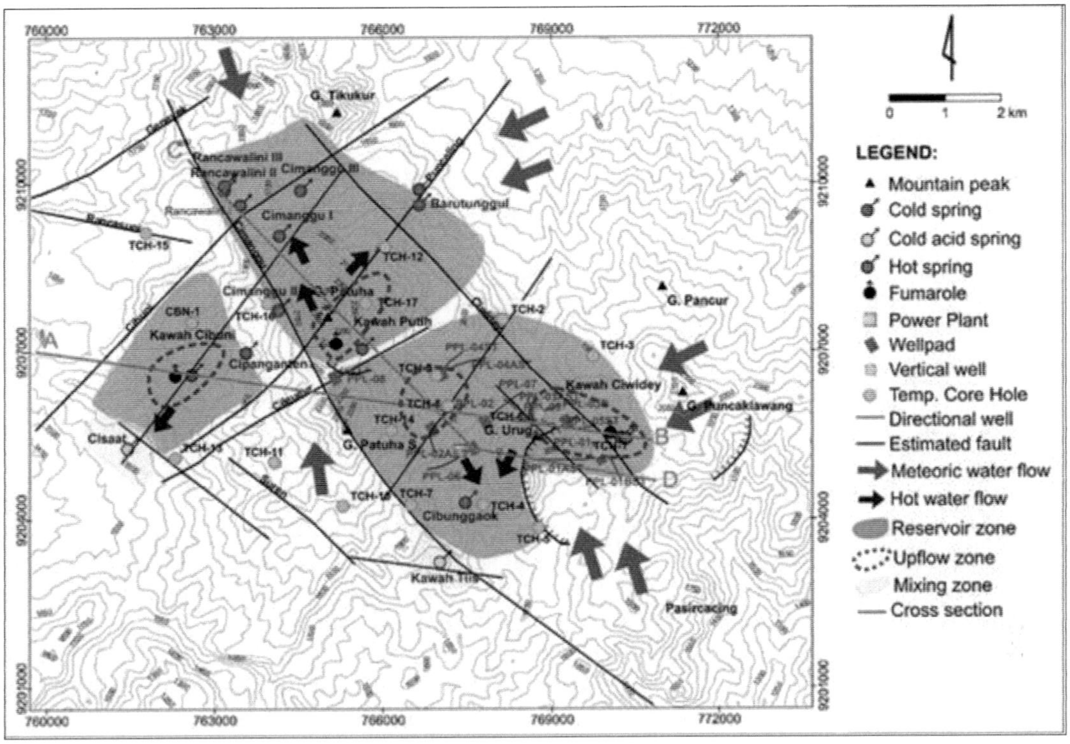

Figure 2. Lateral fluid flow pattern in Patuha Geothermal System [5].

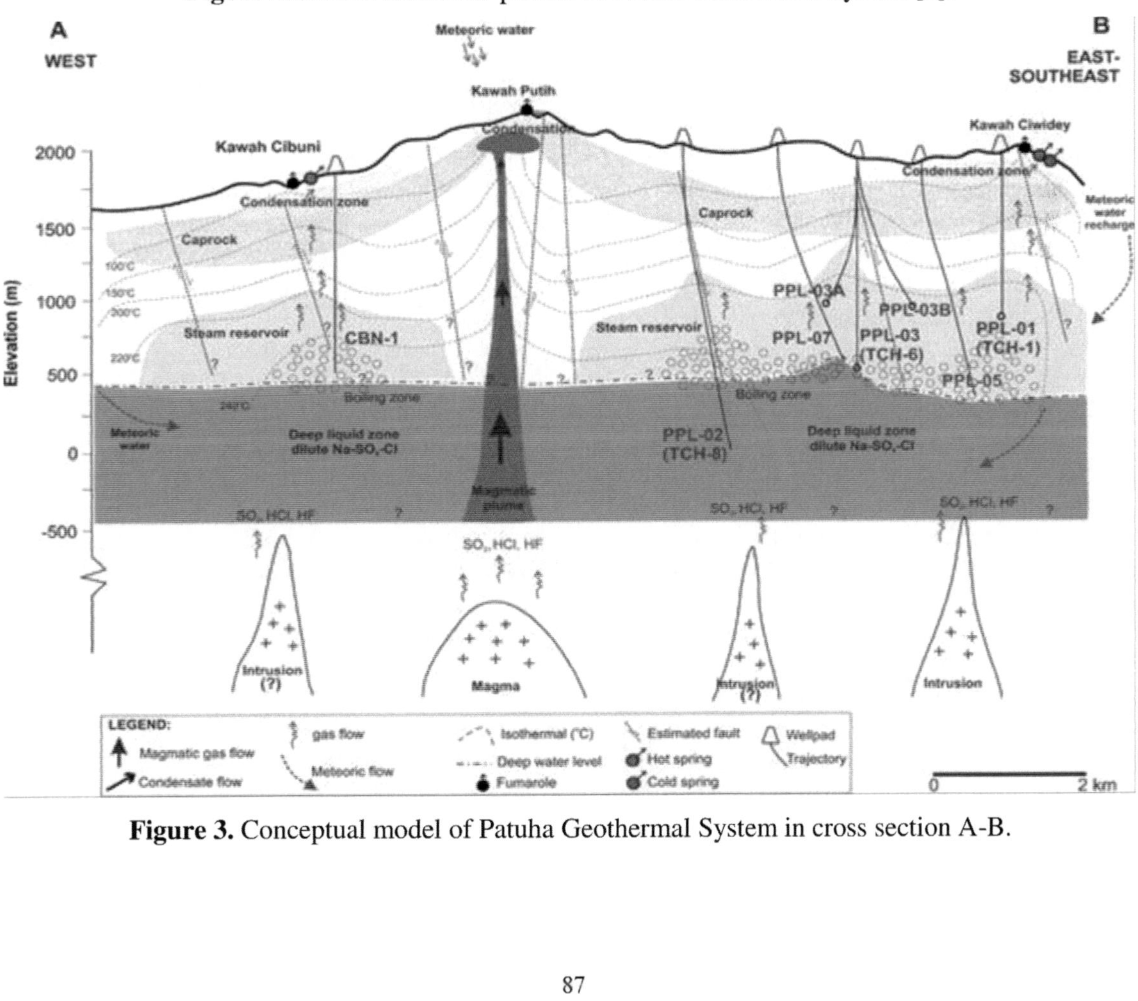

Figure 3. Conceptual model of Patuha Geothermal System in cross section A-B.

The deep reservoir of Patuha system is recharged by meteoric water in elevation 1.900 – 2.400 m asl, which is surround the centre of field, northwest, eastern and southwest part [6] as shown in figure 2. It is associated with area around G. Tikukur, NE of G. Patuha, G. Puncaklawang, Pasircacing and southern part of G. S. Patuha. Layman (2003) also reported that periperal meteoric water penetrate near G. S. Patuha and possibly quench the steam reservoir near this area.

Conceptual model of Patuha system is constructed by W-ESE cross section (A-B) as describe in figure 2. The cross section is reflected the reservoir characteristic of area of Kawah Cibuni, Kawah Putih and Kawah Ciwidey (figure 3).

Boiling condition within deep liquid reservoir lead the steam separated and give rise to vapor-dominated zone in the Patuha system. Steam condensation occurs and percolates downward the deep liquid zone, which may formed dilute and neutral Na-SO4-Cl waters. Steam reservoir in Kawah Ciwidey zone flows vertically toward area of G. Urug following Ciwidey Fault, and discharge fumarole in eastern part of Kawah Ciwidey. Hot spring also present in this area, whose sulfate acid water type, as steam-heating water due to groundwater heated by steam of reservoir near shallow depth (figure 3). The steam of production well in eastern such as PPL-01, PPL-05, PPL-03 and PPL-07 are supplied by reservoir which is associated with G. Urug. While the well of PPL-02 in the western part is connected with potential heat source in western part (figure 3). Corresponding to chemical analysis of production well, some of well such as PPL-01, PPL-01B, PPL-03, PPL-03A, and PPL-03B has same characteristic of gases composition, but different with PPL-02 and PPL-02A. It suggests that a different heating system is affected it.

Well of PPL-06 is presumed that supplied by the same source as PPL-02. On other hand, PPL-01A is a deepest well drilled in pad 1 and is not produced now. Based on chemical analysis, steam condensation occurs within this well due to meteoric recharge infiltrated through crater rim of G. Urug, as the trajectory direction of PPL-01A closed to this location [3]. Whereas, PPL-01B is situated in east corner, is purposed as injection well of PT. Geo Dipa. Reversal temperature occurs within this well due to groundwater influx in depth at elevation of 500 – 700 m asl. It is expected to recharge of meteoric water around G. Puncaklawang and Pasircacing [3].

Fluid of Kawah Putih crater lake is mainly related to magmatic gases derived from magmatic vapor plume that rising toward surface at the central of Kawah Putih, as condensation of steam into shallow aquifer or surface water affect the lake water. Meanwhile, the occurrence of fumarole in the shore of lake direct discharge of steam reservoir, based on chemical composition, no magmatic gases contained in it. The existence of the Kawah Putih reservoir zone is indicated by the TCH-17 drilling wells that penetrates the vapor zone at elevation of about 1,400 masl with a maximum temperature of 240°C. Magmatic steam condensates in Kawah Putih flows lateral to the NW slopes of G. Patuha and probably mixture with steam condensate from vapor dominated reservoir.

4. Ciwidey Numerical Model

A numerical model is constructed with conceptual model. Therefore it is set of interconnected elements which each of them represents of actual subsurface. The thermal and rock properties of the rock materials are represented as the thermodynamic properties of fluids of reservoir. To confirm that the constructed numerical model represents the Ciwidey-Patuha reservoir, it is subjected to two steps of calibration. The first is a natural state and the second step is a history matching calibration. In this study, after natural state is attained therefore the model run for experimental design (table 1). Material data for computer model.

The numerical model in these study is limited to ciwedey reservoir because GeoDipa generated 55 MW production from steam cap. The numerical model rotated parallel with Cwidey fault as a main fault (**Figure 4.** Ciwidey grid/block of numerical model.). The computer model has grid size of 5.5 x 5.5 km with rectangular mesh (250 x 250 m). The model is dived into 15 layers corresponding to 3.01 km in total thickness (2,010 msal to -1,000 msal). The layers have different thickness depending on the region (**Figure 5**). The atmosphere, reservoir and bottom layer are 10 m, 200 m and 500 m respectively. The total block from Ciwidey computer model is 7,260 blocks. The top layer as atmosphere was set homogenously at 1 bar and 25°C for initial pressure and temperature respectively. Bottom layer, heat source, was set 120 bar and 317°C it is based on the conceptual model with inflow around Mt. Patuha and Mt. Urug. The side boundary using hydrostatic pressure and normal temperature gradient to represent the surrounding environment condition. At side boundary of the model is in impermeable condition which is there are no flows of both heat and mass during natural state condition.

Table 1. Material Data.

Material	Porosity	kxy (m^2)	kz (m^2)
RESV1	0.10	1.00E-13	5.00E-14
RESV2	0.10	8.00E-14	4.00E-14
RESV3	0.10	8.00E-14	4.00E-14
RESL1	0.10	5.00E-14	2.00E-14
RESL2	0.07	4.00E-14	2.00E-14
RESL3	0.05	2.00E-14	1.00E-14
RSVG	0.08	2.00E-15	1.00E-15
FCIWI	0.15	3.00E-14	3.00E-14
FBDTL	0.125	2.50E-14	2.50E-14
FCIMA	0.125	2.00E-14	2.00E-14
FBLTG	0.10	3.00E-15	3.00E-15
SEAL1	0.05	8.00E-15	8.00E-15
UPZ07	0.10	8.00E-14	4.00E-14
UPZ02	0.10	8.00E-13	4.00E-13
VERTI	0.10	8.00E-13	4.00E-13
SEAL2	0.05	1.00E-17	1.00E-17

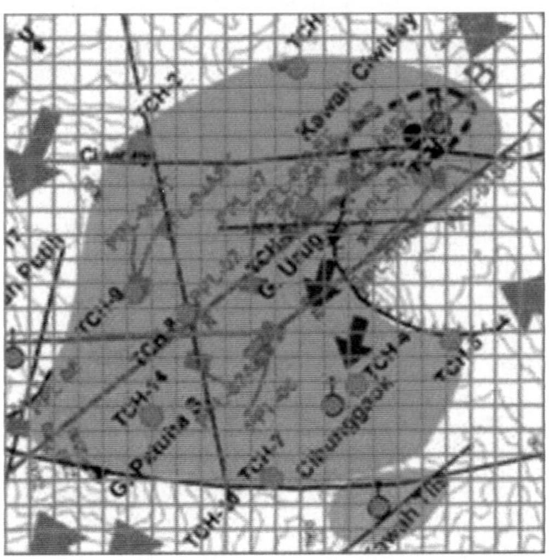

Figure 4. Ciwidey grid/block of numerical model.

The numerical model should represent actual conditions based on the information from conceptual model. Thus the material properties that represent rock type for the model should be assign into the computer model. It consist of density, porosity, permeability, thermal conductivity and specific heat. Additionally the relative permeability corey and relative liquid saturation 0.3 applied in these model (Permeability and porosity are considered as most important parameter in order to have steam cap underlying brine reservoir natural state model (**Table 1**).

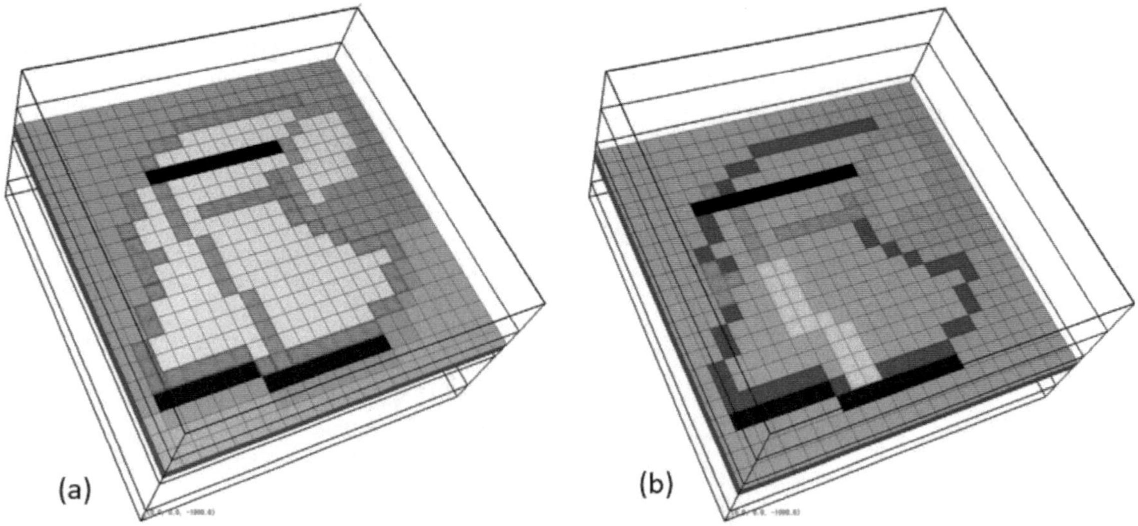

Figure 5. Ciwidey numerical block at steam cap layer.

5. Ciwidey Natural State

The main objective of the natural state calibration is to verify the pressure and temperature distribution and the heat and mass flow of the model. Pressure and temperature profile at natural state shown at Apendix A. It shows a good match of pressure and temperature between a model output and actual well data for PPL and TCH Wells. The simulated pressure and temperature profiles at existing production wells, PPL-02 and PPL-06 for example (**Figure 6.** Matching on the pressure and temperature profile of PPL-02 and PPL-06 Well.), has reproduced the steam dominated at top of the reservoir underlying liquid dominated reservoir. While the TCH well shown a conductive heat transfer because it through impermeable layer.

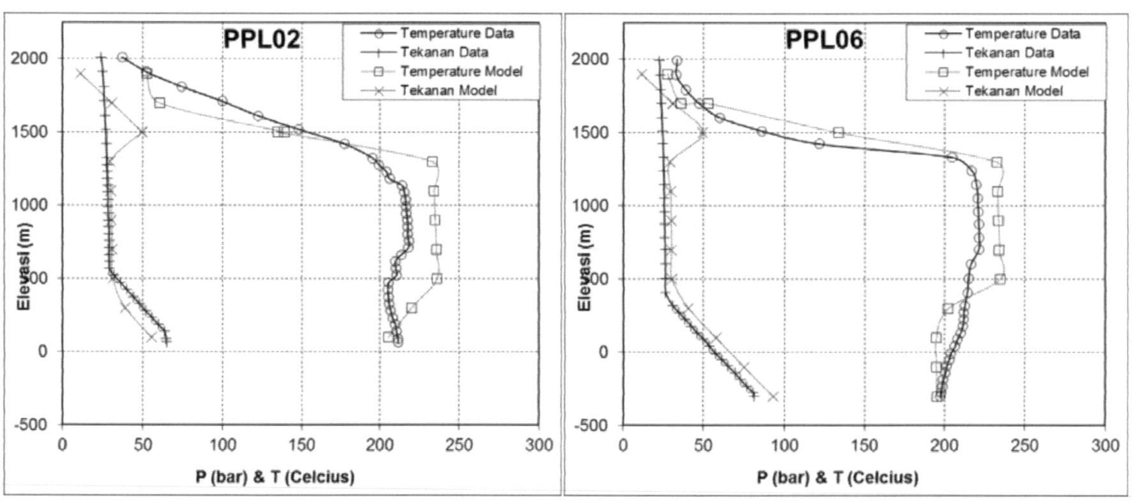

Figure 6. Matching on the pressure and temperature profile of PPL-02 and PPL-06 Well.

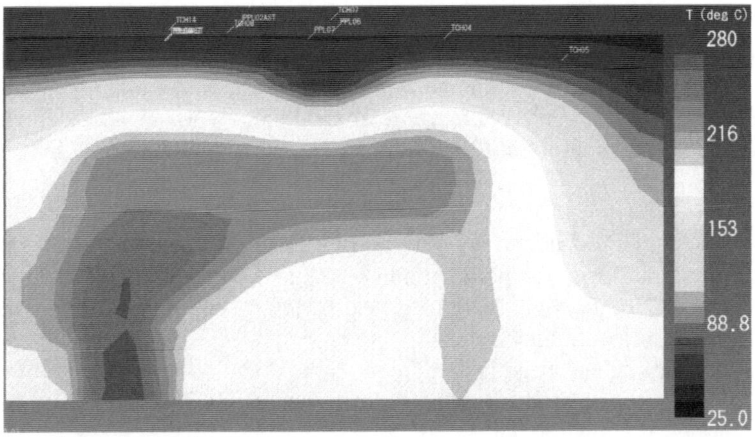

Figure 7. The temperature cross-section of natural sate Ciwidey Model.

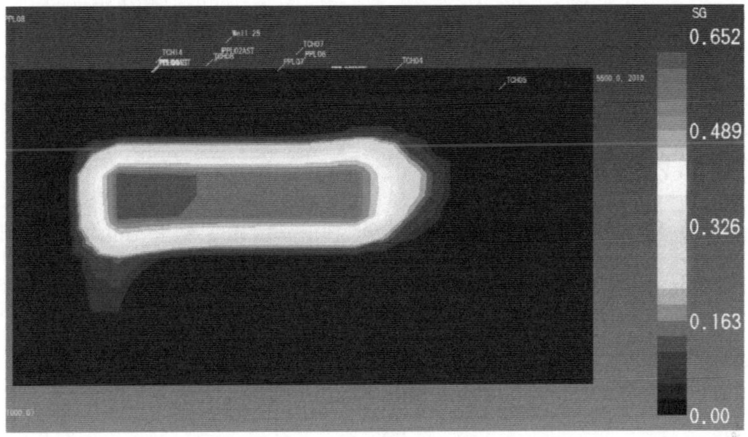

Figure 8. The gas saturation cross-section of natural sate Ciwidey Model.

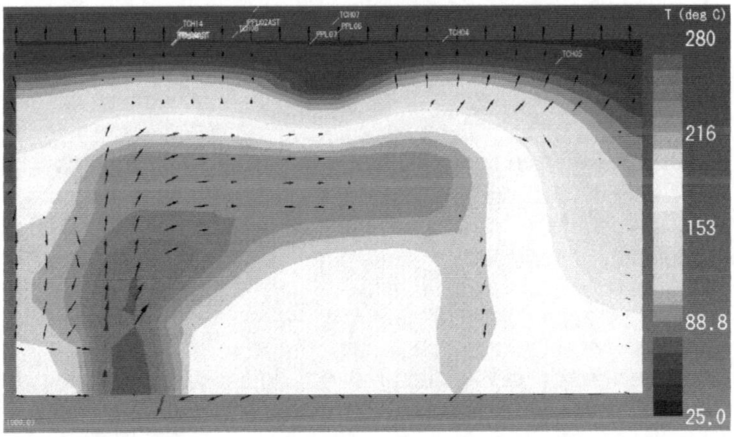

Figure 9. The heat transfer at natural sate Ciwidey Model.

The cross-section of natural state model shown at **Figure 7** and **Figure 8** for temperature and gas saturation. These has similarity with Ciwidey conceptual model shown in **Figure 3**. The heat source in these model is Kawah Putih as main heat source and intrusion at the bottom of Urug Mountain. Hot water is heated by a hot reservoir rock and will go up through good vertical permeability because high temperature fluids has a lower density. The boiling zone will segregate vapor phase and liquid phase then the steam cap will form and will be moved laterally the reservoir and spread (**Figure 9**). The gas saturation on the steam cap has a maximum value around 65%. The PPL-02 well is affected by a heat source indicated by higher temperature. PPL-06 indicated has a main heat source as PPL-02.

6. Sustainability

The experimental design (ED) concept were applied in Ciwidey reservoir model in order to study the dependency from uncertain reservoir parameters of to know the amount of production (MW) for 30 years. It also has a purpose to derive a proxy model (polynomial equation) to the reservoir simulation in the probabilistic Monte Carlo simulation.

There are 3 main parameters were investigated: liquid saturation (SL), matrix porosity (ϕ), and permeability (k_{xyz}). The low and high value from matrix porosity and permeability are ±20% deviations from material properties as shown in **Table 2**.

Table 2. Parameter for ED

Parameter	Low (-1)	High (+1)
Liquid saturation (SL)	0.3	0.5
Matrix Porosity (ϕ)	-20% from base	+20% from base
Permeability (k_{xyz})	-20% from base	+20% from base

A two level (low and high) of full factorial with three factor for ED was implemented. The total run in these study is 8 runs for 30 years of production. In the design table (**Table 3**), the standard order column is an experimental design idea to do the experiments in a random order (run order) and avoid run-dependent effects.

Table 3. A two level of full factorial design.

StdOrder	RunOrder	SL	ϕ	k_{xyz}
7	1	-1	1	1
2	2	1	-1	-1
4	3	1	1	-1
1	4	-1	-1	-1
5	5	-1	-1	1
3	6	-1	1	-1
6	7	1	-1	1
8	8	1	1	1

The reservoir simulation were run equally distributing using well on deliverability method with PI 2.0E-12 m^3 and WHP 10 bar. The total production into 37 wells with targeting at steam dominated reservoir. The power capacity (MW) estimated by steam production over time project times specific steam consumption shown in Equation 2.

$$MW_e = \frac{\sum_{m=1}^{L} m \times \Delta t}{L \times SSC} \tag{2}$$

Where MWe is the power capacity, m is steam produced (kg/s), Δt is delta time at simulator (years), L is time project (years) and SSC is specific steam consumption (kg/s/MW).

The results of Ciwidey ED for 30 years of steam production with 2.016 kg/s/MW of SSC shown in **Table 4**. The minimum power capacity is 124.5 MW and the maximum is 131.1 MW.

The Ciwidey numerical simulation are approximated by polynomial equation to generate the power capacity (MWe). The polynomial equation of power capacity as a function of three independent parameters are the results in these study. A two level proxy model using -1 and +1 parameters to generate the power capacity responses for a Ciwidey model shown in these Equation 3.

Table 4. The results of Ciwidey ED.

RunOrder	MWe
1	128.4
2	126.5
3	126.3
4	125.3
5	129.9
6	124.5
7	131.1
8	130.4

$$MWe = 127.8 + 0.7670*SL - 0.4074*\phi + 2.132*k + 0.1789*SL*\phi + 0.02135*SL*\phi - 0.1459*\phi*k + 0.01660*SL*\phi*k \quad (3)$$

The next step are running Monte Carlo simulation based on generated polynomial equation of power capacity with 50,000 iterations using spreadsheet. The result is a probability distribution function of the power capacity covering the full range of possible outcomes. The cumulative distribution for 30 years steam production of the power capacity shown in **Figure 10**. The P10, P50 and P90 are 126 MW, 128 MW and 129 MW respectively.

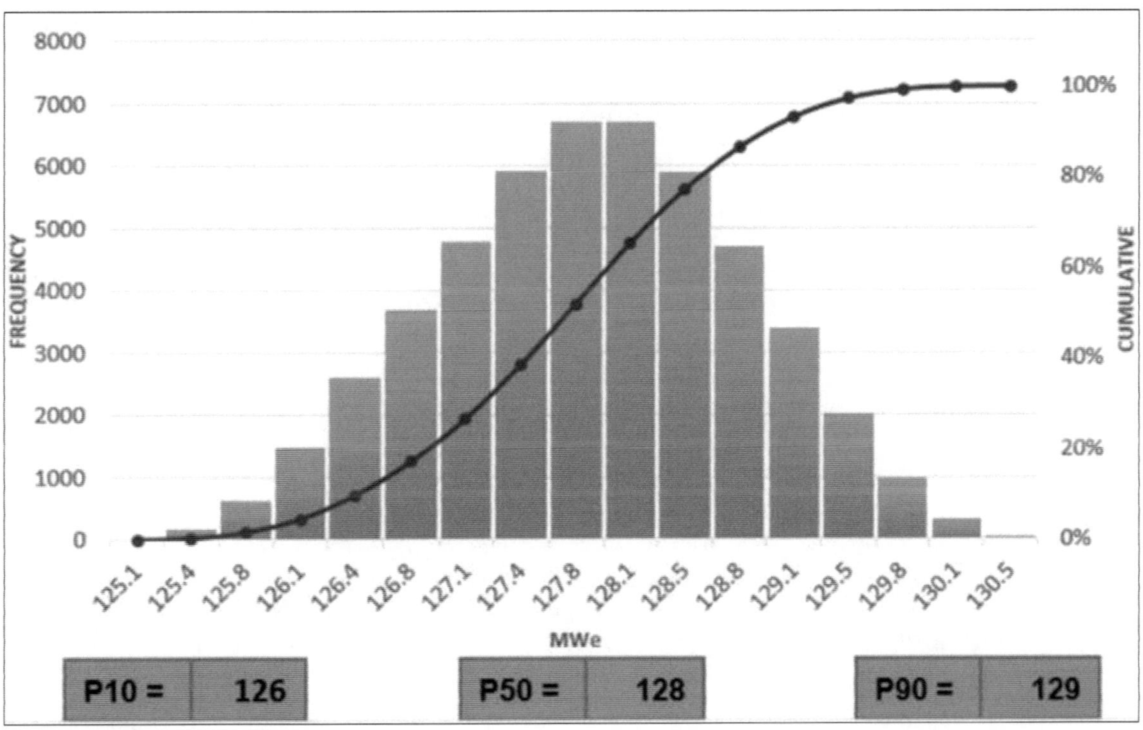

Figure 10. Monte Carlo simulation results with histogram and cumulative distribution function with P10, P50 and P90 indicated.

7. Conclusion

1. The model of Ciwidey-Patuha geothermal field with 5.5 km x 5.5 km area has successfully developed and it showed a good match with actual PPL well and conceptual model.
1. The numerical simulation with Experimental design has successfully applied for Ciwidey-Patuha geothermal field.
2. The probabilistic power capacity using Monte Carlo based on experimental design for 30 years production, P10-P50-P90, are 126 MW, 128 MW and 129 MW respectively.

Acknowledgement

The authors would like to sincerely thank my supervisor Prof. Dr. Ryuichi ITOI, Kyushu University and Ir. Nenny M. Saptadji Ph.D., Institut Teknologi Bandung. The reaserch was supported by PT. Geo Dipa and Ruly H. Ridwan (Engineering Manager) for supporting the study and allowing us to access the data.

References

[1] Alhamid, I. (1989): The Resources Potential of the G. Patuha Geothermal Area, West Java, Proceedings Indonesian Petroleum Association, 89-14.01.

[2] Pradipta, R. A. (2016): Penentuan Zona Permeabilitas Menggunakan Citra ALOS Palsar, Pemetaan Geologi, dan Geokimia Tanah dan Udara Tanah (Hg-CO 2), Studi Kasus Lapangan Panasbumi Ciwidey, Kabupaten Bandung, Thesis of Magister, Institut Teknologi Bandung, Unpublished.

[3] Amelia, Y. (2014): Studi Pola Aliran Fluida Sistem Panas Bumi Patuha, Thesis of Magister, Institut Teknologi Bandung, Unpublished.

[4] Layman, E. B., and Soemarinda, S. (2003): The Patuha Vavor-Dominated Resource West Java, Indonesia, Twenty-Eighth Workshop on Geothermal Reservoir Engineering (p. SGP-TR-173). Stanford: Stanford University.

[5] Suswati, Mulyana, A. R., and Sutawidjaja, L. S. (2000): Pemetaan Geologi Komplek Gunungapi Patuha Kabupaten Bandung, Jawa Barat. Bandung, Center of Volcanology and Geological Hazard Mitigation.

[6] Fahlevi, I. (2009): Penentuan Daerah Resapan Lapangan Panas Bumi Patuha, Thesis of Geology Engineering, Institut Teknologi Bandung, Unpublished.

IOP Conf. Series: Earth and Environmental Science **103** (2017) 012010 doi:10.1088/1755-1315/103/1/012010

Make-up wells drilling cost in financial model for a geothermal project

Fitri Oktaviani Purwaningsih[1], **Ruly Husnie**[2], **Waldy Afuar**[1], **Gugun Abdurrahman**[1]

[1]Study Program of Geothermal Engineering, Faculty of Mining and Petroleum, Bandung Institute of Technology, Bandung, Indonesia
[2] PT Geo Dipa Energy, Jakarta, Indonesia

Email : Feetre@gmail.com

Abstract. After commissioning of a power plant, geothermal reservoir will encounter pressure decline, which will affect wells productivity. Therefore, further drilling is carried out to enhance steam production. Make-up wells are production wells drilled inside an already confirmed reservoir to maintain steam production in a certain level. Based on Sanyal (2004), geothermal power cost consists of three components, those are capital cost, O&M cost and make-up drilling cost. The make-up drilling cost component is a major part of power cost which will give big influence in a whole economical value of the project.

The objective of this paper it to analyse the make-up wells drilling cost component in financial model of a geothermal power project. The research will calculate make-up wells requirements, drilling costs as a function of time and how they influence the financial model and affect the power cost. The best scenario in determining make-up wells strategy in relation with the project financial model would be the result of this research.

1. Life cycle of a geothermal field

A geothermal field has a life cycle that is different with other energy resource. It consists of four periods: (1) Developing, (2) Sustaining, (3) Declining and (4) Renewable [2]. In developing period, the increments of a plant capacity come in line in steps. In sustaining period, the output of the field remains steady over an extended period of time. The declining period begins when the underlying declines in the capacity of individual wells become evident in the actual output of the entire facility. The renewable period would achieved when the field reach a state in which net mass withdrawals in production stage and the operations are still profitable at this level.

A geothermal power plant is built to generate a specific capacity that requires a certain supply of geothermal steam. One of the consequence in a continuing operation of a geothermal field is decreasing in production of steam. A lowering production to below the limit would cause a reduction in plant output. A continued reduced output would reducing the economic value of the project along with inability to meet previous contract with the electricity buyer.

Typically, a developer will drill enough wells at the early period of the project to be able to meet plant capacity requirement, even set a limit of minimal surplus capacity at the wellhead to anticipate declines in well productivity. Later, when previous production wells have decreasing production, constant output is sustained by drilling make-up wells [2].

IOP Conf. Series: Earth and Environmental Science **103** (2017) 012010 doi:10.1088/1755-1315/103/1/012010

2. Flowchart of this paper

The objective of this paper is to analyze make-up well cost component in a financial model of a geothermal power project. In this research, synthetic data is used to calculate make-up wells requirement to maintain sustainability electricity production of 55 MW plant with project lifetime 30 years. The data is obtained from sample case in a geothermal field in Indonesia. The project is assumed to starts in 2017 with four years allocation time for exploration, feasibility study stage and plant development. The field is estimated to start producing electricity by 2021. The complete flow chart of this research is in figure 1.

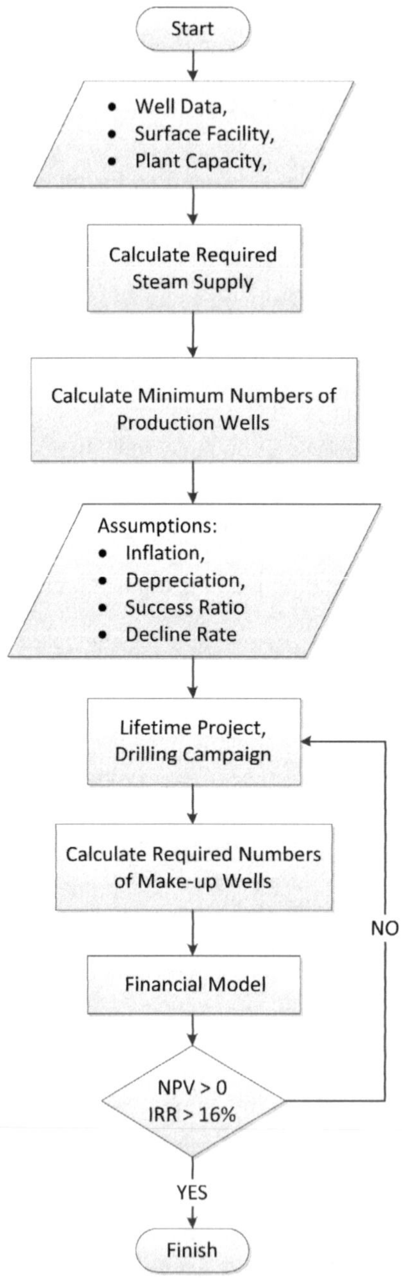

Figure 1. Working flowchart of the research

Data collection is the first step done from this research. The required data is:

1. Well data retrieved from exploration and production wells
2. Surface facilities data, that include separator pressure, turbine characteristics and other surface facilities components

Those data is required to determine required minimum steam supply to generate total capacity of 55 MW and accommodate 10% of excess steam supply and to calculate required number of production wells. Then the data is used to calculate required number of make-up wells. The number of make-up wells and its drilling cost then put into a geothermal financial model.

The financial model will calculate make-up well cost together with other parameters, including exploration, well drilling, O&M and other costs, as well as revenue, tax, interest and all parameters that built the model. The model then will used to evaluate value of the project using some financial parameters as NPV (Net Present Value), IRR (Internal Rate of Return) and PBP (Payback Period).

This research will study make-up wells cost as one of component in a financial model. Evaluation of the project with NPV, IRR and PI calculation would be done only to describe variation of those financial indicators related with different strategy in make-up wells drilling. Calculation will be done using basic financial model with expected result NPV > 0 and IRR > 16%. The IRR number is retrieved from Abadi Purnomo (2007) stated that ideal IRR of a geothermal project is 16%.

3. Required parameters to calculate make-up wells

Miller stated that the term 'make-up' wells are applied to all wells drilled after plant start-up [3]. The timing of the drilling of make-up wells and the number of those wells can make the difference between project economic success and failure. The required number for make-up wells and the timing of drilling those wells is directly dependent on some aspects as: plant capacity, life time of the project, capacity of production wells, decline rates of the field and the minimum steam allowance.

Make-up wells are drilled to accommodate steam required by the geothermal plant to be able to produce certain capacity. There are some parameters that need to be determined to calculate the number of make-up wells required in a geothermal field for a specific lifetime project. The most important parameters are plant capacity, lifetime of a project, average flow rate of geothermal well and steam consumption rate to produce electricity. All calculation in this research use data from a synthetic field of a steam dominated geothermal plant in Indonesia.

Table 1. Parameters of the field and calculation of number of wells required as production wells in the beginning of the project.

No.	Parameter	Value	Simbol
1	Plant Capacity	55	MWe
2	Lifetime project	30	year
3	Turbin Input Pressure	6.5	bar
4	Turbin Efficiency	85%	ηt
5	Steam flow rate	371	ton/h
6	Specific steam consumption	6.7	ton/h/MW
7	Average well production rate	20	kg/s
8	Steam fraction in well	90%	
9	Steam flow rate from a well	18	kg/s
10	Total mass flow rate (required)	114.6	kg/s
11	Well drilling success ratio	80%	
12	Average well capacity	9.6	MW
13	Required production wells	8	unit
Make-up wells calculation parameters			

No.	Parameter	Value	Simbol
14	Excess Steam	10%	
15	Minimum Steam Supply	60.5	MW
16	Decline Rate	3%	Per Year
17	Average Production Capacity	9.6	MW/well

Some parameters in table 1 determined based on data on synthetic field in a steam dominated geothermal field in Indonesia. Some others are calculated based on previous data. Calculation in this paper use specific steam consumption 6.7 ton/hour/MW and plant capacity 55 MW. Based on the calculation, the requirement of productions wells that have to be drilled in the beginning of the project is 8 wells with each well have flow rate 20 kg/s or able to produce 9.6 MW in average. It means that the geothermal plant need steam supply similar to 114.6 kg/s to be able to produce 55 MW in electricity. The requirement is fulfilled by drilling 8 production wells with flow rate 18 kg/s and turbine efficiency 85%, or capacity 9.6 MW/well.

At initial production, there are 8 production wells that produce steam to fulfil plant requirements. Concomitant with time, the field will have production decline due to lowering pressure, scaling or other factor and the production will decline. Then the make-up wells need to be drilled to fulfil steam production requirement.

4. Calculating make-up wells

To be able to calculate the number of make-up wells, there are some other parameters required. Those are: excess steam, minimum steam supply and decline rate. Excess steam is a level of surplus in steam production to maintain supply to geothermal plant above minimum requirement. In this research the calculation for make-up wells will use excess steam 10%.Minimum steam supply is a number of minimum steam production supplies to geothermal power plant. With excess steam 10% then the minimum supplies value is 110% of plant requirement.

Decline rate is directly related to wells capacity to supply steam to geothermal power plant. Well productivity decline rates may vary widely in different geothermal fields. In this paper decline value is determined in 3% per year as stated in Adiprana (2015). Those parameters are calculated in order to determine the required time to drill make-up wells and number of wells to be drilled. Specific values of each parameter are in table 1.

Based on table 1, minimum flow rate to be applied to supply a 55 MW geothermal plant is 114.6 kg/s or equal to 55 MW. But ideal requirement is about 60.5 MW considering 10% excess steam requirement. Thus, when production rate almost reach 60.5 MW or below, the field needs addition supply, which can be accommodated by drilling make-up wells.

However, there are some limitations that need to be considered in make-up wells drilling procedure as:

1. The real decline rate is vary through the year and wells, but in this research we use similar decline rate for the field to forecast the production decline in whole period of plant (30 years)
2. The minimum number of wells drilled in a drilling campaign period is 2 or 3, considering the high rig mobilization cost. This research will use different number of minimum wells drilled in a drilling campaign period as part of determining make-up wells strategy.
3. Success rate of well make-up well drilling will affect the number of required well drilling in each drilling campaign period. Calculation in this paper will use success rate 80%.

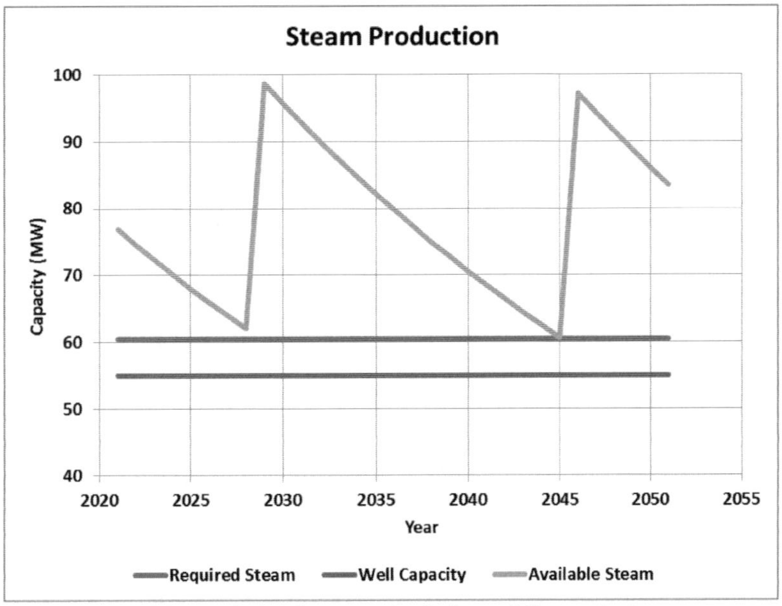

Figure 2. Make-up calculation of Scenario 1

Figure 2 contains make-up wells strategy for the whole 30 years of project. The field needs 8 make-up wells to maintain production level. Those wells need to be drilled in two drilling campaigns, four wells each, considering the 80% success ratio. The drilling time would be in year 8 (2029) and year 25 (2046) with assumption that the project start in 2017 (for exploration and feasibility stage) and electricity production start in 2021.

5. Make-up well cost in financial model

Make-up well drilling is treated as an occasional operating cost allocated to the years in which is anticipated that additional drilling will be required [1]. Inserting the make-up wells drilling into financial model must consider some important parameters, as inflation and depreciation.

Inflation is defined as a sustained increase in the general level of prices and usually measured as an annual percentage increase (Investopedia). In make-up wells cost component, the cost is increasing with a certain degree of inflation. The escalation cost is influenced by initial drilling cost in the start of the project, inflation rate and year, the longer drilling time from year start of the project and the higher inflation rate, the more costly the drilling price would be.

The calculation of make-up well cost is using initial cost of make-up well drilling as USD 8 million per well in start year of the project (2017), and inflation rate based on US PPI (United States Producer Price Index) 2.5% in January 2017.

The more detailed calculation of present make-up well cost is in APPENDIX A. Based on previous calculation, the additional make-up wells are required in year 8 (2029) and year 25 (2046) of the project. Considering inflation, the required cost to fund 4 make-up wells drilling in 2029 is USD 38.98 million and USD 59.33 million in 2046. The total cost to drill make-up wells for the whole project is USD 98.3 million.

6. Geothermal financial model

Every geothermal project is different with others because each of them has their own resource characteristics with different risks and opportunity. The cost driver also behaves differently under different conditions. Sanyal stated that geothermal power cost consists of three components, those are capital cost, O&M cost and make-up drilling cost [5]. The make-up drilling cost component is a major part of power cost because it requires big amount of money that must be spent during the plant project.

A financial model is anything that is used to calculate, forecast or estimate financial numbers. A valuable financial model integrates accurately all the costs throughout all phases of development and presents the resulting information in a manner to help various users make the appropriate decision (Ngugi, 2014). One of the objectives of this paper is to provide analysis on technical aspect of a geothermal project, focusing on make-up wells cost to apply in a proper financial model of the project.

The research will not focus on detailed data on specific process and determined cost in a financial model, yet it will explain about how to put the make-up well cost properly in the structure of general geothermal financial model.

6.1. Make-up well cost in geothermal financial model

A financial model in general is designed to meet the information need of investor and decision maker with objective to consider whether the investment for the project is viable. The model is consists of four financial statements that prepared annually in most cases, present the result and financial position of a project (or a company) by a certain date (Ngugi, 2014). Those are:

1. Income statements
   The objective of income statement is to establish the profitability of the project. The main elements in this statement are revenue, expenses, depreciation, interest and tax.
2. Balance sheets
   This statement detailed financial position and consists of two major grouping, the assets and the equity and liability.
3. Cash flow statement
   This statement mainly account for the cash availability to the company or project owner. Amount of the cash accounted will define how much it can meet its financial obligation. The statement is divided into three major segments: cash flow from operation, cash flow from investment activities and cash flow from financing activities.
4. Statement of owner's equity

The make-well costs are included in a geothermal financial model in some statements as:

- Operating expenses in Cash Flow Statements, that encompass all cost directly and indirectly associated with the generation of the sold energy
- Depreciation expenses, because the make-up well cost is applied depreciation with a straight line or fixed percentage deduction or other depreciation method applied

6.2. Make-Up Well Cost in Cash Flow Statement

Make-up wells cost is occasional operating cost which allocation years and amount is retrieved from previous calculation in table 2. Forecasting the year of required drilling time needs technical calculation while the amount of required cost would be affected by inflation rate. If there is more than one power plant unit in the project, calculating the requirement of make-up well and forecasting their cost must be done for each unit.

6.3. Make-up Well Cost in Depreciation Expense

Depreciation is an accounting method of allocating the cost of tangible asset over its useful life. A business financial model will depreciate long-term asset for both tax and accounting purposes. For tax purpose, businesses can deduct the cost of the tangible assets as business expense. For accounting purpose, depreciation expense does not represent a cash transaction but it indicates how much of an asset's value has been used up over time.

Calculation depreciation of make-up well cost will provide better basis for investment decision. The life of a make-up well would be calculated based on the time it was drilled. There are some methods to calculate depreciation: the Straight Line and the Double Declining method. Besides choosing the method, calculating depreciation also must consider the depreciation period time.

6.4. Make-Up Well Cost in Balance Sheet

Make-up Wells are considered as fix asset with book value calculated by taking the cost minus the accumulated depreciation. Book values of make-up wells cost are carried in the balance sheet with other fix assets as exploration book value and power plant.

7. Make-up well cost in determining investment feasibility of a project

Make-Up well cost will determine the feasibility of a geothermal project. Make-up wells drilling cost is varying with range USD 3.2 Million (SKM, 2009) to 6 Million (Sudarman, 2012). The cost will increase along the time and will be allocated as operating cost for the whole project. The huge cost of make-up wells drilling along the project would affect the investment feasibility.

Some financial indicators as Nett Present Value (NPV), Internal Rate of Return (IRR) and Payback Period (PBP) determine investment feasibility of a geothermal project.

7.1. Nett Present Value (NPV)

NPV is equal to present value of cash flow substract with initial investment, which represented by following formula:

$$NPV = \sum_{t=1}^{n} \frac{C_t}{(1+r)^t} - C_0 \quad (1)$$

With r is internal rate of return (IRR), C0 is initial investment, Ct is cash flow in Year t and n is period of investment in year. Positive NPV indicate that the present value of cash flow is more than initial investment. Investor would need positive NPV that represent they have benefit from their investment.

7.2. Internal Rate of Return (IRR)

IRR is the discount rate that makes the present value of investment's income stream total to zero. Following formula represents the IRR:

$$\sum_{t=0}^{N} \frac{(CF_t)}{(1+IRR)^t} = 0 \quad (2)$$

With CFt is cash flow at year t. The IRR is expected rate of return of the project that can be used to forecast margin. When the IRR is larger than the cost of capital, the project will generate a positive margin for the shareholder. Therefore a project with IRR greater than the cost of capital will increase the project worth.

7.3. Payback Period (PBP)

The payback period is the length of time required to recover the cost of an investment. This indicator is important to determine whether to undertake the position or project, as longer payback periods are usually not desirable for investment position. The amount of make-up wells cost along the whole time of the project will affect those financial feasibility projects. The huge cost that allocated for some periods of the project will influence NPV, IRR and Payback Period.

Calculation of those financial indicators would not be done in this research. That calculation would need further research and more complex parameters and project strategy that would not cover in this paper.

However, there are some factors that influence make-up wells calculation in cash flow that will generate different value of IRR, NPV and Payback Period. Those are:

1. Amount of initial well drilling cost
2. Inflation rate
3. Minimum well drilled in a drilling campaign period
4. Make-up well drilling success ratio

5. Allocated drilling time

Considering the time value of money, the bigger cost in the beginning of the project will generate smaller NPV and IRR and longer PBP. Further research must be done to find out how different strategy of make-up wells drilling strategy influences those indicators. The sensitivity analysis also needs to be done to determine how different strategy of make-up well drilling impact NPV, IRR and PBP.

8. Make-up well cost affect project financial

Determining when and how much make-up wells need to be drilled is one of production strategy. In technical view, the make-up wells should be drilled soon after production rate decreasing to the minimum requirement. In financial consideration make-up wells should be drilled in cost efficient strategy.

In this paper, we only present four scenarios. Actually, in the real world there are many possible scenarios regarding the number of required make-up wells to be drilled and when they are drilled. The research choose only four possible scenarios only based on minimum steam requirement and minimum wells drilled at one drilling period. In this research the limitation of minimum wells drilled in one drilling period is 3 wells, based on economical reason and experience in real case in geothermal field.

To be able to know how big make-up well cost affects the feasibility of a geothermal project, this paper will compare financial indicators from four different make-up wells strategy. Both are in the same field and use similar data as previously calculated. The differences between those scenarios are drilling time and the number of wells drilled in a drilling campaign period.

All the scenarios were analysed using basic geothermal financial model with constant cost and general assumption. The objective of this analysis is to compare the result of NPV, IRR and PBP indicator for different make-up wells strategies

8.1. Scenario 1

Based on figure 2, total make-up wells = 8 wells (4 wells in 2029 and 4 wells in 2046), total make-up well cost = USD 98.3 Million (APPENDIX A).

8.2. Scenario 2

Based on scenario 2 in figure 3, total make-up wells = 8 wells, all drilled in 2029, total make-up well cost = USD 77.97 Million (APPENDIX B).

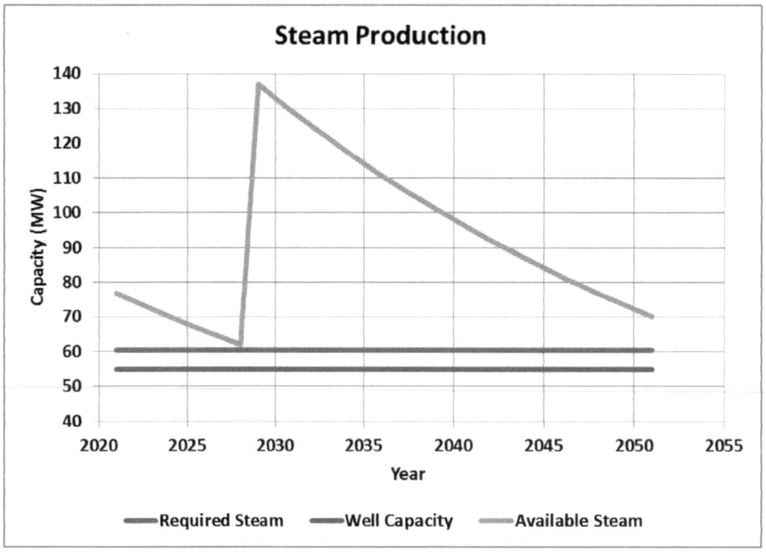

Figure 3. Make-up calculation of Scenario 2

8.3. Strategy 3

Based on scenario 3 in figure 4, total make-up wells = 8 wells (4 wells in 2029 and 4 wells in 2038), total make-up well cost = USD 87.68 Million (APPENDIX C).

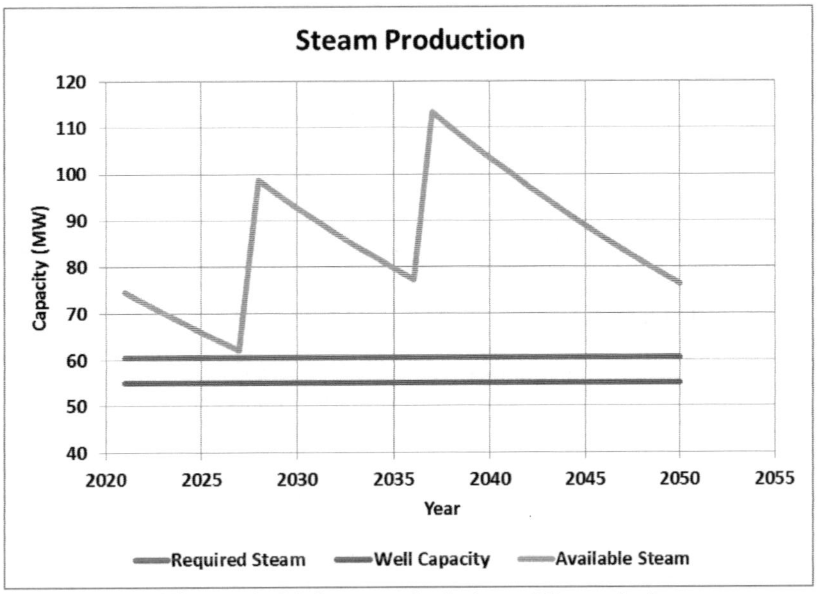

Figure 4. Make-up calculation of Scenario 3

8.4. Strategy 4

Based on scenario 4 in figure 5, total make-up wells = 6 wells (3 wells in 2029 and 4 wells in 2042), total make-up well cost = USD 69.55 Million (APPENDIX D).

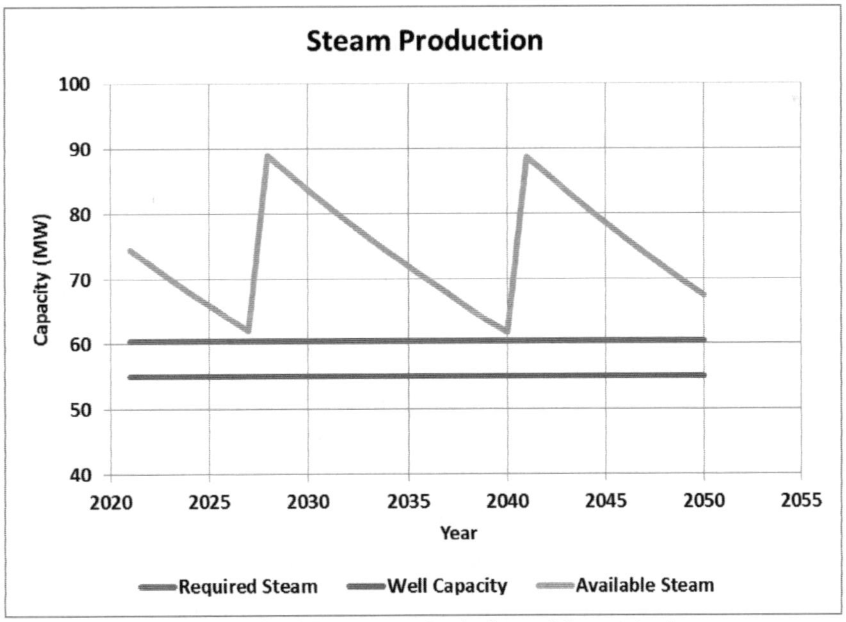

Figure 5. Make-up calculation of Scenario 4

Using scenario 1, 2, 3 and 4, financial indicators of the project are description in table 2.

Table 2: NPV and IRR and Payback.

Scenario	NPV	IRR	Payback Period
1	74 Million	17.5%	11 years
2	66 Million	16.5%	11 years
3	72 Million	17.4%	11 years
4	77 Million	17.8%	11 years

9. Resume of 4 make-up wells strategies

Resume of analyses from previous make-up wells drilling strategies are as in Table below:

Table 3. Resume of financial model for all scenario

Resume	Drilling Time		Total Make-Up Wells Cost (USD Million)	NPV (USD Million)	IRR (%)	Payback Period (Year)
Strategy 1	4 Wells	2029	98.32	74	17.5	11
	4 Wells	2046				
Strategy 2	8 Wells	2029	77.98	66	16.5	11
Strategy 3	4 Wells	2029	87.68	72	17.4	11
	4 Wells	2038				
Strategy 4	3 Wells	2029	69.55	77	17.8	11

Based on calculation that had been done using financial model for 4 different make-up wells drilling strategies, there are some conclusions as:

- Different strategies in drilling make-up wells will require different cost
- Different time of drilling and number of make-up wells drilled will affect NPV and IRR of the project, but not significant
- Degree of IRR difference is between 0.1 to 6% or 0.16 to 1.2 point for average IRR Project 17.3%
- Amount of NPV difference is between 2.7 to 6% or USD 2 to 5 Million for average NPV project USD 73.75 Million
- The Payback Period is similar for all strategies

10. Conclusion

From this study, there are some points that need to be considered in calculating make-up wells requirement in a geothermal project and development. Calculations make-up well was not only based on required power plant and decline rate, but also financial factors of the developer and drilling vendor into the calculations that the optimum strategy. All calculation in those scenarios using electricity price USD 12 ¢/kWh

- The best IRR (17.76%) is obtained with make-up wells strategy that used fewest make-up wells
- The best NPV USD 98.32 Million is obtained with make-up wells strategy that used 8 wells in 2 drilling

Based on technical requirement, the make-up wells strategy could vary as long as the steam production not below the minimal production level. Then applying a make-up wells strategy should consider not only technical requirement, but also the financial consideration.

Suggestion for further work is to develop more accurate model to find the most optimum strategy of make-up wells scenario in a geothermal field.

11. References

[1] Randle, James B. (2005), "Financial Modelling of Geothermal Project," Proceeding World Geothermal Congress, Antalya, Turki

[2] Lovekin, James W. (1998), "Sustainable Geothermal Power: The Life Cycle of a Geothermal Field," Geothermal Resource Council Transaction, Vol. 22

[3] Miller, Richard J. (1982), "The Influence of Declines Rates and Pressure Interference Effects of the Economic Viability of Vapor-Phase Geothermal Reservoir Development," New Zealand Geothermal Workshop

[4] Sanyal, Subir K. (2004), "Cost of Geothermal Power and Factors that Affect it," Proceeding Twenty-Ninth Workshop on Geothermal Reservoir Engineering, Stanford University, California

[5] Sanyal, Subir K. and Morrow, James W. (2004), "An Investigation of Drilling Success in Geothermal Exploration, Development and Operation," GRC Transaction, Vol. 35

IOP Conf. Series: Earth and Environmental Science **103** (2017) 012010 doi:10.1088/1755-1315/103/1/012010

APPENDIX A. Calculation of required time, which is anticipated that additional make-up wells drilling is required and make-up well (Scenario 1)

Production Year	Well Capacity (MW)	Required (MW)	Available before make up (MW)	Additional Make-up Wells	Available After Make-up (MW)	Make Up Well Cost (USD)
2021	55	60.5	76.82	0	76.82	0
2022	55	60.5	74.51	0	74.51	0
2023	55	60.5	72.28	0	72.28	0
2024	55	60.5	70.11	0	70.11	0
2025	55	60.5	68.01	0	68.01	0
2026	55	60.5	65.97	0	65.97	0
2027	55	60.5	63.99	0	63.99	0
2028	55	60.5	62.07	0	62.07	0
2029	55	60.5	60.21	4	98.62	38,988,893
2030	55	60.5	95.66	0	95.66	0
2031	55	60.5	92.79	0	92.79	0
2032	55	60.5	90.00	0	90.00	0
2033	55	60.5	87.30	0	87.30	0
2034	55	60.5	84.68	0	84.68	0
2035	55	60.5	82.14	0	82.14	0
2036	55	60.5	79.68	0	79.68	0
2037	55	60.5	77.29	0	77.29	0
2038	55	60.5	74.97	0	74.97	0
2039	55	60.5	72.72	0	72.72	0
2040	55	60.5	70.54	0	70.54	0
2041	55	60.5	68.42	0	68.42	0
2042	55	60.5	66.37	0	66.37	0
2043	55	60.5	64.38	0	64.38	0
2044	55	60.5	62.45	0	62.45	0
2045	55	60.5	60.57	0	60.57	0
2046	55	60.5	58.76	4	97.17	59,326,211
2047	55	60.5	94.25	0	94.25	0
2048	55	60.5	91.42	0	91.42	0
2049	55	60.5	88.68	0	88.68	0
2050	55	60.5	86.02	0	86.02	0
2051	55	60.5	83.44	0	83.44	0
Total				**8**		**98,315,104**

IOP Conf. Series: Earth and Environmental Science **103** (2017) 012010 doi:10.1088/1755-1315/103/1/012010

APPENDIX B. Calculation of required time, which is anticipated that additional make-up wells drilling is required and make-up well (Scenario 2)

Production Year	Well Capacity (MW)	Required (MW)	Available before make up (MW)	Additional Make-up Wells	Available After Make-up (MW)	Make Up Well Cost (USD)
2021	55	60.5	76.82	0	76.82	0
2022	55	60.5	74.51	0	74.51	0
2023	55	60.5	72.28	0	72.28	0
2024	55	60.5	70.11	0	70.11	0
2025	55	60.5	68.01	0	68.01	0
2026	55	60.5	65.97	0	65.97	0
2027	55	60.5	63.99	0	63.99	0
2028	55	60.5	62.07	0	62.07	0
2029	55	60.5	60.21	8	137.02	77,977,785
2030	55	60.5	132.91	0	132.91	0
2031	55	60.5	128.93	0	128.93	0
2032	55	60.5	125.06	0	125.06	0
2033	55	60.5	121.31	0	121.31	0
2034	55	60.5	117.67	0	117.67	0
2035	55	60.5	114.14	0	114.14	0
2036	55	60.5	110.71	0	110.71	0
2037	55	60.5	107.39	0	107.39	0
2038	55	60.5	104.17	0	104.17	0
2039	55	60.5	101.05	0	101.05	0
2040	55	60.5	98.01	0	98.01	0
2041	55	60.5	95.07	0	95.07	0
2042	55	60.5	92.22	0	92.22	0
2043	55	60.5	89.45	0	89.45	0
2044	55	60.5	86.77	0	86.77	0
2045	55	60.5	84.17	0	84.17	0
2046	55	60.5	81.64	0	81.64	0
2047	55	60.5	79.19	0	79.19	0
2048	55	60.5	76.82	0	76.82	0
2049	55	60.5	74.51	0	74.51	0
2050	55	60.5	72.28	0	72.28	0
2051	55	60.5	70.11	0	70.11	0
Total				**8**		**77,977,785**

IOP Conf. Series: Earth and Environmental Science **103** (2017) 012010 doi:10.1088/1755-1315/103/1/012010

APPENDIX C. Calculation of required time, which is anticipated that additional make-up wells drilling is required and make-up well (Scenario 3)

Production Year	Well Capacity (MW)	Required (MW)	Available before make up (MW)	Additional Make-up Wells	Available After Make-up (MW)	Make Up Well Cost (USD)
2021	55	60.5	76.82	0	76.82	0
2022	55	60.5	74.51	0	74.51	0
2023	55	60.5	72.28	0	72.28	0
2024	55	60.5	70.11	0	70.11	0
2025	55	60.5	68.01	0	68.01	0
2026	55	60.5	65.97	0	65.97	0
2027	55	60.5	63.99	0	63.99	0
2028	55	60.5	62.07	0	62.07	0
2029	55	60.5	60.21	4	98.62	38,988,893
2030	55	60.5	95.66	0	95.66	0
2031	55	60.5	92.79	0	92.79	0
2032	55	60.5	90.00	0	90.00	0
2033	55	60.5	87.30	0	87.30	0
2034	55	60.5	84.68	0	84.68	0
2035	55	60.5	82.14	0	82.14	0
2036	55	60.5	79.68	0	79.68	0
2037	55	60.5	77.29	0	77.29	0
2038	55	60.5	74.97	4	113.38	48,691,784
2039	55	60.5	109.98	0	109.98	0
2040	55	60.5	106.68	0	106.68	0
2041	55	60.5	103.48	0	103.48	0
2042	55	60.5	100.37	0	100.37	0
2043	55	60.5	97.36	0	97.36	0
2044	55	60.5	94.44	0	94.44	0
2045	55	60.5	91.61	0	91.61	0
2046	55	60.5	88.86	0	88.86	0
2047	55	60.5	86.19	0	86.19	0
2048	55	60.5	83.61	0	83.61	0
2049	55	60.5	81.10	0	81.10	0
2050	55	60.5	78.67	0	78.67	0
2051	55	60.5	76.31	0	76.31	0
Total				**8**		**87,680,677**

APPENDIX D. Calculation of required time, which is anticipated that additional make-up wells drilling is required and make-up well (Scenario 4)

Production Year	Well Capacity (MW)	Required (MW)	Available before make up (MW)	Additional Make-up Wells	Available After Make-up (MW)	Make Up Well Cost (USD)
2021	55	60.5	76.82	0	76.82	0
2022	55	60.5	74.51	0	74.51	0
2023	55	60.5	72.28	0	72.28	0
2024	55	60.5	70.11	0	70.11	0
2025	55	60.5	68.01	0	68.01	0
2026	55	60.5	65.97	0	65.97	0
2027	55	60.5	63.99	0	63.99	0
2028	55	60.5	62.07	0	62.07	0
2029	55	60.5	60.21	3	89.01	29,241,670
2030	55	60.5	86.34	0	86.34	0
2031	55	60.5	83.75	0	83.75	0
2032	55	60.5	81.24	0	81.24	0
2033	55	60.5	78.80	0	78.80	0
2034	55	60.5	76.44	0	76.44	0
2035	55	60.5	74.15	0	74.15	0
2036	55	60.5	71.92	0	71.92	0
2037	55	60.5	69.76	0	69.76	0
2038	55	60.5	67.67	0	67.67	0
2039	55	60.5	65.64	0	65.64	0
2040	55	60.5	63.67	0	63.67	0
2041	55	60.5	61.76	0	61.76	0
2042	55	60.5	59.91	3	88.72	40,309,964
2043	55	60.5	86.05	0	86.05	0
2044	55	60.5	83.47	0	83.47	0
2045	55	60.5	80.97	0	80.97	0
2046	55	60.5	78.54	0	78.54	0
2047	55	60.5	76.18	0	76.18	0
2048	55	60.5	73.90	0	73.90	0
2049	55	60.5	71.68	0	71.68	0
2050	55	60.5	69.53	0	69.53	0
2051	55	60.5	67.44	0	67.44	0
Total				**6**		**69,551,634**

IOP Conf. Series: Earth and Environmental Science **103** (2017) 012011 doi:10.1088/1755-1315/103/1/012011

Probabilistic approach: back pressure turbine for geothermal vapor-dominated system

Angga Alfandi Ahmad[1], **Fransiscus Xaverius Guwowijoyo**[1], **Heru Berian Pratama**[1]

[1]Geothermal Engineering Study Program, Faculty of Mining and Petroleum, Bandung Institute of Technology, Jl. Ganesha No.10 Bandung, Indonesia 40132

Email: anggaalfandi@gmail.com

Abstract. Geothermal bussiness nowadays needs to be accelerated in a way that profit can be obtained as soon as reasonable possible. One of the many ways to do this is by using one of geothermal wellhead generating unit (GWGU), called backpressure turbine. Backpressure turbine can be used in producing electricity as soon as there is productive or rather small-scale productive well existed after finished drilling. In a vapor dominated system, steam fraction in the wellhead capable to produce electricity based on each well productivity immediately. The advantage for using vapor dominated system is reduce brine disposal in the wellhead so it will be a cost benefit in operation. The design and calculation for backpressure turbine will use probablistic approach with Monte Carlo simulation. The parameter that will be evaluated in sensitivity would be steam flow rate, turbine inlet pressure, and turbine exhaust pressure/atmospheric pressure. The result are probability for P10, P50, and P90 of gross power output which are 1.78 MWe, 2.22 MWe and 2.66 Mwe respectively. Whereas the P10, P50, and P90 of SSC are 4.67 kg/s/MWe, 5.19 kg/s/MWe and 5.78 kg/s/MWe respectively.

1. Introduction

Geothermal vapor dominated system, as the name implies, consist steam fraction dominantly. In an early geothermal development plan, many wells can be drilled and remain idle as the plan continue. Surface facilities in early development are still not complete and steam that already produced in a drilled well can be wasted unwantedly. In order to utilize the produced steam, even just a little, geothermal wellhead generating unit (GWGU) can be used to utilize the produced steam and turn it into electricity.

In Kenya, KenGen lost a potential of 11,211 GWh by not utilizing steam in the idle wells, which is approximately one billion usd lost in revenue [6].

Table 1. Example of GWGU application worldwide

Year	Field	Country	Type	Capacity
1978	Kamojang	Indonesia	Backpressure Turbine	1 x 0.25 MW
1982	Los Azufres	Mexico	Backpressure Turbine	5 x 5 MW
1997	Miravalles	Costa Rica	Backpressure Turbine	1 x 5 MW
2012	Eburru	Kenya	Mini Condensing Unit	1 x 2.5 MW

Kamojang Geothermal Field can be a representative data for vapor dominated system in this calculation and design, which had more than 40 years of production history and still operating in present time. It will become good practice for vapor dominated system near Kamojang which still in the development process and able to use GWGU for increasing their negative income in the beginning of the project.

2. Back Pressure Turbine

Geothermal Wellhead Generating Unit (GWGU) has short pipelines because of its location which is close to the production well and steam supplied from one production well or more in a single wellpad, usually able to generate electricity less than 10 MW [2]. Generally, GWGU can be divided into two distinguished types namely non-condensing type and condensing type. In time, for condesing type GWGU can be utilized for many technologies especially related to binary technology.

Backpressure turbine is the most simple GWGU type. This type does not require condensor, cooling system, gas removal system, and separator (for vapor-dominated system only). This type only needs scrubber, turbine-generator, and electric transmission. Steam from production well flow into scrubber to purify the steam. Then from scrubber, steam used to rotate turbine which will generate electrical power. The steam from turbine will goes out from atmospheric exhaust and flow to the atmosphere/environment directly.

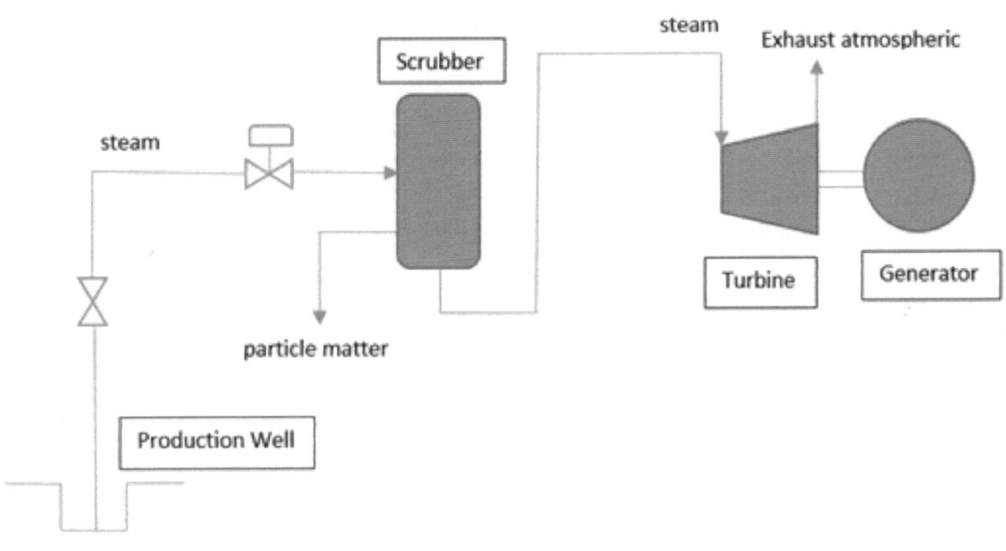

Figure 1. Simplify Schematic Diagram of Steam Flow

Backpressure turbine is ideal for new developed project. The productive well that being used for backpressure turbine is ideally the well that will be used for the big project in the next phase of development. From the components itself, backpressure turbine components are easily to relocate to any available and possible well. It has simple yet effective components. One concern from backpressure turbine is atmospheric turbine exhaust system utilization. The system would produces noisy sound because higher pressure steam out from turbine directly goes to atmospheric with lower pressure.

In GWGU, steam can driving turbine-generator system directly to produce electricity immediately. The gross power output produced by turbine-generator system [1] is given by:

$$\dot{W_e} = \frac{\dot{m}_s \; x \; \eta_t \; x \; \eta_m \; x \; \eta_g \; x \; (h_{HP} - h_{LPisen})}{1000}$$

The performance of GWGU application for vapor dominated field determined by specific steam consumption of turbine as termed SSC. SSC is the ratio of steam consumed by turbine per unit gross power output. Unit of SSC basically is kg/s/MWe or commonly it uses kg/MWh [1]. The formula of SSC is given by:

$$SSC = \frac{\dot{m}_{steam}}{\dot{W}_e}$$

Where:

$\dot{W}_e$: gross power output [MWe]
$\dot{m}_s$: steam flow rate [kg/s]
η_t : isentropic turbine efficiency [%], which is ratio of the actual work to the isentropic work
η_m : mechanical losses of turbine-generator system [%]
η_g : generator efficiency [%]
h_{HP} : specific enthalpy of the steam at the turbine inlet pressure [kJ/kg]
h_{LPisen} : specific enthalpy of the steam [kJ/kg]

3. Methodology

This study used probabilistic approach, so the calculation might be estimated by using a polynomial response or proxy equation for full factorial of 3 factors, as follows:

$$Y = \beta_0 + \beta_1 A + \beta_2 B + \beta_3 C + \beta_4 AB + \beta_5 AC + \beta_6 BC + \beta_7 ABC$$

Where β_0 is the average value of the response and β_1, β_2,..., β_7 is regression linear. The polynomial describes the behavior of the responses between parameter including the interaction effects between parameters.

Experimental design concept were applied in GWGU application to calculate the performance with probablistic approach based on several main parameters. There are three main parameters investigated in this paper: **steam flow rate** ($\dot{m}_{steam}$), **turbine inlet pressure** (TIP), and **turbine exhaust pressure** (TEP) or in this case we use atmospheric pressure as turbine exhaust pressure. The parameter of back pressure unit are summarized in table 2 where the low and high value are categorized based on typical value of average vapor dominated field in Kamojang as representative data [7].

Table 2. Parameter for Sensitivity

Parameter	Low (-1)	High (+1)
Steam flow rate [kg/s]	8	15
Turbine inlet pressure [bar abs]	3	5
Atmospheric pressure [bar abs]	0.8	1

A two level (low and high) of full factorial with three factor for ED was implemented using Minitab 17 free trial software. The total run in these study is 8 runs. In the design table (table 3), the standard order column is an experimental design idea to do the experiments in a random order (run order) and avoid run-dependent effects.

Table 3. A two level of full factorial design

StdOrder	RunOrder	Msteam	TIP	TEP
7	1	-1	1	1
2	2	1	-1	-1
4	3	1	1	-1
6	4	1	-1	1
8	5	1	1	1
1	6	-1	-1	-1
5	7	-1	-1	1
3	8	-1	1	-1

IOP Conf. Series: Earth and Environmental Science **103** (2017) 012011 doi:10.1088/1755-1315/103/1/012011

The polynomial equation of gross power output and specific steam consumption as a function of three independent parameters are the results in these study. A two level proxy model using -1 and +1 parameters to generate the gross power output (GPO) and SSC responses GWGU back pressure are represented below:

$$\boldsymbol{GPO} = 2.253 + 0.6857\,M_{steam} + 0.3933\,TIP - 0.1571\,CIP + 0.1197\,M_{steam} * TIP$$

$$\boldsymbol{SSC} = 5.295 + 0.000000\,M_{steam} - 0.9344\,TIP + 0.3946\,CIP - 0.000000\,M_{steam} * TIP$$

4. Data, Results, and Analysis

The basic data design for vapor dominated fields is obtained from Kamojang geothermal field [7] which are presented below:

Table 4. Data Design

Parameter	Value
Well Head Pressure (WHP)	15 bar abs
Quality of steam in well head	100%
NCG content	1%
Isentropic turbine efficiency	85%
Mechanical losses of turbine-generator system	95%
Generator efficiency	98%

The results of GWGU back pressure model calculation shown in table 5. The minimum and maximum gross power output is 1.2 MWe and 3.7 MWe, and the minimum and maximum SSC value is 4.1 and 6.8 kg/s/MWe.

Table 5. The result of GWGU back pressure performance

RUN ORDER	Gross Power Output	SSC
	MWe	*kg/s/MWe*
1	1.7	4.6
2	2.6	5.7
3	3.7	4.1
4	2.2	6.8
5	3.2	4.6
6	1.4	5.7
7	1.2	6.8
8	1.9	4.1

The next steps are running Monte Carlo simulation based on generated polynomial equation of gross power output and SSC with 50,000 iterations using spreadsheet. The result is a probability distribution function of the gross power output covering the full range of possible outcomes. The cumulative distribution results of the gross power output shown in figure 2, the P10, P50 and P90 are 1.78 MWe, 2.22 MWe and 2.66 MWe respectively. In other side, The cumulative distribution results of the SSC shown in figure 3, the P10, P50 and P90 are 4.67 kg/s/MWe, 5.19 kg/s/MWe and 5.78 kg/s/MWe.

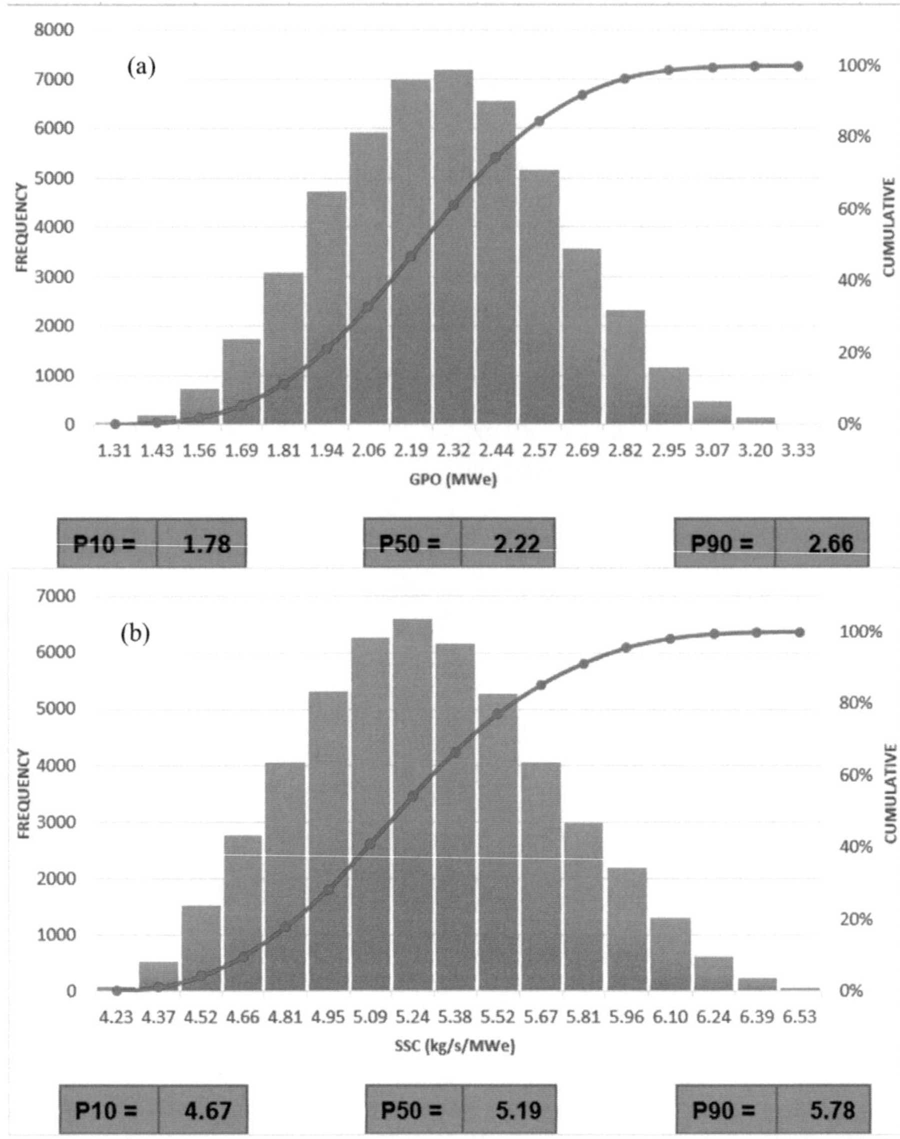

Figure 2. Monte Carlo simulation results for Gross Power Output (left side) and SSC (right side) with histogram and cumulative distribution function with P10, P50 and P90 indicated

In the Monte Carlo simulations, we also know about the most sensitive parameters of calculation based on Pareto Chart. As shown in figure 4 and 5, the highest effect for gross power output calculation is steam flow rate whereas for SSC calculation is turbine inlet pressure. It is important things for the next development project of back pressure unit in some vapor dominated field to select the best parameter in order to obtain the optimum performance.

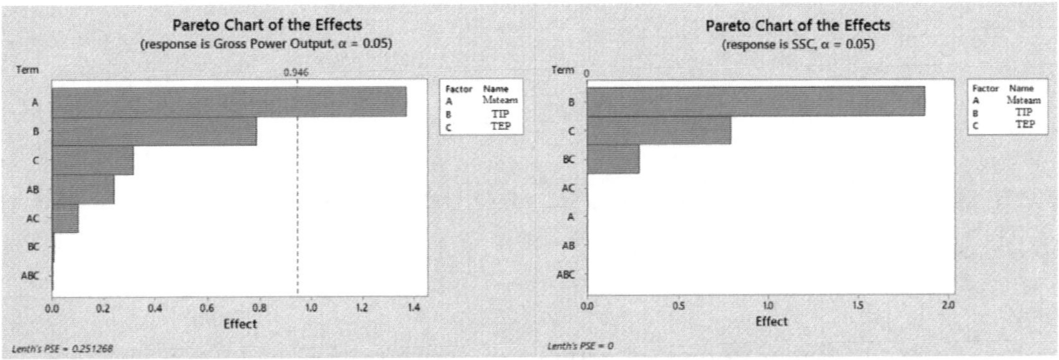

Figure 3. Pareto Chart for Gross Power Output (left side) and SSC (right side) Results

5. Conclusion

GWGU back pressure model is the best model for green field development project uses a well to generate electricity meanwhile drilling campaign still on going until the end of project. The probablistic approach using Monte Carlo simulation for gross power output and SSC based on 50,000 iteration has successfully represented and applied of GWGU experimental design. The P10, P50, and P90 of gross power output are 1.78 MWe, 2.22 MWe and 2.66 MWe. Whereas the P10, P50, and P90 of SSC are 4.67 kg/s/MWe, 5.19 kg/s/MWe and 5.78 kg/s/MWe respectively.

References

[1] DiPippo, R. (2008). *Geothermal Power Plant Second Edition : Principles, Applications, Case Studies and Environmental Impact.* Elsevier.

[2] Hudson, R. B. (1988). Technical and Economic Overview of Geothermal Atmospheric Exhaust and Condensing Turbines, Binary Cycle and Biphase Plant. *Geothermics*, 51-74.

[3] McHugh, J., & Bodylski. (1985). A Generic Overview of Geothermal Wellhead Generating Units. *GRC Bulletin, May.*

[4] Mendive, D. L., & Green, L. (2012). Wellhead Geothermal Power Plant at Eburu, Kenya. *GRC Transaction Vol. 36.*

[5] Moya, P., & DiPippo, R. (2011). Miravalles PGM-29 Wellhead Unit, Guanacaste, Costa Rica : Technical and Environmental Performance Assessment. *Short Course on Geothermal Drilling, Resource Development and Power Plants.*

[6] Saitet, D., & Kwambai, C. (2015). Wellhead Generating Plants: KenGen Experience. *World Geothermal Congress.* Melbourne.

[7] Suryadarma, Dwikorianto, T., Zuhro, A., & Yani A. (2010). Sustainable Development of the Kamojang Geothermal Field. *Geothermics*, 391-399.

[8] Sutter, J., Kipyego, E., & Mutai, D. (2012). The Use of Portable Geothermal Wellhead Generators as Small Power Plants to Accelerate Geothermal Development and Power Generation in Kenya. *Thirty-Seventh Workshop Geothermal Reservoir Engineering Stanford University.* California: Stanford University.

A new idea: The possibilities of offshore geothermal system in Indonesia marine volcanoes

Teguh Rahat Prabowo[1], Fithriyani Fauziyyah[2], Suryantini[1], Sutikno Bronto[3]

[1]Faculty of Mining and Petroleum Engineering, Bandung Institute of Technology (ITB)
[2]Faculty of Earth Science and Technology, Bandung Institute of Technology (ITB)
[3]Pusat Survei Geologi, Ministry of Energy and Mineral Resource, Bandung

Email : teguhprabowo_23@yahoo.com

Abstract. High temperature geothermal systems in Indonesia are commonly associated with volcanic systems. It is believed that volcanoes are acting as the heat source for a geothermal system. Right now, most of the operating geothermal fields in the world are assosiating with volcanic settings which known as the conventional geothermal system. Volcanoes are created in active tectonic zone such as collision zone and MOR (mid oceanic ridge). The later is the one which formed the marine volcanoes on the sea floor. The advances of today's technology in geothermal energy has created many ideas regarding a new kind of geothermal system, including the ideas of developing the utilization of marine volcanoes. These marine volcanoes are predicted to be hotter than the land system due to the shorter distance to the magma chamber. Seamounts like NEC, Banua Wuhu, and Kawio Barat in Indonesia Sea are good spots to be studied. Methods such as remote sensing using NOAA images, sonar, and MAPR are commonly used, eventhough these would be more accurate with more detailed techniques. This has become the challenge for all geothermal scientists to overcome for a better study result.

1. Introduction

The offshore geothermal has begun to intrigue some of the scientist minds as a result of out-of-the-box thinking in geothermal. Even though there are a lot of issues that needs to be handled, this doesn't stop them to research more about the possibilities of offshore geothermal system. The research has begun in several countries such as Italy, known as the Marsili Project, and Iceland. Some said there is also an underway project for offshore geothermal project in Gulf of California. So far Italy leads in offshore geothermal energy development with its Marsili Project which was planned to be finished around 2015. Many studies have been conducted, from the study of the marine volcano as the heat source to the technologies to exploit the energy in the future, especially the power plants and the pipes.

As we know volcanoes are not only located on land but there are still a lot more in the ocean, which called seamount. Some of the examples are Mt. Emperor of China and Mt. Niewerkerk, or known as NEC volcanic complex. Both of these mountains located in Banda Sea, although they are not much known since there was only one expedition conducted in 1934 by Snellius. There are also Mt. Banua Wuhu, Mt. Submarin, and Mt. Kawio Barat which located around Sangihe Island. Unlike NEC volcanic complex, these seamounts are more known. There are several records of their eruption which starts around 1800 and some still considered active.

Located around more than 3000 m in depth, these underwater volcanoes are predicted that it could be acting as the heat source of the offshore geothermal system. With the surrounding sea water it means the system has literally unlimited supply water for the system. The cracks in the sea floor are allowing the water to go in and out the system becoming the recharge and the discharge zone. Even though all of these are still possibilities, the idea of offshore geothermal is interesting to be discussed and studied further.

2. Seamounts

Marine volcanoes are believed to be a result of sea floor spreading and mostly found in Mid Oceanic Ridge (MOR). The spreading caused the magma to move towards surface. The usual type of eruption is effusive thus creating a shield type volcano which spread several times larger than the stratovolcano type. Magma is rapidly cooled from the impact with the sea water, causing the magma to form an igneous rock with pillow lava structure. The rapid cooling could also create autobreccia, a type of breccia where the component and the matrix are the same. The magma type is different than the magma in on land volcanoes. In marine volcanoes the magma is basaltic due to the magma path to the surface is shorter thus minimizing the contact with surrounding rocks.

According to the volcano height, underwater volcanoes can be divided into 3 kinds; (1) Seamounts, marine volcanoes which can stand up to 1000 m from the sea floor, (2) Hills, a smaller volcano with the height below 1000 m from the sea floor, (3) Pinnacle, which looks like a small pillar. In this paper, seamount is the main focus because it is predicted to have a significant heat to be considered a heat source.

2.1. Seamounts Types

Usually the research for the study of seamounts is conducted using remote sensing images taken from ships or close up videos and pictures taken by a drone. This equipment can be found in National Oceanic and Atmospheric Administration (NOAA). These remote sensing images are limited to the shipping line around the world. Up close images and videos taken by a drone or Remotely Operated Vehicle (ROV) gave detailed information regarding the seamounts. These seamounts can be located several thousand under the mean sea level, like Mt. Kawio Barat and Mt. Submarine 1922. Some are not even having enough information such as Mt. Emperor of China and Mt. Niewerkerk.

Figure 1: Illustration of seamount type (Seamount, Sea Knoll, and Pinnacle)

Seamount type's elevation can vary from 1000 m to a smaller elevation feature which has the same function as the larger feature. Sea knolls, hills, and pinnacle can even be as low as 100 m[1]. Pinnacles are important in marine ecological system, often isolated and can reach near the surface.

Some seamounts can erupt above the sea level somewhere in the ocean so the peak will submerge above the surface. After a long time waves then will eroded the top portion thus making it flat. The sea floor will gradually move away from the oceanic ridge thus making the seamounts submerged. This is known as guyots or tablemount. Guyots located around 200 m below the surface and the width can exceed 10 km. Seamount never reach the surface thus maintained its volcanic shape.

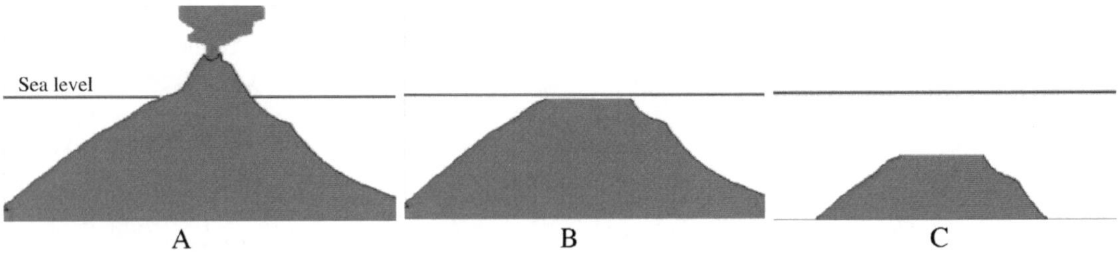

Figure 2. Illustration of Guyots establishments steps. Seamounts which erupt above the sea level will be eroded, then will be submerged due to the sea floor movement

2.2. Seamounts Profile

Although there are quite a lot of seamounts in Indonesia, only some of the seamounts are written here as examples. The reasons are these seamounts are widely known around the world and sparked some interests for the researcher. Informations about the seamount, such as height and eruption record, are already available from earlier researches. The unique characteristics are also one of the other reasons. But some are still very much unknown but still debatable about their existences. The seamounts mentioned below are some of the most infamous seamounts in Indonesia Sea.

2.2.1. Kawio Barat

This seamount is pressumed to be the biggest seamount in Indonesia. Research was conducted by a joint operation between Indonesia-United States in Sangihe Sea in 2010. The image and sonar taken from NOAA's Okeanos showed that this seamount is approximately has height around 3000 m and located around 1800 m below the sea level. A video was taken by the *Little Hercules* ROV during its second dive in June 2010 showed active hydrothermal vents (white smoker and black smoker) on the west side of the seamount[2]. The volcanic rocks around the hydrothermal vent was covered in white sulfur deposits and became the home of many underwater organism, such as shrimp and stalked barnacles. Along with the barnacles, imagery disclosed a field of sulphide chimneys, pressumably precipitating from the hot, mineralized vent fluids as they cooled and mixed with seawater. This hydrothermal vents proved that Kawio Barat Seamount is still active and the size is indicating a large heat source underneath.[3]

Figure 3. Pictures taken from the Little Hercules ROV in June 2010. A picture of sulphide chimneys in Kawio Barat Seamount (above). An active hydrothermal vent was found on the summit (below, left). Sulphide-rich hydrothermal vents was also found created home for many underwater organism (below, right).[2]

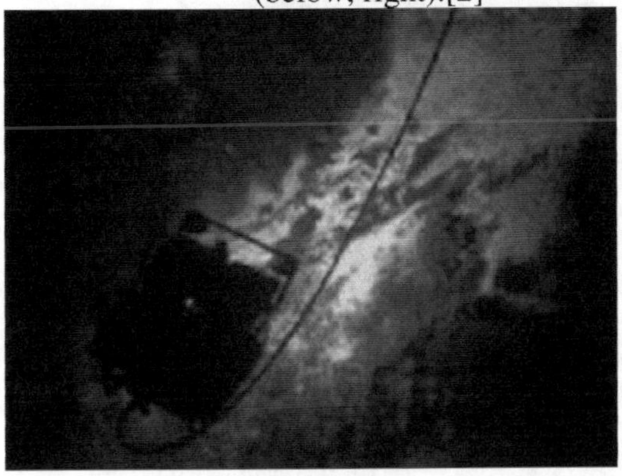

Figure 4. An active hydrothermal vent in Kawio Barat Seamount taken from the "Little Hercules" ROV, Jne 2010[2]

Figure 5. A sulphide-rich qhydrothermal vent in Kawio Barat Seamount taken from the "Little Hercules" ROV, Jne 2010.[2]

2.2.2. Submarine

There are not much record of this seamount except that it was erupted in 1922 after a series of earthquakes since 1912 in between Mindanau and Sangir Talaud Island. Geographically this seamount is located between 4°-6° LU and 124°-126° BT. The nearest city is Tahuna. The eruption in 1922 created a big wave along the Sulawesi northern beach. After that there were no eruption recorded. The eruption were dated around February 1, 1922.[4]

2.2.3. Banua Wuhu

Banua Wuhu Seamount, or Banua Buaya, located southwest of Manghetang Island in North Sulawesi province. Geographically it is located in 3°08'16" LU and 125°29'26" BT. The only access to this seamount is by using small boats from Tagulandang[5]. The summit depth is estimated only around 5 m from the sea level. The eruption type is considered an explosive type and the history is a prove. In 1835 the summit was around 90 m above the sea level and in 1919 it was around 12 m above the sea level. It is predicted that the eruption destroyed the summit making it smaller each time. Right now it's still considered ative due to the presence of the bubbles from the cracks in the seamount. It is composed mostly by fall pyrocalstic and lava. The eruption was predicted around April 16 – Agustus 27, 1904[6].

The eruption on August 27 is treted by the Catalogue of Active Volcanoes as an independent event[6]. Banua Wuhu eruption in 1889 was an extrusion of a lava dome and catagorized as a submarine eruption. This eruption created tsunami. While in 1904 was described that the eruption was located in the central crater as a submarine eruption which created a volcanic island[3][4].

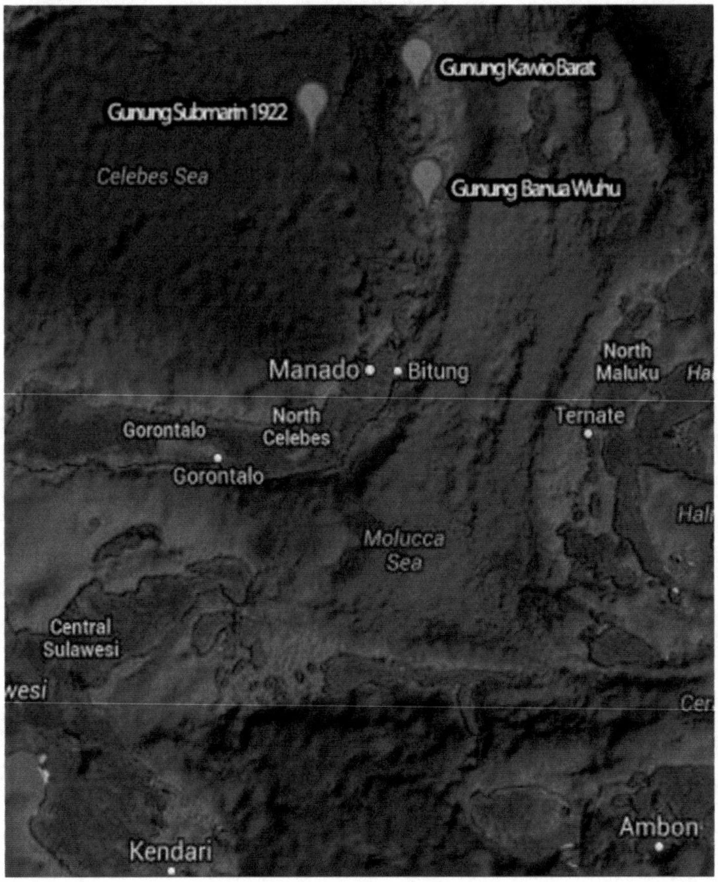

Figure 6. Location of Banua Wuhu (south), Kawio Barat (north) and Submarine 1922 Seamounts (west of Sangihe Islands). All of them are located in Sangihe Ridge, to the north of Sulawesi Island.

2.2.4. Emperor of China – Nieuwerkerk (NEC)

Both of this seamount presences were still questioned by some scientist and has become an interesting debate topic wether they are real. There was only one expedition regarding these seamounts, and that was Snellius Expedition in 1934. Some believed that these seamounts were mere reefs. In 1929 there was a phenomena where many dead sea turtles and fished washed ashore, which usually caused by underwater eruption. Emperor of China was a seamount rising nearly 1500 m above its surroundings, having a summit which was 2850 m below the sea level. Nieuwerkerk was a NNW – SSE stretching seamount with two tops at a mutual distance of more than 7 km, rising about 1800 m and 1900 m above the bottom of the sea. These seamounts created a ridge called NEC. Both these seamounts are predicted to have typical shield-volcanic topography.[4]

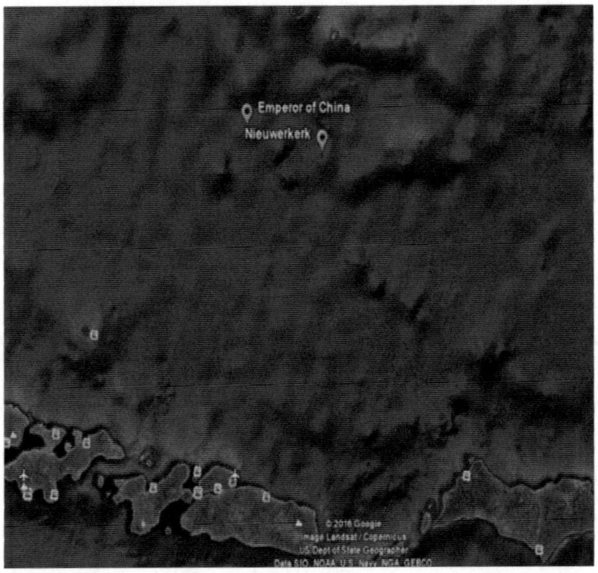

Figure 7. Location of Emperor of China and Niewerkerk Seamounts. Also known as NEC Seamounts Complex. Both are located in the Banda Sea.

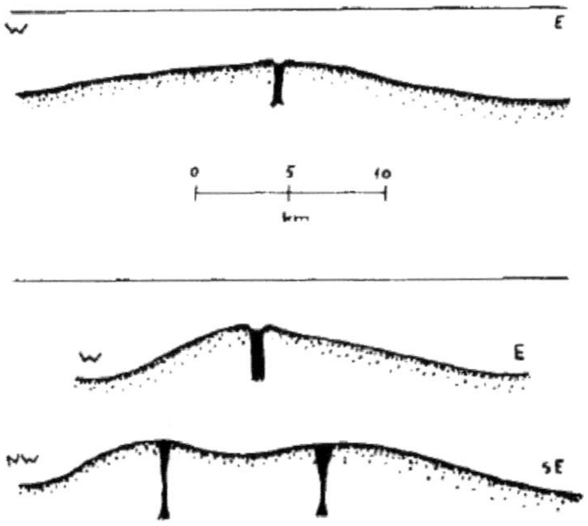

Figure 8. An estimated cross section of Emperor of China (above) and Nieuwerkerk Seamount (below) according to Neumann van Padang.[4]

3. Offshore Geothermal System

Until now there hasn't been many real practices regarding the utilization or even research on offshore geothermal system. The main reason being is that the cost is still too high comparing the systems on land and there are still much on land system that hasn't been discovered yet. But it still cheaper and more reliable in the long run compared to wave power, according to Engineering Committee on Oceanic Resources (ECOR). It is predicted that most of this system is a liquid dominated system since the excess of sea water around the system and the fluid discharged from the system is in liquid form due to high pressure since most seamounts located several thousand meters deep.

Offshore geothermal project has started for quite a long time. The first attempt to utilize the offshore geothermal is Italy, which conducted by Unione Geotermica Italiana. This project is known as the Marsili Project. The research target area is around Marsili Seamount. Its main target is to have an early production of 600 MW after 10 years and approximately 1200 MW after 15 years. Although until now the technology still under research and not yet published. Other countries have also put interest in offshore geothermal energy and began to study other places with good potential, such as in Iceland and Gulf of California. For Iceland, this research is the result of "out of the box" thinking, to always one step ahead in geothermal utilization technology since Iceland's electricity is depending very much with geothermal power. There is also research on hydrothermal vent in Gulf of California which being conducted by the National University of Mexico. This research also for testing their prototype for submarine power plant, an new technology for utilizing heat from hydrothermal vents.

There are also advantages in using offshore geothermal system. In any resource production, either oil, geothermal, or even mining, there will always a need for land to build the site for drilling and production. In offshore geothermal, no land space is required so there's no need for a detailed visual environmental assessment although it will need some general environmental assessment. The surrounding seawater provides literally an unlimited supply of geothermal fluid and the cold seawater can work as the cooing fluid so there will be no need for condenser or cooling tower

Aside from the cost, there is also another problem that need to be taken care of which is the technologies. Even though the system is not supposed to be so different with the land system geothermal, but the material and production techniques need to be adjusted with the environment. For example the fluid which contain high level of salinity since the fluid used in this system is sea water, the power plant structure that can withstand wave, and pipes that can hold higher pressure or preserve the heat.

There are several concept methods visualization in terms of utilizing the offshore geothermal system; platform based power plant, land based power plant, underwater power plant, binary power plant based on land, and thermoelectricity power station.

3.1. Offshore Geothermal System Components

In general, there are 4 main components for the system to work; heat source, reservoir, recharge, and discharge. Each one of these components holds specific characteristics. Heat source is the main energy in this system which will heat up the fluid. Reservoir is acting as the storage of that fluid and usually composed of rocks with high permeability. Recharge is a zone where the fluid comes into the system to refill the system. The heated up fluid will be moving upwards toward the surface due to it's low density through cracks in the rock to the surface which called the discharge. Both recharge and discharge are contain high permeability.

There are not much differents between on land geothermal and offshore geothermal system. Generally the basic concept is the same, but the components may consisted of different material. As stated before the seamount can become the heat source for the offshore geothermal system. The difference is the magma isn't as acidic as the one on land due to shallow magma kitchen. The temperature is usually hotter too with approximately around 800 – 1200°C. Seamounts are created from seafloor spreading thus located mostly in a ridge. Seamounts are composed mostly by igenous rocks and pyroclastic. This also creates prediction that the reservoir rocks will mostly has the same composition. Seawater is cooling the lava rapidly and making igneous rocks with autobreccia texture, creating cracks in the rocks and increase its permeability for the reservoir. These cracks, together with the underwater geological structure, can also become the fluid pathway to the surface or into the system and become the recharge and discharge zone. One of the seamounts that has both recharge and discharge is Mt. Banua Wuhu. The bubbles emitted from the seamount indicates that there might be cracks which become the path for the fluid to go in and out of the system.

The most common surface manifestation is hydrothermal vent. It starts with the seawater seeps into the sea floor through cracks then get heated up and moved upwards toward the seafloor. This hot fluid also contains minerals from the interaction with surrounding rocks. After discharge to the surface it meets with cold sea water and decrease the temperature. With the temperature decreasing the minerals dissolved are deposited near the discharge area. After a while it will accumulate and create sulphide-rich chimney. Hydrothermal vent can discharge black smoke or white smoker depending on the composition. Black smoke indicates there are mafic minerals involved such as magnesium and iron minerals while white smoke indicates that felsic minerals are the main dissolved minerals in the fluid[7]. These hydrothermal vents can be measured up to ~200 – 400°C. Hot seawater from the vent won't be turned into steam because of extreme pressure underwater. Mineral ore deposits such as copper ore are commonly found near hydrothermal vent and some even become the world's richest ore deposit.

Looking at the surrounding environment, it can be predicted that the system might be a volcanic hydrothermal system with high terrain and it will be liquid dominated. The hot fluid discharge might mostly be in liquid form due to extremely high pressure.

3.2. Exploration Method

The differences in environmental aspect between on land and offshore geothermal have created different methods to explore it. The main difference is that offshore geothermal system isn't as visible as land geothermal since it's located deep underwater. Basically the first thing to do is check wether the system is available or not. Special tools needed to study this subject. Scientists have developed some research methods and have been used in some explorations including Marsili Project and offshore geothermal prospect around Iceland.

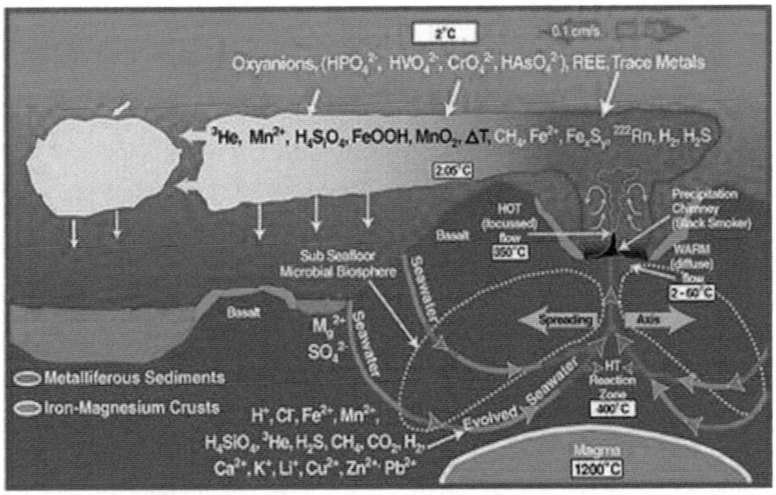

Figure 9. Hydrothermal vent forming process

3.2.1. Sound Navigation and Ranging (SONAR)

This method is often called by multibeam swath bathymetry or echo sounding and used for mapping the ocean floor. It can be useful to locate hydrothermal vents on the sea floor. Basically, hydrothermal vents will emit gas bubbles from within the sea floor and this will create accoustic scatter which is used to locate the hydrothermal vent. The analysis for the accoustic scatter can be seen in figure 11[8].

There are another method called side scan sonar. This method is using the back scattering strength of multibeam data and then analyze it. There are several datas that can be obtained from side scan sonar. It could be used to identify rock types and structures so the ocean floor composition and geological settings could be obtained. It could also identify and distinguish between each lava flow event, landslide, and fault so it can tell the differences in the igneous rocks source and magma characteristic.

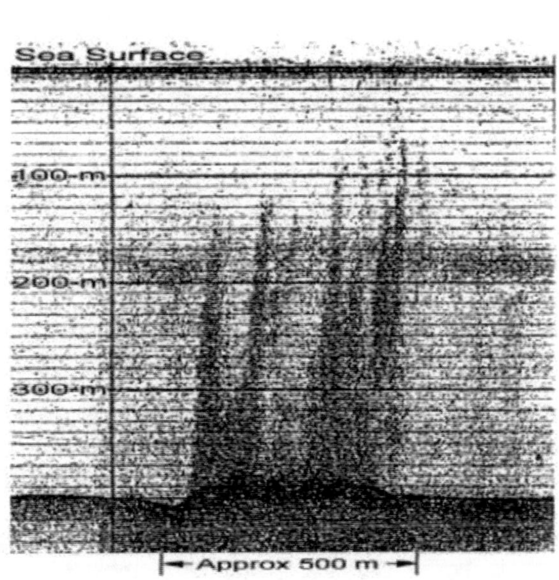

Figure 10. An example of accoustic scatter from hydrothermal vents

There is also method called passive sonar. Its hydrophones array can cover wide areas and it's commonly used for military purpose, such as locating submarines. This method are not yet proven but with further study maybe it could prove useful in the future. The idea is to recognize the particular frequencies from hydrothermal vents using the frequency shift. The array can detect, isolate, and display a wide range of acoustic frequencies in the water. Several requirements are needed for this method to work. Clear and strong signal is a must if one want this method to work. Also clear area is important so the there will not be any disturbance in the data. Things like shipping traffic is a disturbance since it is giving too much noise so the data won't be accurate.

3.2.2. Chemical Analysis
Gas and hot fluid discharged from the hydrothermal vent can be a proof that there are system working underneath it. It means that there is an active heat source below it and there are cracks which allow the seawater to seep into and flow out the system. Gases and hot fluid emitted from hydrothermal vents can give informations regarding the characteristic of the system. For example, analyzing the isotop ratio like $^3He/^4He$ ratio which become indication for hydrothermal activity. Dissolved gas and substances such as CH_4, CO_2, Si, H_2, H_2S, Fe, and Mn are also potential indicator of hydrothermal activity. There are several real-time instruments to measure the chemical components in the fluid like electrochemical redox (eH) and pH sensor[8].

3.2.3. Camera
A close up pictures and videos of hydrothermal vent is the most satisfying and direct way to study it. This method using a towed instruments with camera attached usually called Remotely Operated Vehicles (ROVs) and Autonomous Underwater Vehicles (AUVs). Little Hercules used in Kawio Barat expidition is one of the example of ROV. From the pictures and videos taken, we can study the area more detailed thus giving more accurate informations, providing clues on how big the estimated area and temperature[8].

3.2.4. Dredging
Rock samples from dredging can give information of hydrothermal vent existence because rocks are usually affected by the hydrothermal activity. Sulphide minerals in the rocks are common indicators of hydrothermal activity[8].

3.2.5. Magnetic Survey
The basic concept is not different with on-land geothermal system. This method is searching for the magnetic anomalies caused by the heat thus it can also locate hydrothermal vents. A reduction of magnetic value is usually found near hydrothermal upwelling zones and can reach approximately a few hundred meters across so more detailed magnetic survey is needed.

A rapidy cooled basalt contain titanomagnetites and normally has high value of magnetism. That's why sea floor, especially young mid-ocean basalts are highly magnetized. Hot fluid from hydrothermal vents can alter the mineral in the rocks and the heat could lower the magnetic value of the surrounding area. Hot fluid could also leech out the iron content in rocks making it demagnetized. This lowering value of magnetic attribute could cause anomaly which then later identified as hydrothermal vents[8].

3.2.6. Conductivity, Temperature, Depth (CTD) Sensor
This instrument is usually attached to vehicles towed from the ships and used to measure the conductivity, temperature, and pressure.The measuring usually real-time so the data can be obtained right away. This instrument package can also equipped by other instruments such as optical sensor, camera, pH sensor, chemical detectors, and sampling bottle. There is also similar instrument called

Miniature Autonomous Plume Recorder (MAPR). This intstrument is inexpensive and can be operated by someone with little special training[8].

3.2.7. Remote Sensing

The usage of remote sensing for many exploration has been renown. Many things can be obtained, like lineaments analysis. Lineament analysis are predicted to be a potential geological structure which give a geothermal system permeability although field studies are still need to be done in order to check geological structure availability. Lineaments analysis are conducted using image taken from satelite. Basically the concept between on-land and offshore lineaments analysis is the same, but since satelite imaging can't reach underwater, offshore remote sensing analysis is using underwater image taken by NOAA. These lineaments can give clues about the existence of underwater structure on sea floor. The example of remote sensing for offshore can be seen in figure 18 and figure 19.

3.3. Offshore Geothermal Utilization

Although there are no studies or current development of offshore geothermal energy, there are several ideas of how utilize it[9].

3.3.1. Platform Based Power Plant

This concept of power plant works by installing the power plant in the middle of the ocean. The steam is delivered through the seabed to the power plant using pipeline dircetly from the seabed to the surface. Then the steam will be converted into electricity and delivered to land to be distributed later. The structure for platform based power plant like this need to be able to withstand the wave[9].

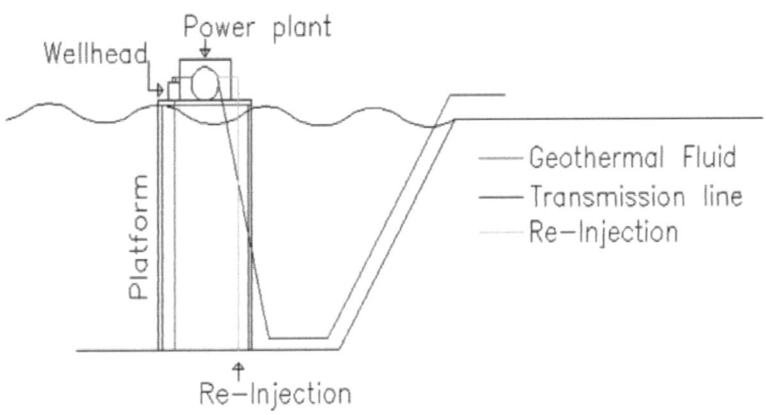

Figure 11. Conceptual model of platform based power plant. All of power plant instrumens are located above the sea. A strong structure is needed in order to withstand the wave impact

3.3.2. Land Based Power Plant

Wave and underwater current may give a promising threat to any platform build in the sea. Another concept was initiated in order to prevent this by building the separator on the seabed, separating the two phased fluid, and then transport the pure steam onto the land. This way the wave won't be a problem. But there is a possibilites of steam changing back to liquid form due to the decreasing of temperature during trasport. To prevent this the pipeline must be calculated so that the steam will stay in the gas form by using certain material or by installing the pipeline under the seabed[9].

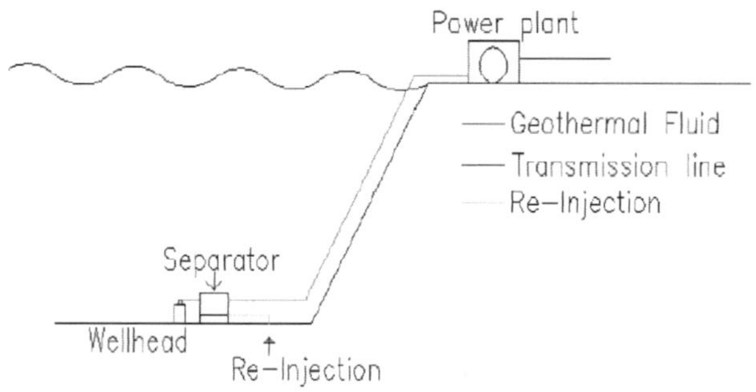

Figure 12. Conceptual model of land based power plant. The separator is installed on the seabed then transport the fluid onto the power plant on land

3.3.3. Underwater Based Power Plant

There are other concept of power plant which is hoped can overcome problems such as wave impact and a decrease in temperature, by building the power plant underwater and distribute the electricity onto the land. This way both the wave and temperature decrease will not be a problem. Although the design for power plant must be carefully calculated to withstand the underwater pressure which of course will be enourmous and to keep the seawater out from all the components. This concept proves to be risky and may need special design for the construction[9].

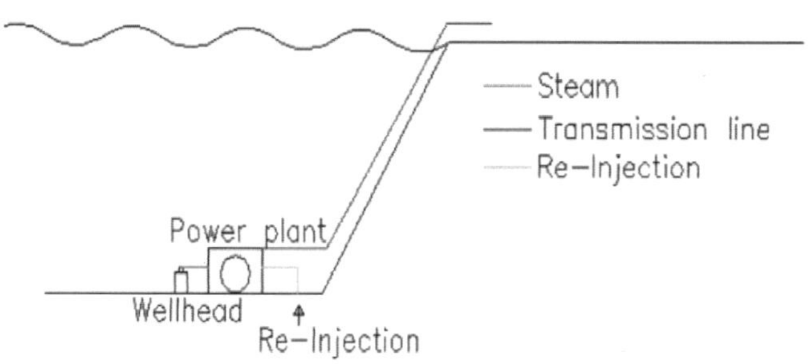

Figure 13. Conceptual model of underwater based power plant. The design must be a ble to withstand the underwater pressure

3.3.4. Binary Power Plant

This concept is similar to the binary power plant concept on land. The only difference is this one is build underwater. The heat exchanger is build on the seabed and the working fluid is transported onto the land.

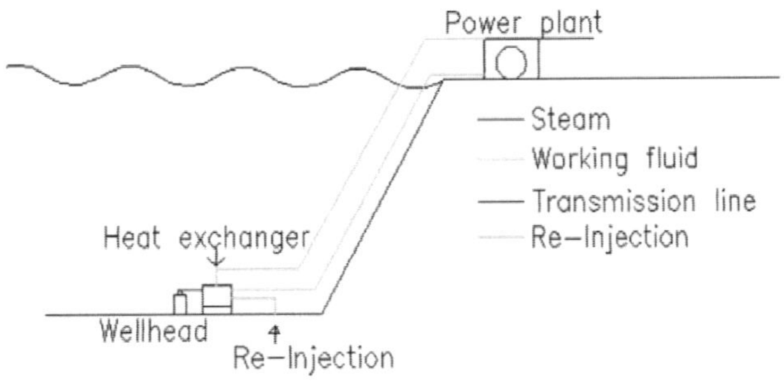

Figure 14. Conceptual model of binary power plant

Recently there is a new concept for this design by building a small submarine where the working fluid, heat exchanger, and generator are placed inside the submarine. This design was developed by The National University of Mexico. The submarine has 2 spiral tubes on the side containing the working fluid. The first spiral is located above the hydrothermal vent where the hot fluid from the vent will heat up the working fluid in the spiral and turn it into steam. Steam will be used to move the turbine in the generator. After this the steam of the working fluid will be streamed into the second tube which will work as the condensor where the tube located in cold seawater, cool down the steam and turn it into liquid phase again[9].

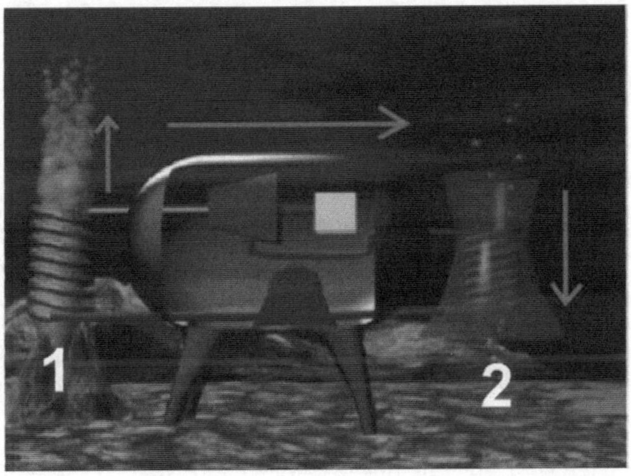

Figure 15. Conceptual model of submarine binary power plant. The hot fluid from vent will heat up the working fluid in tube 1, then transported it to the generator, and cooled in tube 2 by expose the tube to cold seawater changing it back to liquid form

3.3.5. Thermoelectricity
This concept is utilizing the temperature difference between the geothermal fluid and the seawater through a pipeline[9].

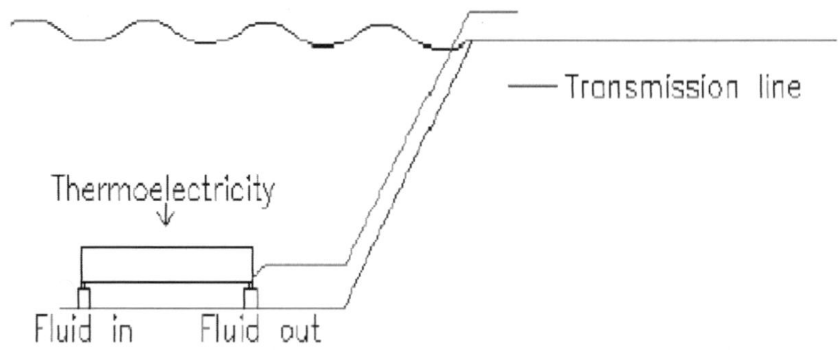

Figure 16. Conceptual model of thermoelectricity power station

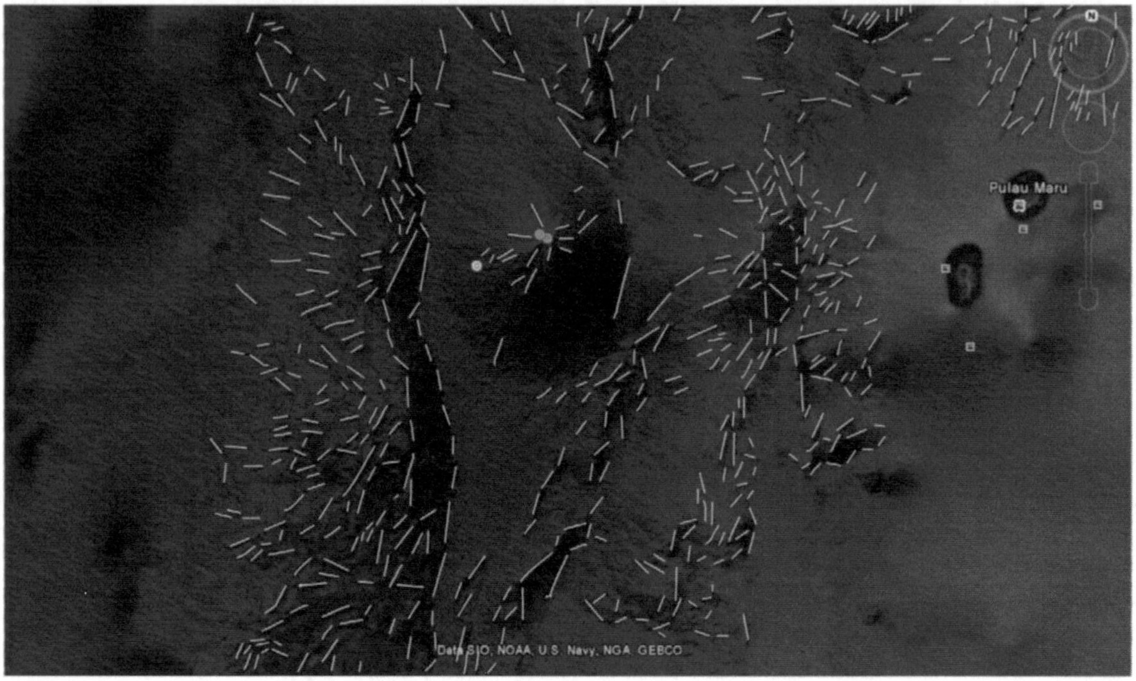

Figure 17. Lineaments from remote sensing image using NOAA image. These lineaments are predicted to be the underwater structures and need to be studied more to prove it

4. Discussion and Conclusion

It is clear from all the information obtained from many experiments around the world that seamounts are indeed can become a great potential for heat source in offshore geothermal system. They are located around spreading sea floors around the world and contain mafic magma. The temperature are possibly hotter than on land volcanoes and not as acidic as one. This is due to the short range for the magma to reach the surface so minimized the contact with the surrounding rocks. There are several seamounts in Indonesia, such as Banua Wuhu, NEC, Submarine, and Kawio Barat. The lava from seamounts will cooled rapidly by the seawater thus creating cracks in the igneous rocks which increase the permeability of the system. These cracks together with underwater structures become the fluid pathway in and out the system.

There hasn't been many detailed researched regarding the offshore geothermal system. Athough some methods for exploration and the technology concepts are available, there are still not enough researched yet to locate all offshore geothermal potential location. Looking from the heat contained in

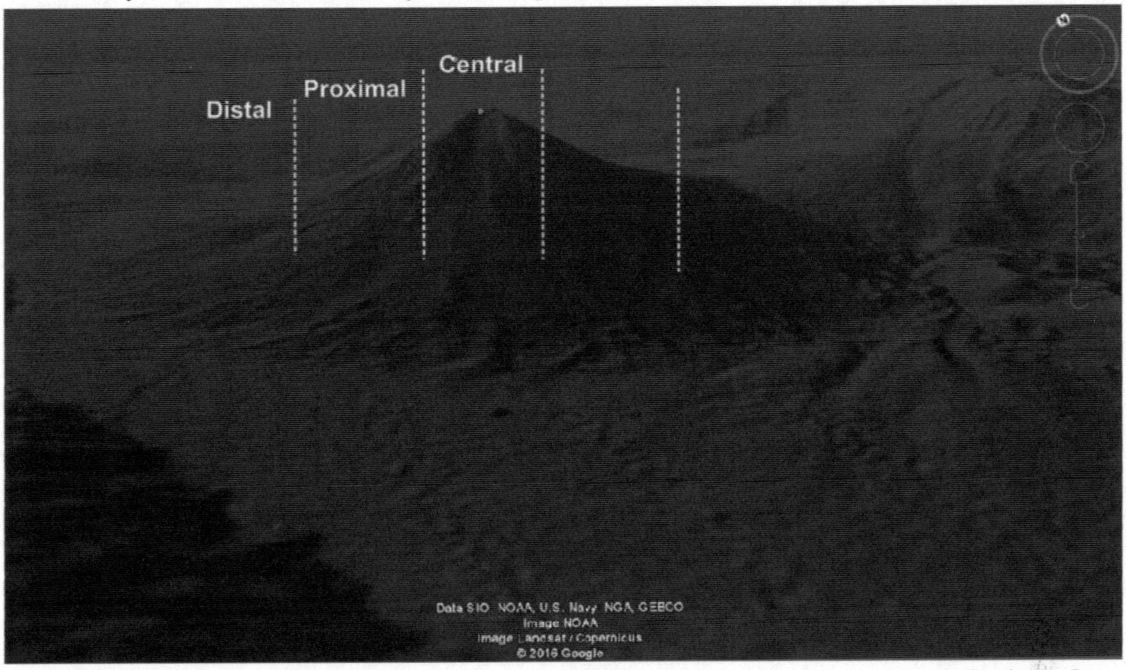

Figure 18. 3D underwater image of Kawio Barat Seamount taken from NOAA image database. It can be divided into 3 parts; central, proximal, distal.

seamounts and the permeability provided, it can be said that offshore geothermal energy is a quite interesting topic to be discussed and studied further. Seawater also provides literally unlimited fluid recharge and can be used as a condenser so there will be no need for cooling tower. Since it's located underwater the land usage for building the facilites can be minimalized.

Even with all those advantages, there are still many disadvantages and quite the vital one too. Some problems such as wave could threat the facilities construction so a strong construction is needed in order to withstand it. For equipments located underwater must be able to hold an enormous pressure in the seafloor. All of these need special materials, constructions design, and routine maintenance which mean an extremely high operation cost. Compared to land geothermal system, of course that offshore geothermal system is not a good choice since it will cost a lot of money. But it will serve as a back up plan for future alternative energy. So further research regarding the best and effective methods and technologies must be conducted. This may give better informations about the offshore geothermal system.

In conclusion, offshore geothermal potential in Indonesia is a very interesting topic since it is located in active rifting area, and as an archipelago state, Indonesia is surrounded by sea which can become the unlimited supply of recharge fluid for offshore geothermal system. Despite its possibility, offshore geothermal potential in Indonesia is not the main development priority due to many reason. It still needs to be studied even more since there are not much data regarding its seamounts, and until now Kawio Barat Seamount maybe the best choice since many researches have been conducted there beside the other seamount, although it is still not enough. The priority to develop conventional on-land geothermal is also one of the reason offshore geothermal system is not to be considered for now. Maybe one day in the future where the demands for geothermal energy has increased, offshore geothermal system can be considered as one of the options.

5. References

[1] Clark, Malcom R., Watling, Les, Rowden, Ashley A., Guinotte, John M., and Smith, Craig R 2010 *A Global Seamount Classification to Aid the Scientific Design of Marine Protected Area Networks* Elsevier

[2] Malik, M., Lobecker, E., Stuart, E., Peters, C., and Verplanck, N 2010 NOAA Ship Okeanos Explorer maps Kawio Barat, latitude: 4.6449°N, longitude: 125.03°E. NOAA Ocean Explorer website

[3] Kusumadinata, K 1979 Data dasar gunungapi Indonesia: *Catalogue of References on Indonesian Volcanoes with Eruption in Historical Time*, (Bandung: Direktorat Vulkanologi) p 726-727

[4] Hedervari, Peter 1984 *Catalog of Submarine Volcanoes and Hydrological Phenomena Associated with Volcanic Events 1500 B.C to Desember 31, 1899. World Data Center A for Solid Earth Geophysics* (United States: National Oceanoc and Atmospheric Administration)

[5] Prasojo, Hadi, Kurniawan, Temmy S, and Syaifulloh, Hanata 2014 *Potensi Panas Bumi Lepas Pantai di Kawasan Gunung Api Bawah Laut Kawio Barat, Banua Wuhu, dan Submarin 1922 di Kepulauan Sangihe, Sulawesi Utara: Energi Masa Depan* Bandung

[6] Siebert, Lee, Simkin, Tom, and Kimberly, Paul 2010 Volcanoes of The World. *The Journal of Geology 91, no. 5 September 1983*: 609-610.

[7] Tivey, Margaret K. 2014 *Black and White Smokers* (Netherlands: Springer)

[8] Atkins, Darren and Audunsson, Haraldur 2013 Exploration Technique for Locating Offshore Geothermal Energy Near Iceland *Proceedings Thirty-Eighth Workshop on Geothermal Reservoir Engineering* Standford California

[9] Karason, Baldur, Gudjonsdottir, Maria S, Valdimarsson, Pall, and Thorolfsson, Geir 2013 Utilization of Offshore Geothermal Resources for Power Production *Proceedings Thirty-Eighth Workshop on Geothermal Reservoir Engineering* Standford California

[10] Engineering Committee on Oceanic Resources 2008 *Report of Working Group on Marine Geothermal Energy*

[11] Hiriart, Gerardo 2010 *Submarine Geothermics: Hydrothermal Vents and Electricity Generation* (Proceeding World Geothermal Congress 2010. Bali, Indonesia)

[12] Encyclopedia Britannica Seamount. accessed on 19 August 2017 https://www.britannica.com/science/seamount

[13] EarthSky 2011 Seamonts and Knolls are Largely Unknown Ocean Oases access on 19 August 2017 http://earthsky.org/earth/seamounts-and-knolls-are-largely-unknown-ocean-oases-2

Acknowledgement

We want to thank many people who has been a great help and support in this research, especially Prof. Sutikno Bronto and Dr.Eng. Suryantini for the endless effort guiding the writting of this paper and all of people who helped us in many ways, also to those who have been giving their times and efforts to helpus in any ways which names cannot be mention one by one.

IOP Conf. Series: Earth and Environmental Science **103** (2017) 012013 doi:10.1088/1755-1315/103/1/012013

Enhancement of subsurface geologic structure model based on gravity, magnetotelluric, and well log data in Kamojang geothermal field

Muhammad Yustin Kamah[1], **Adilla Armando**[1], **Dinda Larasati Rahmani**[1], **Shabrina Paramitha**[1]

[1]Upstream Technology Center, PT Pertamina (Persero), Jakarta

Email: yustinkamah@pertamina.com

Abstract. Geophysical methods such as gravity and magnetotelluric methods commonly used in conventional and unconventional energy exploration, notably for exploring geothermal prospect. They used to identify the subsurface geology structures which is estimated as a path of fluid flow. This study was conducted in Kamojang Geothermal Field with the aim of highlighting the volcanic lineament in West Java, precisely in Guntur-Papandayan chain where there are three geothermal systems. Kendang Fault has predominant direction NE-SW, identified by magnetotelluric techniques and gravity data processing techniques. Gravity techniques such as spectral analysis, derivative solutions, and Euler deconvolution indicate the type and geometry of anomaly. Magnetotelluric techniques such as inverse modeling and polar diagram are required to know subsurface resistivity charactersitics and major orientation. Furthermore, the result from those methods will be compared to geology information and some section of well data, which is sufficiently suitable. This research is very useful to trace out another potential development area.

1. Introduction

Kamojang Geothermal Field is located in Garut, West Java, Indonesia. The volcanism in Kamojang region consists of a quartenary volcanic succession that produces lavas with andesitic and basaltic composition. Geologic structure plays important role in controling the path of fluid flow and circulation of geothermal system. There are numbers of uncertainty in predicting the subsurface condition, especially in identifying the geologic structures which related to the geothermal reservoir and the path of fluid flow for tracing out another potential development area. Therefore, this study provides combination of geophysical methods which consist of gravity and magnetotelluric for identifying the subsurface geololgic structure. Moreover, the result from those methods will be compared with geology information and some section of well data, thus the uncertainty interpretation of subsurface geologic structures will be reduce and another potential development area can be delineated.

2. Geologic Settings

The Kamojang Geothermal Field is located in West Java, around 40 km southeast of Bandung (figure 1). The field has 15 km long and 4.5 km wide. It lies in a big volcanic chain between Rakutak Mountain and Guntur Mountain. The development of this volcanic chain can be observed from the alignment of magmatic center of the volcanoes which began to develop from west to east. The order from oldest to youngest: Rakutak Mountain, Ciharus Lake, Pangkalan Lake, Gandapura Mountain, Masigit Mountain, and Guntur Mountain.

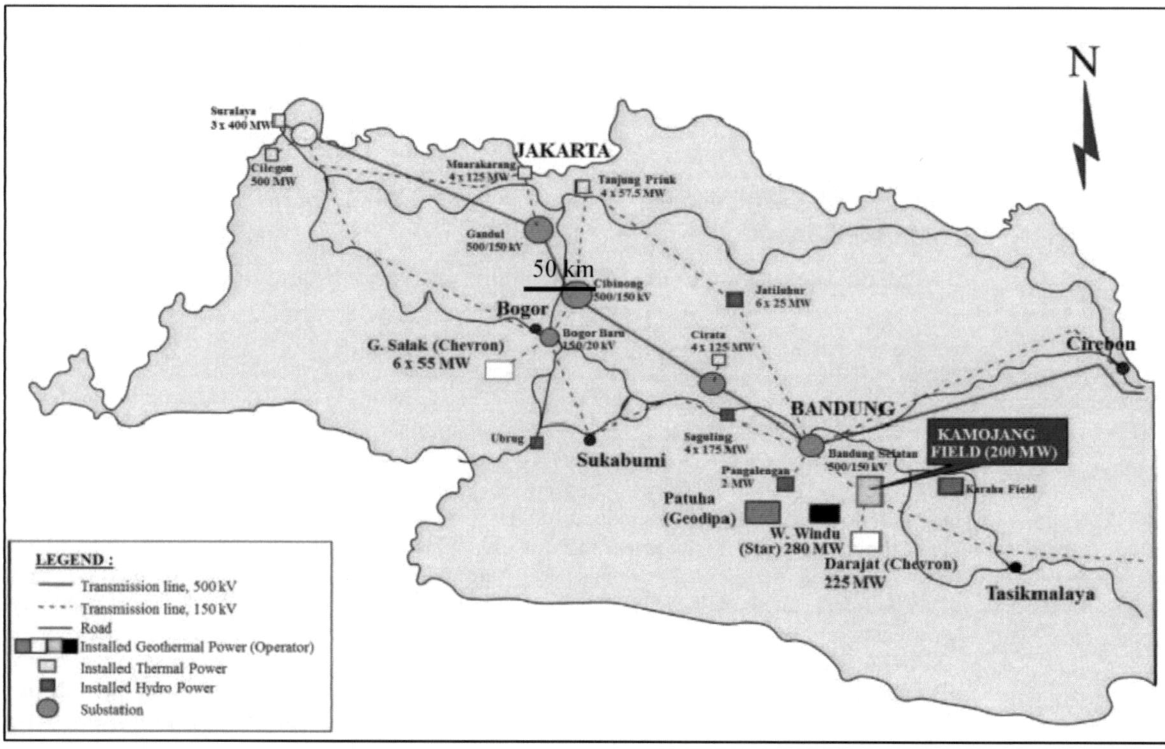

Figure 1. Location of Kamojang Geothermal

Kamojang Area can be divided into two formations (from older to younger): Pangkalan Formation and Gandapura Formation. Pangkalan Formation (1.2 Ma), which consists of weathered andesite, lies to the west of Pangkalan Lake, while Gandapura Formation (0.4 Ma) is formed by pyroxen andesite rocks and situated on the east of Kamojang.

Two important faults in these volcanic series are Kendang Fault and Citepus Fault. Both of them are NE-SW orientation faults. The orientation is in conformity with the tectonic regime in the region where the direction of subduction of Australian Plate under the island in north-south. This opening creates a passage for volcanism. Two known vapor system are found in the vicinity of the fault, Kamojang in the north and Darajat in the south.

These series are also affected by a 6 km wide graben, which is extending from Ciharus Lake to the Kamojang crater. Kamojang Area is located in the Pangkalan Depression Fault that forms a sort of cover which is ideal for the formation of dry steam geothermal systems. The depression which is be able to be observed in the field, can be divided into several sub-structures corresponding to small grabens and horsts. Tectonic activities of this area which have created current Kamojang structures:

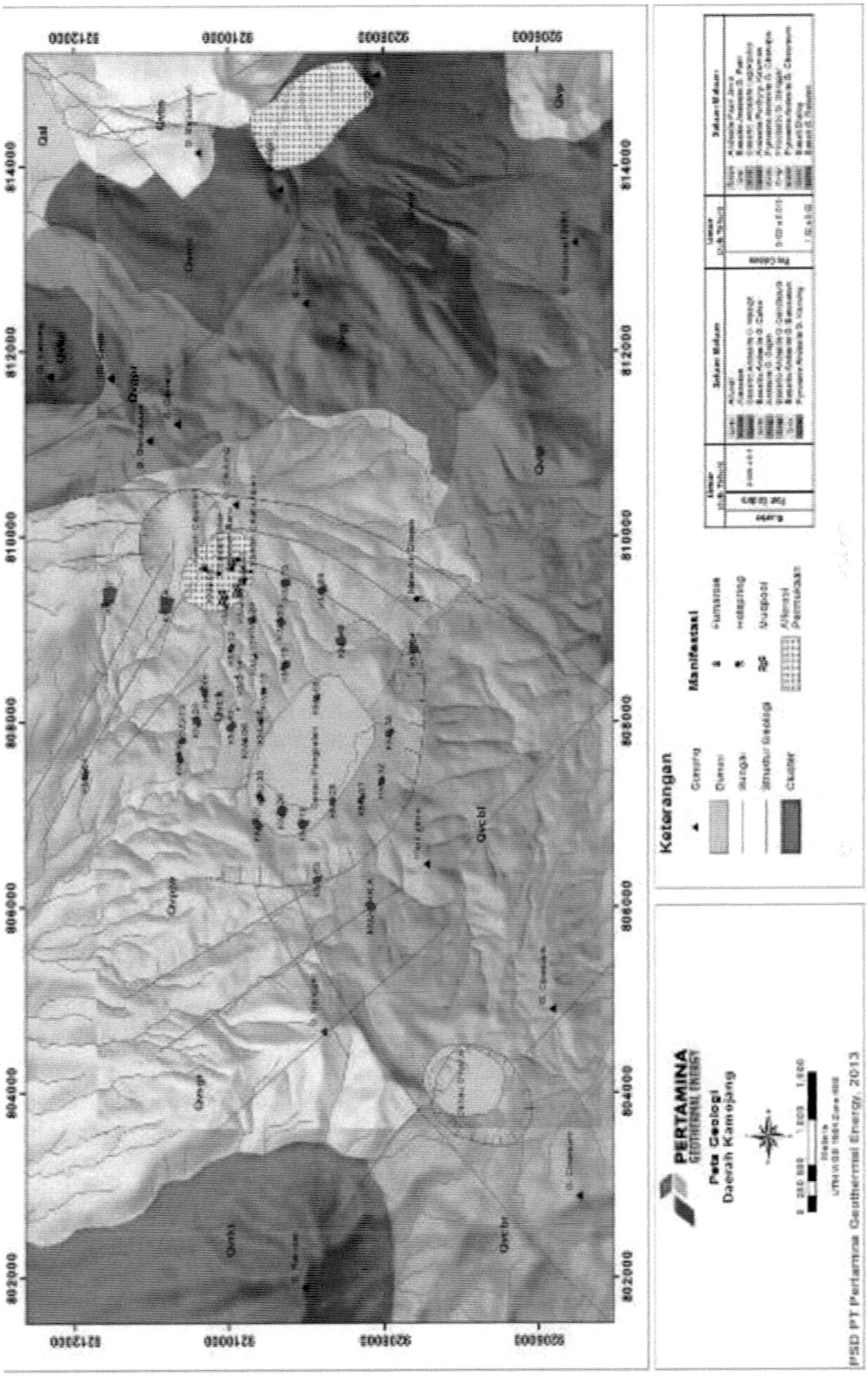

Figure 2. Regional Geological Map of Kamojang [2].

a. The collapse pit of Pangkalan:
This circular structure is located in the western part of Kamojang field. It is limited by a rim which outcrops clearly in the western and southern part. In the northern and eastern part, regional fractures collapse it and young lava flows conceal it. This rim delineates a 2 km long and 1.5 km wide circular area.

b. NE – SW system:
Dominant regional geologic structures in Kamojang have a NE-SW trend, including Kendang Fault and Citepus Fault. This system consists of tensional (lateral origin) faults set, which have been identified as a very important structures, related to the main target of reservoir Kamojang field. Even if the faults and fractures have been altered in the upper part of the surface and form non-permeable seals, the bottom sections may still be highly permeable. This system is parallel to the magmatic axis. After Pangkalan Structure had formed, small volcanic complexes or cones came up alongside the system, which is estimated to present a possible heat source for the geothermal field of Kamojang.

c. NW – SE system:
6 km wide graben is a major expression of NW-SE tensional faults. On the southern part of the field, these normal faults delineate two narrow downthrown structures which clearly affect the rim of Pangkalan Structure. On the northern part, these grabens can be observed too; but on the central part where the NE – SW shear system extends, they cannot be seen at the surface. These tectonic features seem to be slightly younger than the NE – SW system. There is good correspondence between these surface observation and interpretation of electrical surveys carried out by Pertamina (1981).

3. Gravity Techniques

This study was conducted in Kamojang Geothermal Field, around (5233 x 7183) m^2. It used 188 observation stations on gravity survey. The obtained value of Complete Bouguer Anomaly (CBA) based on processing gravity is (9.3 – 21.6) mGal, figure 5(b), and background density 2.61 mGal. Furthermore, it applied spectral analysis, and advance gravity data processing such as First Horizontal Derivative (FHD), Second Vertical Derivative (SVD), and Euler deconvolution to identify boundary contact of body anomalies, type of faults, and depth of body anomalies contact in order to enhance the interpretation of geological structure.

3.1. Spectral Analysis

This method is used to produce a variety of filtered anomaly maps. There are several types of filters in spectral analysis. Low-pass filter is used to remove undesirable high-wave number which associated with small and shallow sources, and to reduce errors in observations and reduction of the gravity anomalies. On the other hand, high-pass filter remove the longer-wavelength components associated with large and deep sources that are usually regarded as the regional anomaly [3]. The objective of wavelength filtering is to focus attention on anomalies derived from specified depth range and to determine the depth slice of residual and regional anomaly.

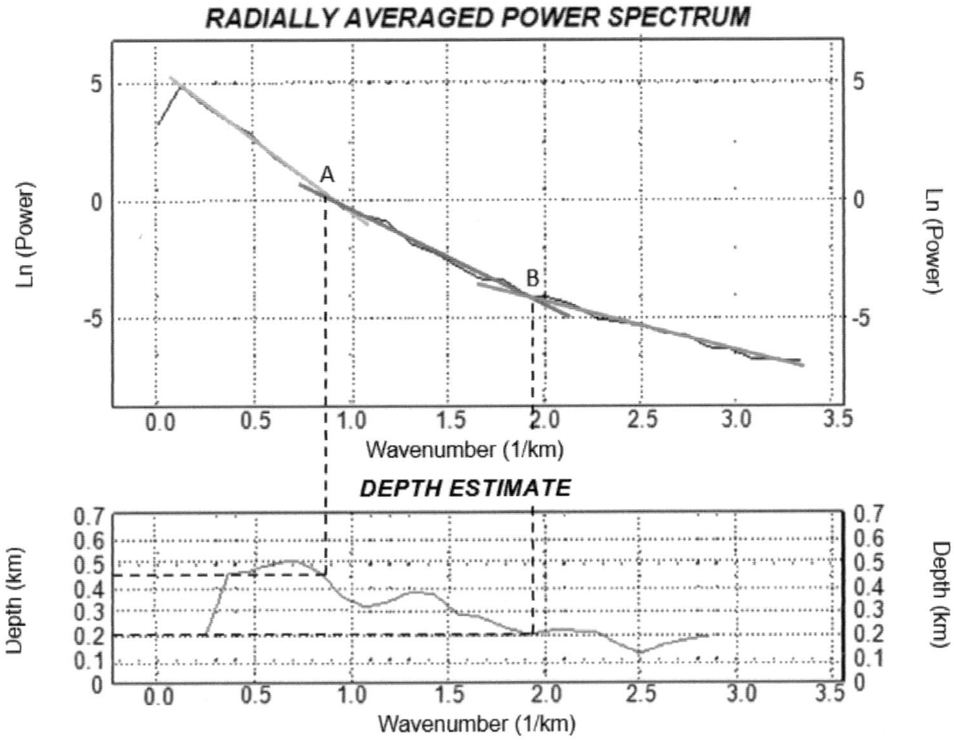

Figure 3. Power spectrum and depth estimate curve.

Based on figure above, there are 2 intersection points hereinafter correlated with depth estimate curve. The A point is located at about 0,45 km depth from surface which is identified as depth of regional body anomalies. While B is located at about 0,2 km depth which is identified as depth of residual body anomalies.

3.2. First Horizontal Derivative (FHD)

This method is used to interpret boundary contact of body anomalies. The accuracy of FHD map is affected by data control and errors, as well as dip variations and depth of body anomalies.

3.3. Second Vertical Derivative (SVD)

SVD can also be used to interpret the boundary contact of body anomalies. Moreover, it could help decide the type of faults, whether it is normal fault or thrust fault based on comparation of maximum absolute SVD value with minimum absolute SVD value.

3.4. Euler Deconvolution

Euler deconvolution method can be used to enhance potential field data interpretation, such as gravity and magnetic, in terms of depth and geological structure trends. By using the Euler deconvolution, the geological features such as faults, contacts, sill, and dykes would be estimated, depends on the structural index used. In this study, the structural index used is 0 (zero) which indicates contact or step of body anomalies at certain depth. The equation of Euler's homogeneity can be shown by equation 1.

$$(x - x_o)\frac{\partial g\,(x,y,z)}{\partial x} + (y - y_o)\frac{\partial g\,(x,y,z)}{\partial y} + (z - z_o)\frac{\partial g\,(x,y,z)}{\partial z} = -N \times g(x,y,z) \quad (1)$$

where:

x,y,z : coordinates of observation station

x_o, y_o, z_o : coordinates of source
g : total gravity values at coordinates (x, y, z)
$\frac{\partial g}{\partial x}, \frac{\partial g}{\partial y}, \frac{\partial g}{\partial z}$: derivatives of g on direction x, y, and z
N : degree of homogeneity, also known as structure index which is correlated with the geomteric structure of the field source

Table 1. Structural Index N for Magnetic Field and Gravity Field.

SI	Magnetic Field	Gravity Field
0	Contact / Step	Sill / Dyke / Ribbon / Step
1	Sill / Dyke	Cylinder / Pipe
2	Cylinder / Pipe	Sphere
3	Sphere / Barrel / Ordnance	N / A

4. Magnetotelluric Techniques

Potential geothermal areas require the presence of water, heat, and high permeability of reservoir. Since the physical parameter of subsurface such as resistivity can changes due to fluid and rock permeability, magnetotelluric (MT) method can be useful and suitable for geothermal exploration. It provides information about lateral and vertical resistivity variation in the subsurface by measuring the magnitude response of electric fields (E) and magnetic fields (H) of the electromagnetic nature.

To provide the MT data, MT-TDEM survey has been done in Kamojang Geothermal area. Instrument used in this survey is MTU-5 Phoenix 24bit satellite synchronized system which has frequency 384 down to 0.001 Hz. It has been applied the robust processing for noise elimination and static shift correction based on TDEM data. There are 3 lines used in MT-TDEM survey which is shown by figure 4.

Moreover, magnetotelluric method can detects the geoelectric strike which is one of the parameter measured by tensor impedance. Principally, when diagonal segment is minimum (segment Z_{xx} and Z_{yy}), it can be represented as geological feature (geological structure).

$$\boldsymbol{Z} = \begin{bmatrix} \boldsymbol{Z_{xx}} & \boldsymbol{Z_{xy}} \\ \boldsymbol{Z_{yx}} & \boldsymbol{Z_{yy}} \end{bmatrix} \tag{2}$$

It's very difficult to set up the MT field layout parallel to the strike trend. So that, for 2D analysing, it takes a proper rotation on the impedance curve tensor by swift angle α, which refer to the strike way in the field. The angle of the swift is:

$$\boldsymbol{\alpha = \frac{1}{4}\tan^{-1}\frac{(Z_{xx} - Z_{yy})(Z_{xy} + Z_{yx}) + (Z_{xx} + Z_{yy})(Z_{xy} - Z_{yx})}{|Z_{xx} - Z_{xx}|^2 - |Z_{xx} + Z_{xx}|^2}} \tag{3}$$

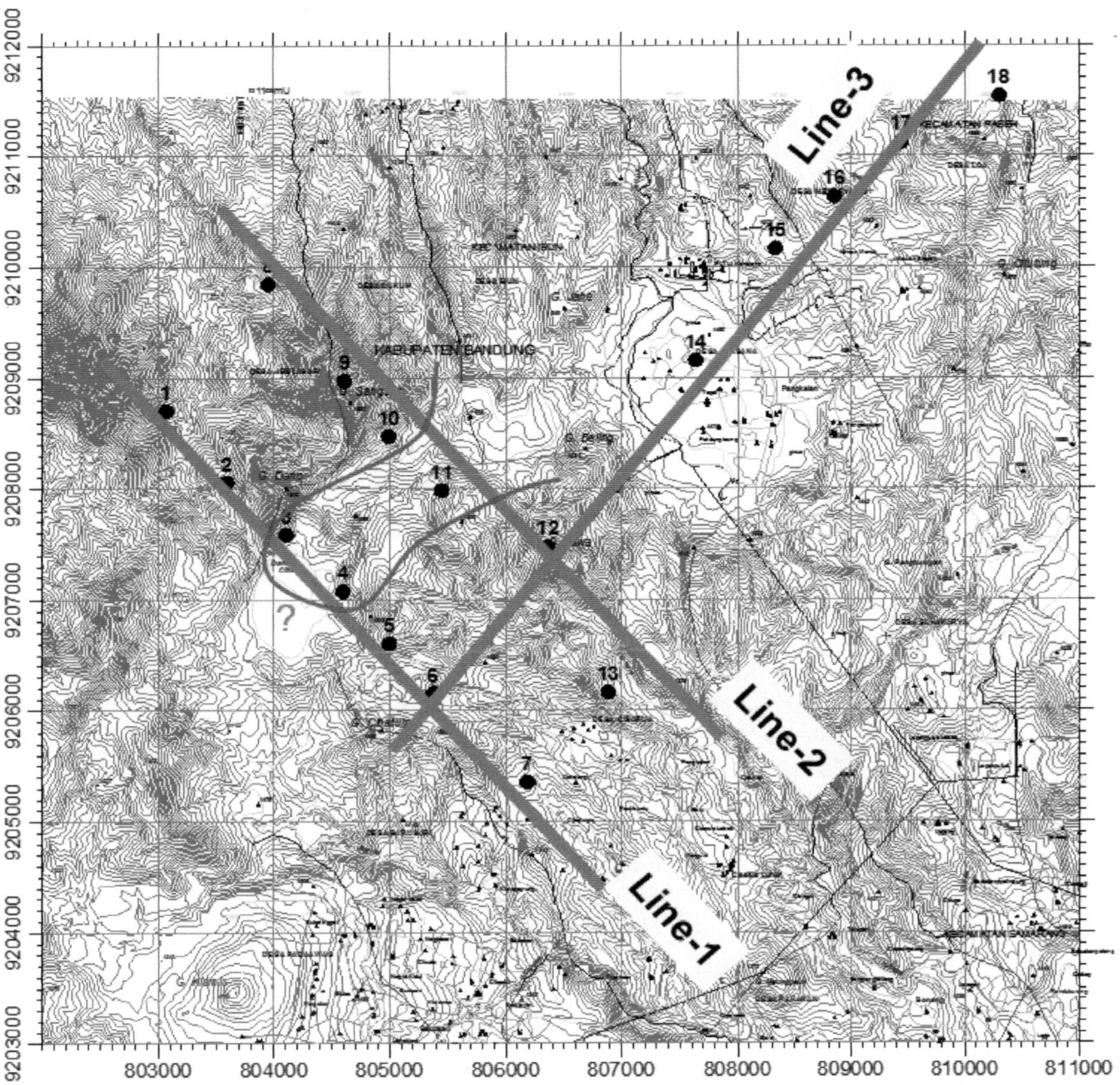

Figure 4. MT-TDEM survey line map. Line-1 crossing Lake Ciharus. Line-2 crossing CHR cluster. Line-3 crossing the entire field to north east.

Identical swift angle obtained by inputting the appropriate impedance values strike direction (by strike angle selection mode when the robust processing) and plotting the impedance curve into the polar diagram. All polar diagram obtained, we can determine the dominant direction trend [2].

5. Result and Discussion

5.1. Gravity

In this study, CBA data is applied spectral analysis in order to dissociate regional and residual anomaly. CBA represents the combination of regional and residual sources. Therefore, it can be used to interpret the existence of regional structure which continous to the surface. On CBA map, Figure 5(b), there are dominated by high gravity anomaly on southeastern area and rounded low gravity anomaly on western area. Low gravity anomaly on CBA and residual map in Figure 5(b) and (d) correlates with Ciharus Lake and fault which has a role as boundary to other rocks adjacent. While regional map, Figure 5(c), tends to represent deeper source of body anomalies (based on regional depth on spectral analysis). There are also high gravity anomaly on southeastern area and low gravity anomaly on western area, the contact between low and high gravity anomalies is identified as regional geological structure which has major trend NE-SW.

Based on FHD map, Figure 5(e), it can be estimated that there are boundary contacts of body anomalies which have trend direction NE – SW as well as NW-SE. Those boundary contacts is identified as geologic structure. SVD map, Figure 5(f), shows geological structure trends which correlate to FHD map. Moreover, it can be seen another gravity anomaly on FHD and SVD maps in northwestern and northeastern survey area which can be estimated as faults or geological structures. This gravity result shows potential existence of geological structures which could be used as additional information to the geological data.

Figure 5(h) and (i) show the FHD and SVD curve toward axis coordinates which can be used for identifying the type of fault in slice area. The normal faults can be identified by the amplitude of absolute maximum SVD value which higher than absolute minimum SVD value, otherwise represented as thrust fault. Based on the curve, it can be determined that the survey area is dominated by normal fault. Based on Euler deconvolution map, Figure 5(g), it can be identified the contact of body anomalies at the certain depth which have NE-SW and NW-SE trends.

5.2. Magnetotelluric

The aim of MT-TDEM inverse modeling is to show the resistivity characteristics of subsurface. It found spot of low resistivity at shallow depth which is associated with surface thermal manifestation (fumarolic) as upflow zone and some alteration area (in middle area). At a shallow-intermediate depth, the conductors merge and become penetrated that has a thickness approximately 1000 m (situated from 1000 to 0 masl). Beyond this level, we find the prophylitic zone contains wairakite and epidote and interpreted from resistivity structure. This zone currently provides the steam zone for the 235 MW power electric generation, especially from the north and west arms [4]. For more enhancing the interpretation of subsurface, MT-TDEM inverse modeling is applied in this study.

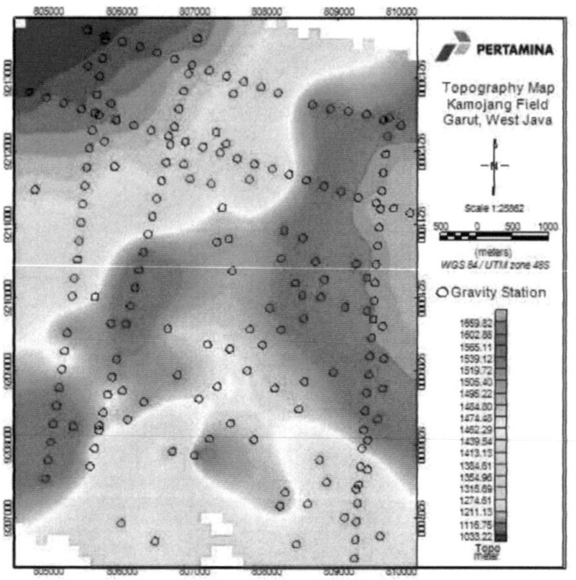

(a)

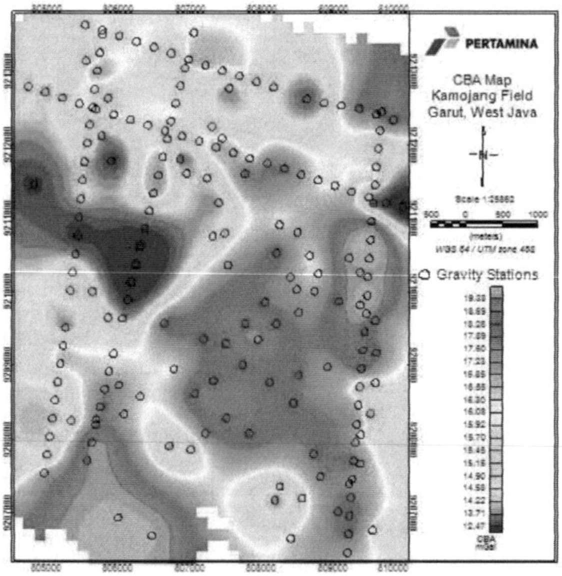

(b)

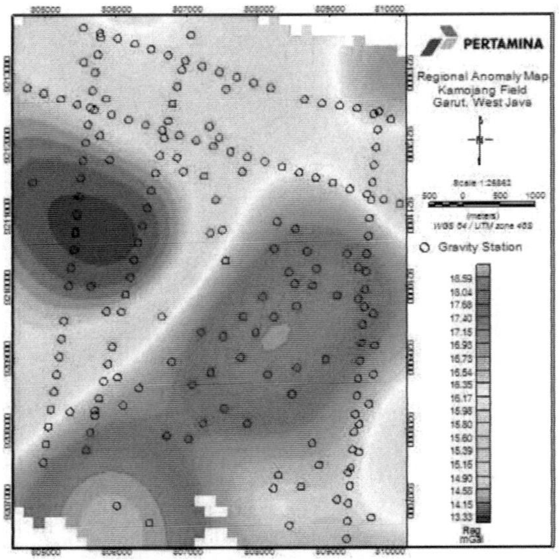
PERTAMINA
Regional Anomaly Map
Kamojang Field
Garut, West Java
Gravity Station

(c)

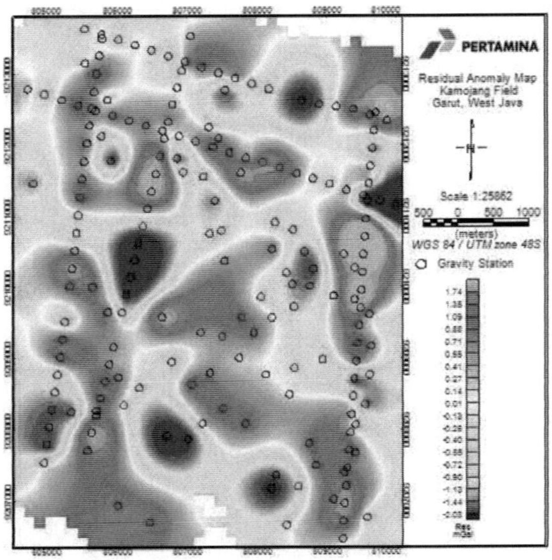
PERTAMINA
Residual Anomaly Map
Kamojang Field
Garut, West Java
Scale 1:25862
WGS 84 / UTM zone 48S
Gravity Station

(d)

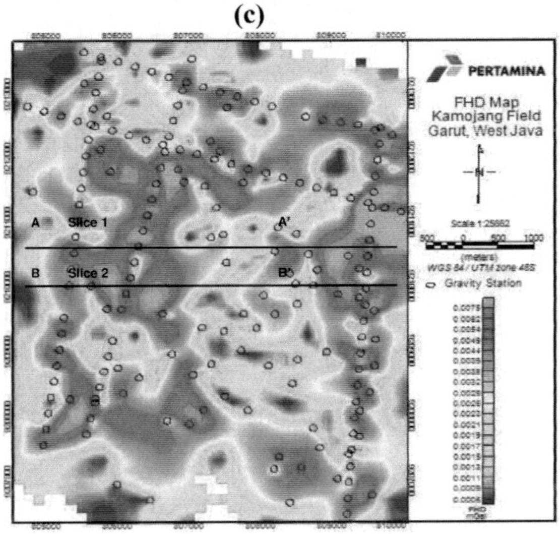
PERTAMINA
FHD Map
Kamojang Field
Garut, West Java
A Slice 1 A'
B Slice 2 B'
Gravity Station

(e)

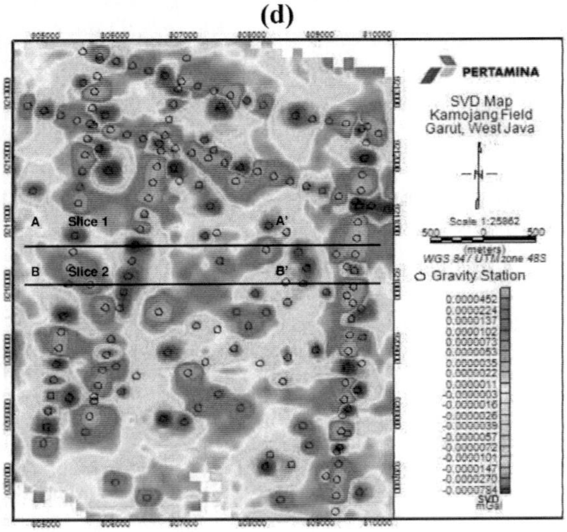
PERTAMINA
SVD Map
Kamojang Field
Garut, West Java
Scale 1:25862
A Slice 1 A'
B Slice 2 B'
WGS 84 / UTM zone 48S
Gravity Station

(f)

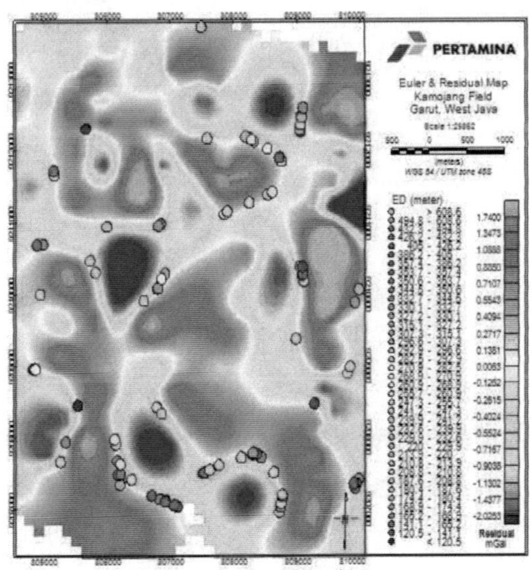
PERTAMINA
Euler & Residual Map
Kamojang Field
Garut, West Java
ED (meter)
Residual mGal

(g)

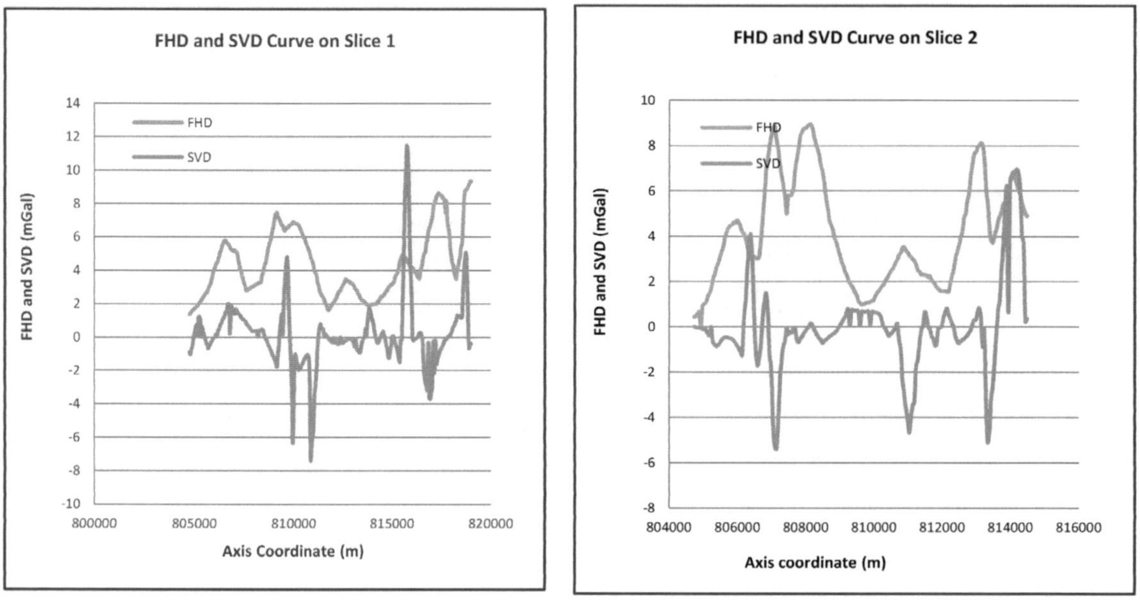

(h) (i)

Figure 5. Processed gravity map and curve result in Kamojang survey area.
(a) Topography map. **(b)** Complete Bouguer Anomaly map. **(c)** Gravity regional map. **(d)** Residual anomaly map. **(e)** First Horizontal Derivative (FHD). **(f)** Second Vertical Derivative (SVD). **(g)** Euler deconvolution map. **(h)** FHD and SVD curve toward axis coordinate on slice 1. **(i)** FHD and SVD curve toward axis coordinate on slice 2.

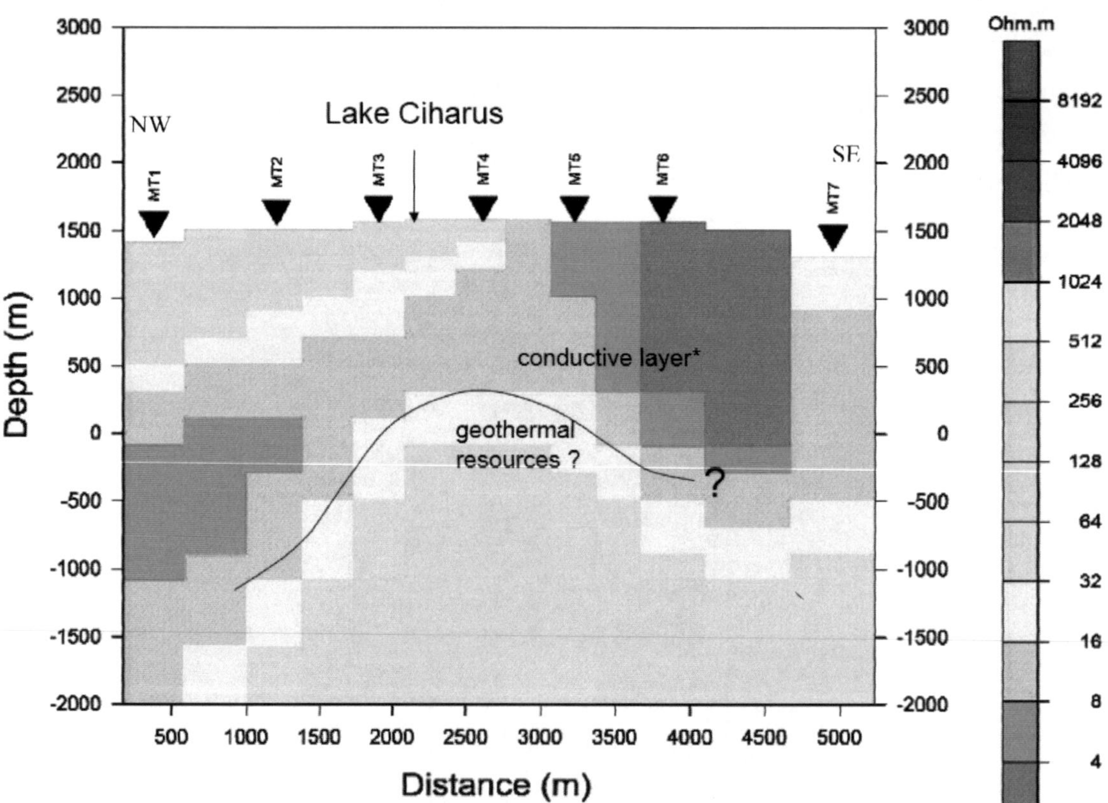

Figure 6. 2D Inversion model in line 1 Kamojang area (SDT Geothermal, 2006)

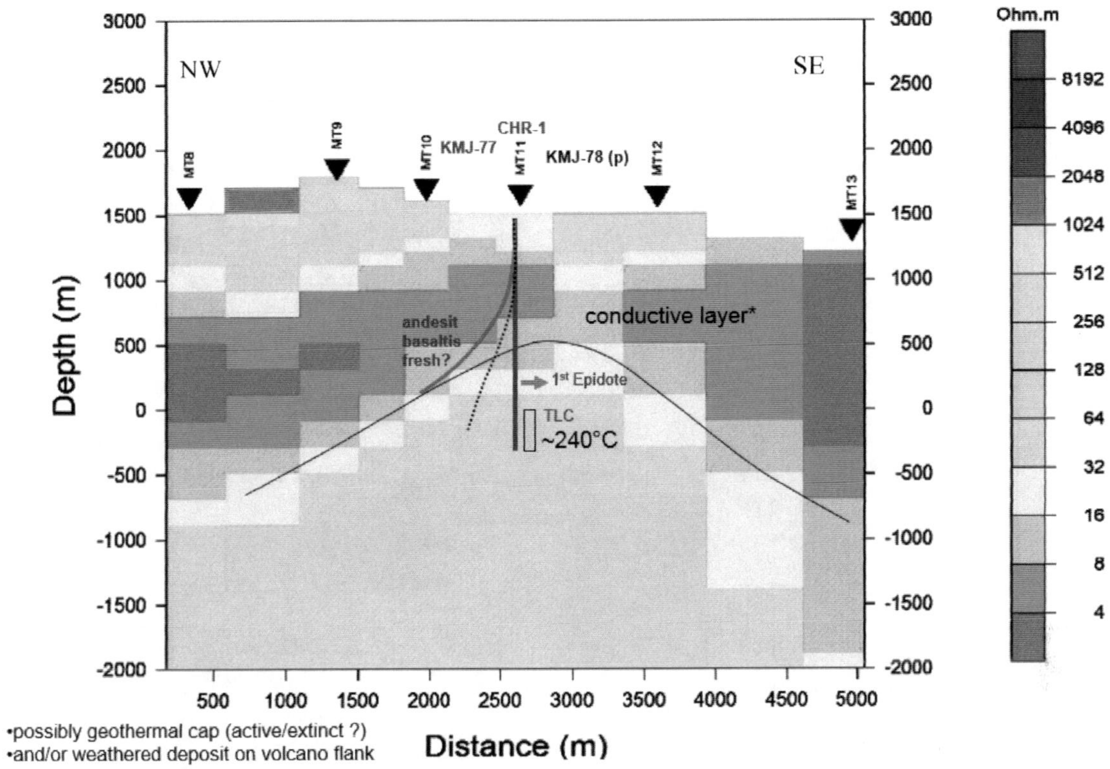

Figure 7. 2D inversion model in line 2 Kamojang area (SDT Geothermal, 2006)

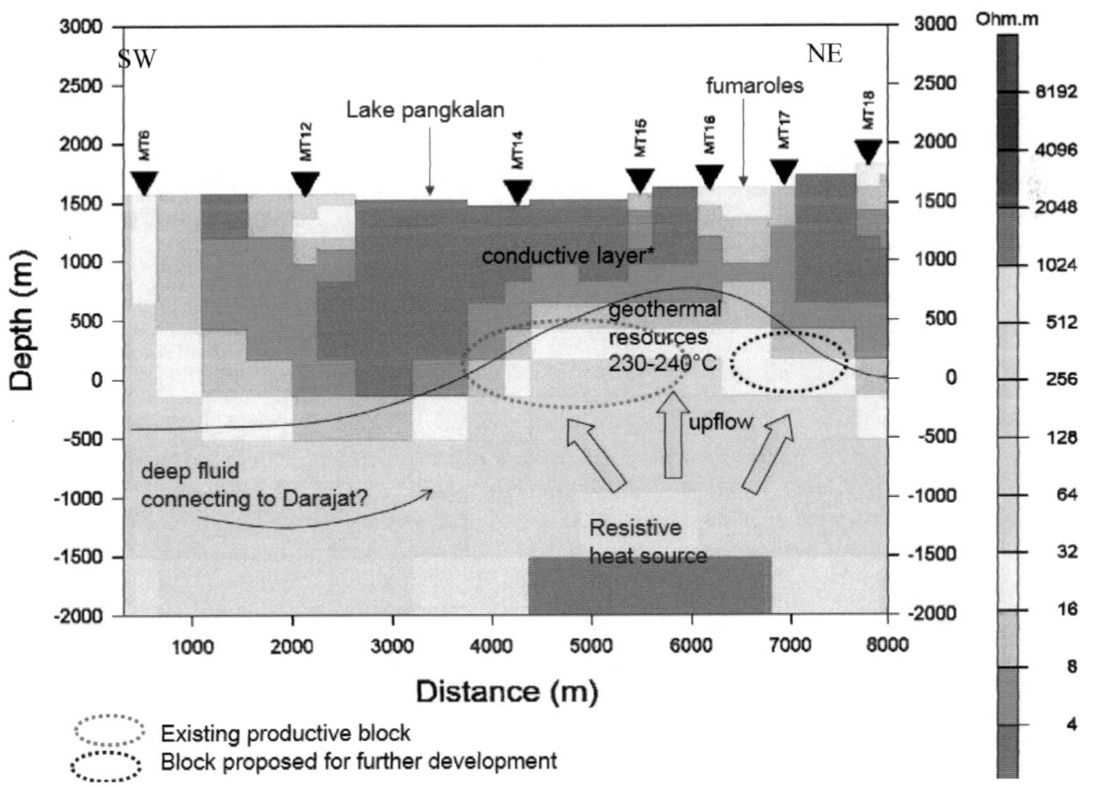

Figure 8. 2D inversion model in line 3 Kamojang area (SDT Geothermal, 2006)

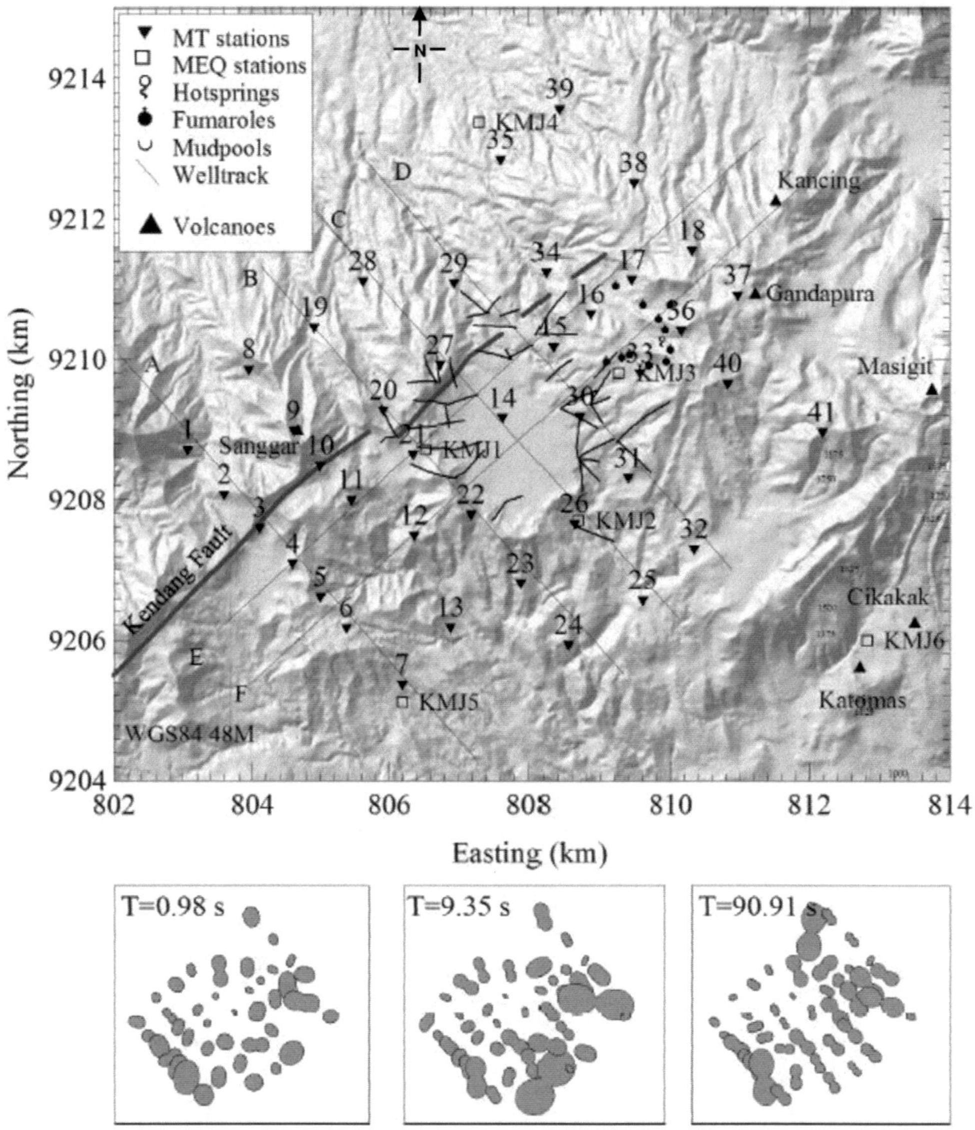

Figure 9. Polar diagram [4].

It has been analyzed 3 panels of Polar Diagram. First Panel 0.98 seconds, shows no obvious alignment except a weak north-south and east west signatures. The second panel shows a transition between the first and third panel where the NW-SE trend becomes apparent. The third panel, at 90.91 seconds, displays an alignment pointing to the NW-SE, where its directions are perpendicular to the Kendang fault [4].

5.3. Well Correlation Data

According to Raharjo[4], FMS (Formation Micro Scanner) logs recorded from Kamojang reservoir are used to confirm the structural alignment. Figure 5 shows two rosette diagrams which reveal the strike of conductive fractures from well KMJ-50 and KMJ-52 (only fractures having a dip greater than 60° are plotted).

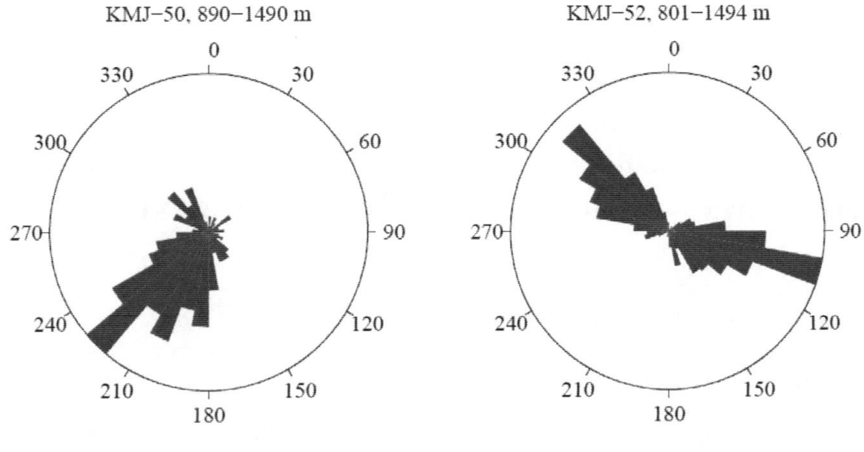

Figure 10. Rosette diagrams of well KMJ-50 and well KMJ-52 [4]

It can be seen clearly from the rosette diagram of well KMJ-50, within the depth of 890-1490 m, fractures dominantly striking N 235° E. This direction is in agreement with the Kendang Fault. In contrast, KMJ-52 encounters major fractures oriented to N 295 ° E – N 315 ° E and N 105 ° E.

Generally, orientation of fractures plotted in the rosette diagrams are suitable to the identification of geologic structures from gravity data, which is interpreted that have trend direction NE–SW and NW–SE.

6. Conclusion

It can be concluded that based on regional geology, gravity, magnetotelluric, and FMS log data, the major structures in Kamojang Geothermal Field have trends NE–SW and NW–SE. From gravity data, we interpret other new potential existence of faults which have trend NW–SE in northwestern part of survey area and trend NE–SW in northeastern part of survey area. Those areas are potentially developed for geothermal field area. There is a good correlation between analysis of regional geology, gravity, magnetotelluric, and FMS log data to the structure interpretation. The result can be used for delineating new potential development areas.

References

[1] Firmansyah, Sujarmaitanto W, Mulyanto H, Hartanto D B and Wicaksono T 2015 Geochemistry Monitoring in the Kamojang Vapor-dominated Geothermal Field from 2010 to 2013 *Proceedings World Geothermal Congress (WGC) 2015*, Melbourne.

[2] Yunus D, Wambra A, Dzil M, Fikri F, Aji P S and Julfi H. 2015 Identification of Subsurface Geological Structure in a Geothermal System using MT Imaging Technology *Proceedings World Geothermal Congress (WGC) 2015*, Melbourne.

[3] Hinze W J, Ralph R B, Von F and Afif H S 2013 *Gravity and Magnetic Exploration* Cambridge University Press, New York.

[4] Raharjo I B 2011 *Geophysical Signatures Of Volcano Hosted Geothermal Systems* Dissertation Geophysics of The University of Utah p 113.

Acknowledgment

We would like to acknowledge to management of PT Pertamina (Persero), Jakarta, Upstream Technology Center, New Energy and Green Technology Department for approving us to publish and present this paper.

Geomorphological classification of post-caldera volcanoes in the Buyan–Bratan caldera, North Bali, Indonesia

Mitsuru Okuno[1], Agung Harijoko[2], I Wayan Warmada[2], Koichiro Watanabe[3], Toshio Nakamura[4], Sachihiro Taguchi[1], Tetsuo Kobayashi[5]
[1] Department of Earth System Science, Faculty of Science, Fukuoka University, Fukuoka, Japan
[2] Department of Geological Engineering, Gadjah Mada University, Yogyakarta, Indonesia
[3] Department of Earth Resources Engineering, Faculty of Engineering, Kyushu University, Fukuoka, Japan
[4] Institute for Space-Earth Environmental Research, Nagoya University, Nagoya, Japan
[5] Emeritus Professor, Kagoshima University, Kagoshima, Japan

e-mail: okuno@fukuoka-u.ac.jp

Abstract. A landform of the post-caldera volcanoes (Lesung, Tapak, Sengayang, Pohen, and Adeng) in the Buyan–Bratan caldera on the island of Bali, Indonesia can be classified by topographic interpretation. The Tapak volcano has three craters, aligned from north to south. Lava effused from the central crater has flowed downward to the northwest, separating the Tamblingan and Buyan Lakes. This lava also covers the tip of the lava flow from the Lesung volcano. Therefore, it is a product of the latest post-caldera volcano eruption. The Lesung volcano also has two craters, with a gully developing on the pyroclastic cone from the northern slope to the western slope. Lava from the south crater has flowed down the western flank, beyond the caldera rim. Lava distributed on the eastern side from the south also surrounds the Sengayang volcano. The Adeng volcano is surrounded by debris avalanche deposits from the Pohen volcano. Based on these topographic relationships, Sengayang volcano appears to be the oldest of the post-caldera volcanoes, followed by the Adeng, Pohen, Lesung, and Tapak volcanoes. Coarse-grained scoria falls around this area are intercalated with two foreign tephras: the Samalas tephra (1257 A.D.) from Lombok Island and the Penelokan tephra (ca. 5.5 kBP) from the Batur caldera. The source of these scoria falls is estimated to be either the Tapak or Lesung volcano, implying that at least two volcanoes have erupted during the Holocene period.

1. Introduction

The Buyan–Bratan caldera on the island of Bali, Indonesia, is approximately 6 km × 11 km in area and contains three lakes (Tamblingan, Buyan, and Bratan). The post-caldera volcanoes (Lesung, Tapak, Sengayang, Pohen, and Adeng) are distributed from the center to the south of the caldera (figure 1). This caldera was previously thought to be less active than the neighboring volcanoes to the east, the Batur caldera and the Agung volcano [1]. The eruptive history of the caldera may provide important information to determine the development of the geothermal system [2]. To reconstruct a detailed Holocene eruptive history of this caldera volcano, we observed its geomorphological features and conducted a field survey on the tephra stratigraphy.

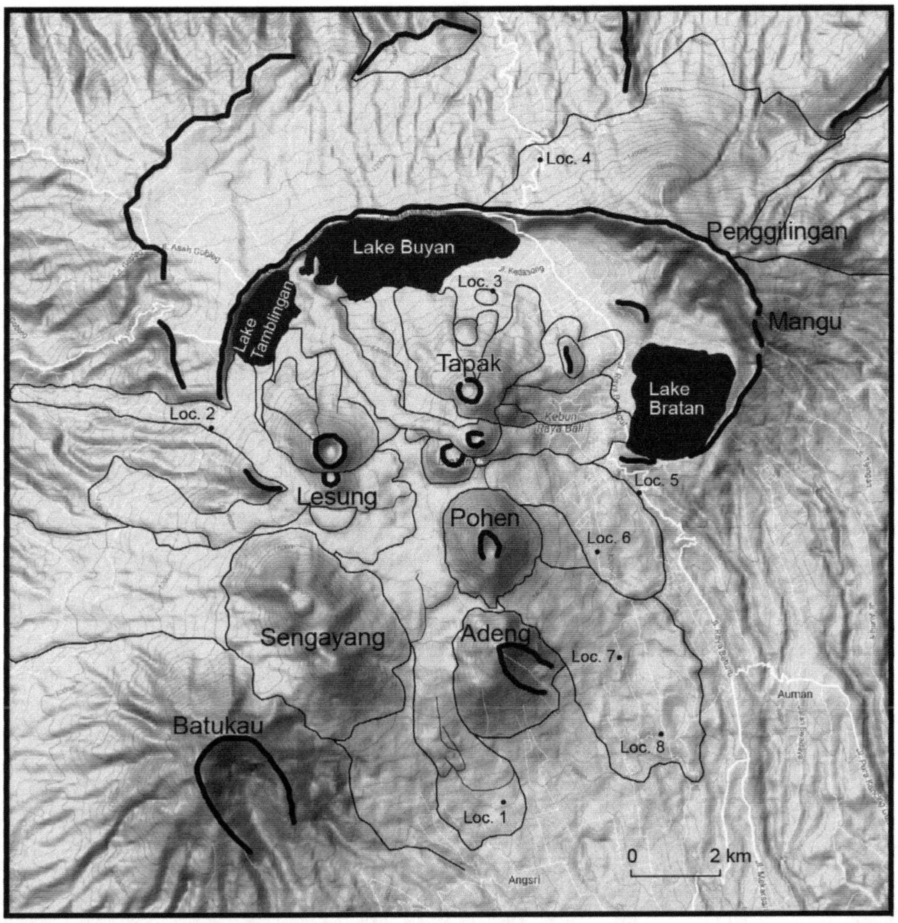

Figure 1. Geomorphological map of the Buyan–Bratan caldera, showing locations of selected outcrops. Thick lines indicate caldera cliffs and the crater rim. Moderately thick lines indicate geological unit boundaries. Thin lines indicate flow boundaries within units. Google Maps (terrain view) was used as the base map.

2. Geomorphology

Based on the topographic relationships (figure 1), the Sengayang volcano is the oldest of the post-caldera volcanoes, followed by the Adeng, Pohen, Lesung, and Tapak volcanoes. The Tapak volcano has three craters, aligned from north to south, which can be correlated with each subdivided edifice. Lava from the central crater has flowed downward to the northwest, dividing the Tamblingan and Buyan Lakes. This lava has covered the tip of the lava flow from the Lesung volcano. The central edifice of the Tapak volcano covers a part of the south crater, and the direction of eastward lava flow is controlled by the north edifice. Therefore, the central edifice is a product of the latest eruption among the post-caldera volcanoes. Southward lava flow from Tapak has reached the south foot of the Adeng volcano (figure 2). This topography has been built by a series of eruptions during the Holocene. However, the lower edifice of this volcano may predate the Holocene. Some parts of the old lava flows have been exposed on the northern slope, such as Location 3 (figure 3).

The Lesung volcano has two craters, and a gully developing on the pyroclastic cone from the northern to the western slope. Lava from the south crater has flowed down to the western flank beyond the caldera rim (figure 4). The horizon of this lava flow is estimated to be near the Penelokan tephra [3] from the Batur caldera [4]. The debris avalanche deposits (figure 5) distribute on the southeastern foot of the Pohen volcano. However, no clear collapsed landform is remained on this volcano. These topographic features indicate that the sector collapse and rebuilt were occurred in Pohen volcano before the Holocene activities of Tapak volcano.

Figure 2. An occurrence of the youngest lava flow on the southern foot of the Adeng volcano (Location 1). Only the Samalas ash (1257 A.D.) covers this lava.

Figure 3. Outcrop of lower lava flow on the northern slope of the Tapak volcano (Location 3). The stratigraphic position of this lava remains unclear.

Figure 4. Lava flow and scoria falls on the western slope (Location 2) of the Buyan–Bratan caldera. (a) Four coarse-grained scoria falls above the lava flow. The scale of the lava is 1 m. (b) Penelokan tephra, which erupted from the Batur caldera (ca. 5.5 kBP), is distributed only outside of the lava.

Figure 5. Outcrop of a debris avalanche deposit on the southeastern slope of the Adeng volcano (Location 8).

3. Tephra stratigraphy and AMS radiocarbon dates

Ashfall layers from the Samalas volcano eruption (1257 A.D.) on the island of Lombok (figure 6) [5, 6, 7, 8] and the Penelokan tephra are useful in establishing the tephra stratigraphy around this caldera. The Penelokan tephra corresponds to the Blingkan ignimbrite at ca. 5.5 kBP [9]. At the foot of the post-caldera volcanoes, coarse scoria falls [10, 11] are intercalated with two foreign tephras (figure 2).

Figure 6. Occurrence of the Samalas tephra (1257 A.D.) on the southeastern slope of the Pohen volcano (Location 7).

Four accelerator mass spectrometry (AMS) radiocarbon dates (figure 7) from Location 4 on the northern slope of the caldera indicate that the two most recent scoria eruptions occurred at ca. 1.1 and

2.5 kBP, respectively [12]. One more AMS radiocarbon date, 1175 ± 25 BP [12], was obtained from charcoal fragments immediately below southward lava flow from Tapak volcano at Location 1 (figure 2). It is consistent with stratigraphic relation with the Samalas tephra. Based on the thickness and grain size of these scoria falls (Locations 5, 6, and 7) above the Penelokan tephra, we estimate that they originated from the Tapak or Lesung volcano. Therefore, at least two volcanoes erupted during the Holocene period.

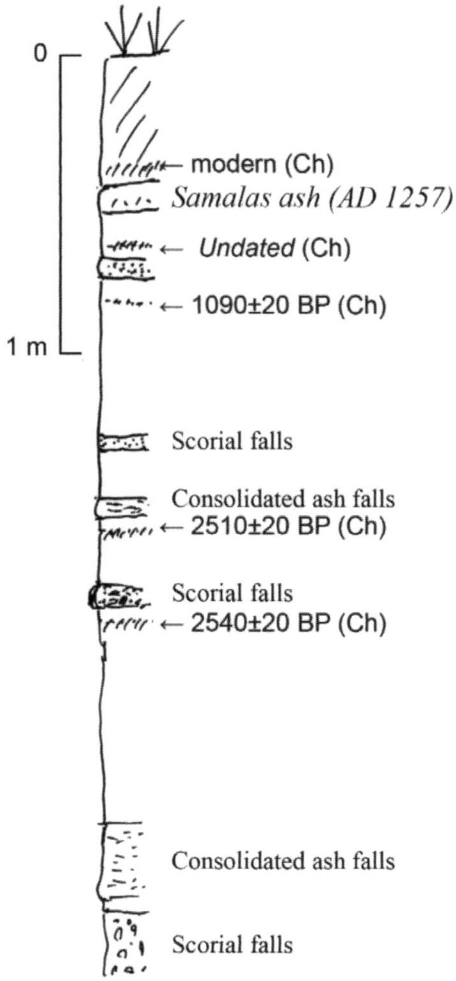

Figure 7. Columnar section and AMS radiocarbon dates [12] from the northern slope of the caldera (Location 4).

4. Conclusions

The geomorphological classification map can provide framework of eruptive history of the post-caldera volcanoes in the Buyan–Bratan caldera. Combination with tephra stratigraphy as well as AMS radiocarbon dates implies that at least Tapak and Lesung volcanoes have erupted during the Holocene period. The Tamblingan and Buyan Lakes were divided by the youngest lava flow from the Tapak volcano. Furthermore, the other youngest lava flow (ca. 1.1 cal kBP) has reached the south foot of the Adeng volcano (figure 2). More detail tephra-stratigraphy and radiocarbon chronology should be determined to refine their eruptive history during Holocene period.

Acknowledgement
This work was partly supported by JSPS KAKENHI (B) Grant Number 24360153 and funds from the Central Research Institute of Fukuoka University (Number 167002). The comments by the reviewers and editor improved the manuscript.

References

[1] Ryu S Kitagawa H Nakamura E Itaya T and Watanabe K 2013 *J. Volcanol. Geotherm Res* **264** 107–16

[2] Purnomo B J and Pichler T 2015 *J. Volcanol. Geotherm. Res.* **304** 349–58

[3] Kobayashi T Harijoko A Warmada I W Watanabe K Nagata T Nakamura T Taguchi S and Okuno M 2016 Eruptive sequence of the Penelokan eruption from Batur caldera, North Bali, Indonesia. *Proc. Int. Meeting Eruptive History Informatics 2015-2* (Fukuoka: Fukuoka University) 35–9

[4] Wheller G E and Varne R 1986 *J. Volcanol. Geotherm. Res.* **28** 363–78

[5] Fontijn K Costa F Sutawidjaja I Newhall C G and Herrin J S 2015 *Bull. Volcanol.* **77** doi:10.1007/s00445-015-0943-x

[6] Lavigne F Degeai J-P Komorowski J-C Guillet S Robert V Lahitte P Oppnheimer C Stoffel M Vidal C M Surono Pratomo I Wassmer P Hajdas I Hadmoko D S and de Belizal E 2013 *Proc. Natl. Acad. Sci. USA* **110** 16742–7

[7] Vidal C M Komorowski J C Metrich N Pratomo I Kartadinata N Prambada O Michel A Carazzo G Lavigne F Rodysill J Fontijn K and Surono 2015 *Bull. Volcanol.* **77** 73

[8] Vidal C M Metrich N Komorowski J C Pratomo I Michel A Kartadinata N Robert V and Lavigne F 2016 *Sci. Rept.* doi: 10.1038/srep34868

[9] Sutawidjaja I S 2009 *J. Geol. Indonesia* **4** 189–202

[10] Watanabe K Yamanaka T Harijoko A Saitra C and Warmada I W 2010 *J. SE Asian Appl. Geol.* **2** 283–90

[11] Yamanaka T Harijoko A Warmada I W Itaya T and Watanabe K 2009 Petrochemical and geochronological characterization of Bratan Caldera, Bali, Indonesia. *Proc. Int. Symp. on Earth Science and Technology 2009* 249–52

[12] Okuno M Harijoko A Warmada I W Nakamura T Taguchi S and Kobayashi T 2017 AMS Radiocarbon dating of Holocene tephra from post-caldera volcanoes of Buyan-Bratan caldera, North Bali, Indonesia *Abst. 14th Int. Conf. on Accelerator Mass Spectrometry* (Ottawa: University of Ottawa) #183

IOP Conf. Series: Earth and Environmental Science **103** (2017) 012015 doi:10.1088/1755-1315/103/1/012015

Cl/B ratio of geothermal fluid around Slamet Volcano, Jawa Tengah, Indonesia

Agung Harijoko[1], Saefudin Juhri[1]

[1] Department of Geological Engineering, Faculty of Engineering, Universitas Gadjah Mada Jl. Grafika No. 2, Kompleks Kampus Fakultas Teknik UGM, Yogyakarta, E-mail: aharijoko@ugm.ac.id

Abstract. Geothermal manifestations occurred in four areas surrounding Slamet Volcano, such as Guci, Baturraden, Paguyangan, and Bantarkawung. These areas are located of about 7.5 km, 8 km, 25 km and 33 km from the summit of Slamet volcano, respectively. We analyzed the chemical composition of cold and hot hater in order to understand the genesis and hydrological the relationship of the hot springs. The plot on HCO_3-Cl-SO_4 ternary diagram classified the hot water into four water types i.e. chloride-bicarbonate water (Bantarkawung), chloride water (Paguyangan), sulfate-chloride water (Baturraden), and bicarbonate water (Guci). The Cl/B ratio values indicate that the southern part of the Slamet volcano (Baturaden) hot springs have high Cl/B ratio compared to that of the northern hot springs (Guci area). While the hot springs in the western part (Paguyangan and Bantarkawung) are classified into high and low Cl/B ratio. This indicates that the hot springs in Paguyangan and Bantarkawung are the outflow of Baturraden and Guci.

1. Introduction

The occurrence of geothermal system beneath Slamet volcano is featured by hot springs discharged in four areas (Baturraden, Guci, Paguyangan and Bantarkawung) surrounding the Slamet volcano (figure 1). The Baturraden area is located about 8 km from the summit at southern flank, while the Guci area is located about 7.5 km away from the summit at northern flank of Slamet volcano. The Paguyangan and Bantarkawung both are located at western area at the distance about 25 and 33 km from the summit, respectively. The hot springs in Paguyangan and Bantarkawung discharge from the fracture of sedimentary rock (figure 2).

The Government of Indonesia divides the geothermal prospect surrounding Slamet volcano into two Geothermal Working Areas (WKP) of Baturraden and Guci (figure 1). The hot springs in Paguyangan and Bantarkawung are located out of those WKP. Until now, those two WKP's are still in exploration stage. For further exploration therefore it is important to understand the characteristic of the geothermal fluid of Baturraden and Guci hot springs. In addition it also important to understand the relation of hot springs in Paguyangan and Bantarkawung to the geothermal system surrounding Slamet volcano based on its chemical composition especially chloride and boron concentration.

Chloride and boron content of geothermal fluids have been commonly used to interpret the origin of the fluid [1]. Besides, it may give a hint to evaluate the degree of geothermal water mixing and to assess other characteristics of geothermal systems [1][2][3][4][5][6].

2. Geology of Study Area

Slamet volcano can be divided into two stages, the old Slamet and young Slamet [7]. Slamet Volcano stands on tertiary sedimentary rocks as its basement, consists of Rambatan, Halang, and Tapak Formation [8] shown in figure 2. The composition of Rambatan Formation is quite similar in Bantarkawung and Paguyangan compared with that around Slamet Volcano. It is composed mainly by shale, marl, calcareous sandstone and conglomerate [8][9]. Around Slamet Volcano, Halang Formation is slightly affected by volcanic product, shown in the composition of andesitic sandstone, tuffaceous conglomerate and marl which is interbedded with sandstone [8], while around Bantarkawung Halang Formation consists of sedimentary rocks show typical structure of turbidite product. The appearance of several oil seepages around Slamet volcano [10][11] can be the evidence for the existence of organic-rich sedimentary rocks and probably formation water. Even though, the exact formation which contains those organic-rich rocks are still uncertain.

Generally, in the western part of Slamet Volcano is dominated by normal fault trending NW-SE and NE-SW surroundings this volcano, while in the eastern part strike-slip fault can be found [7]. Further to the west of Slamet Volcano where the lithology is dominated by tertiary sedimentary rocks, the geological structure changes into fold and thrust fault trending NE-SW and gradually change into NW-SE [9], indicating the change of physiographical zone into Bogor Zone.

3. Samples and Laboratories Analytical Method

The water samples were collected from hot springs, cold springs, river, and dug well. Water sampling procedure follow the procedure described by Nicholson [12]. The sampling applied water filtration using 0.45 μm pore size of cellulose membrane. The acid treatment was applied for the samples used for cation and silica analysis.

Laboratory analyses were conducted in Department of Earth Resources Engineering, Kyushu University. Three methods were used to determine cation-anion and neutral components, which are titration, ion chromatography (using Dionex ICS-90 instrument) and Inductively Coupled Plasma - Atomic Emission Spectrometry (ICP-AES) of Optima 5300 DV series.

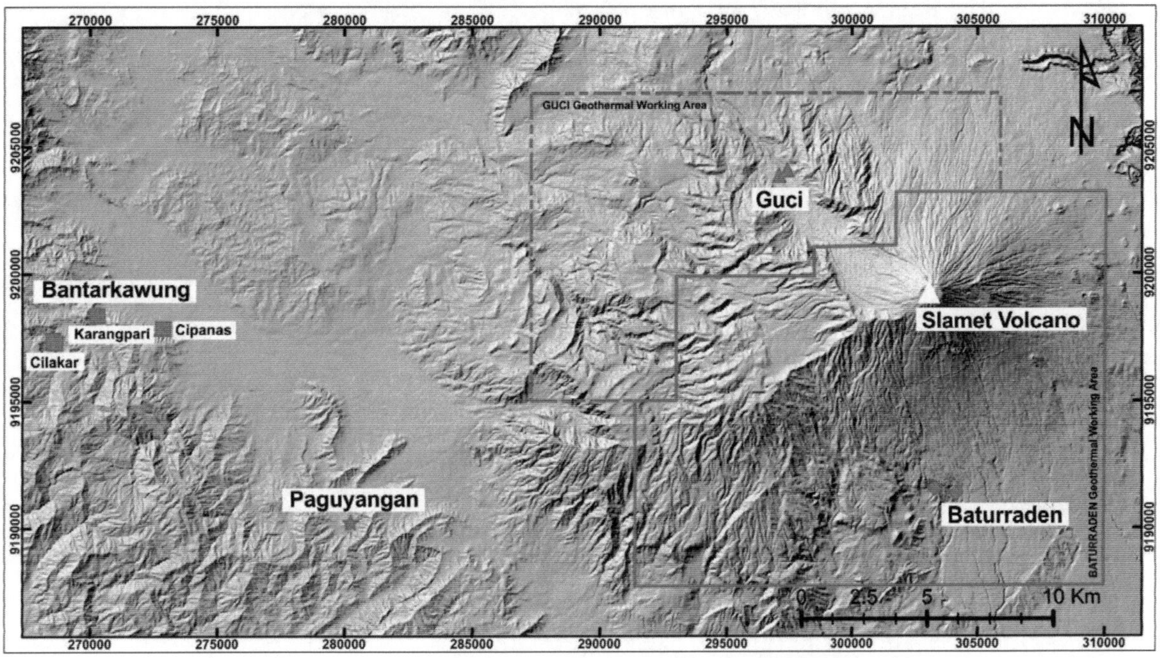

Figure 1. Map of Guci and Baturraden geothermal working area [13]

3.1. Water Type

The type of geothermal water is defined based on the main anions concentration of Cl^-, SO_4^{2-} and HCO_3^- [14]. Plot on the ternary diagram of those anion (figure 3) reveals that the type of geothermal waters discharged surrounding Slamet volcano are different. The hot springs from Baturraden are classified into chloride-sulfate water, while hot springs from Guci are bicarbonate water. The hot springs in Bantarkawung are chloride-bicarbonate water. The Paguyangan hot spring is classified as chloride water. All of the cold waters collected from all the areas are classified as bicarbonate water. The cold watrs have bit high chloride and sulfate concentration. The Bantarkawung cold waters, Guci and Baturraden hot springs seem having a mixing trend of bicarbonate and chloride-sulfate waters.

The high sulfate and chloride concentration in hot waters may be produced due to condensation of volcanic steam. Therefore the Baturraden hot springs can be interpreted as a result from mixing of surface water with volcanic condensate water. Possibly, there is a condensate of volcanic steam flowing from the summit to the southern flank of Slamet volcano. In the passage of flowing the condensate mixed with the surface water forming hot spring in Baturaden.

IOP Conf. Series: Earth and Environmental Science **103** (2017) 012015 doi:10.1088/1755-1315/103/1/012015

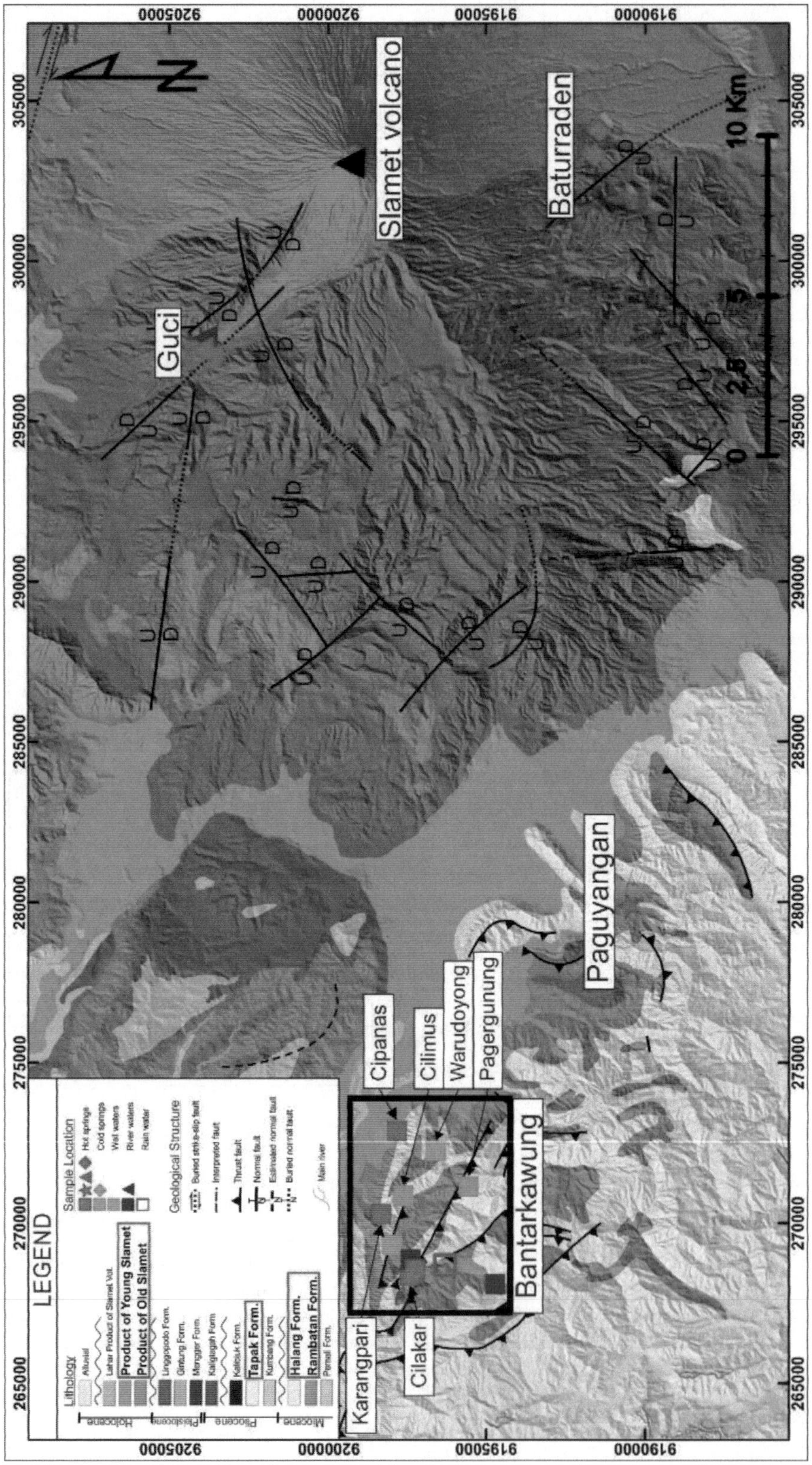

Figure 2. Regional geology map of Slamet Volcano and surrounding area [8][9][7]

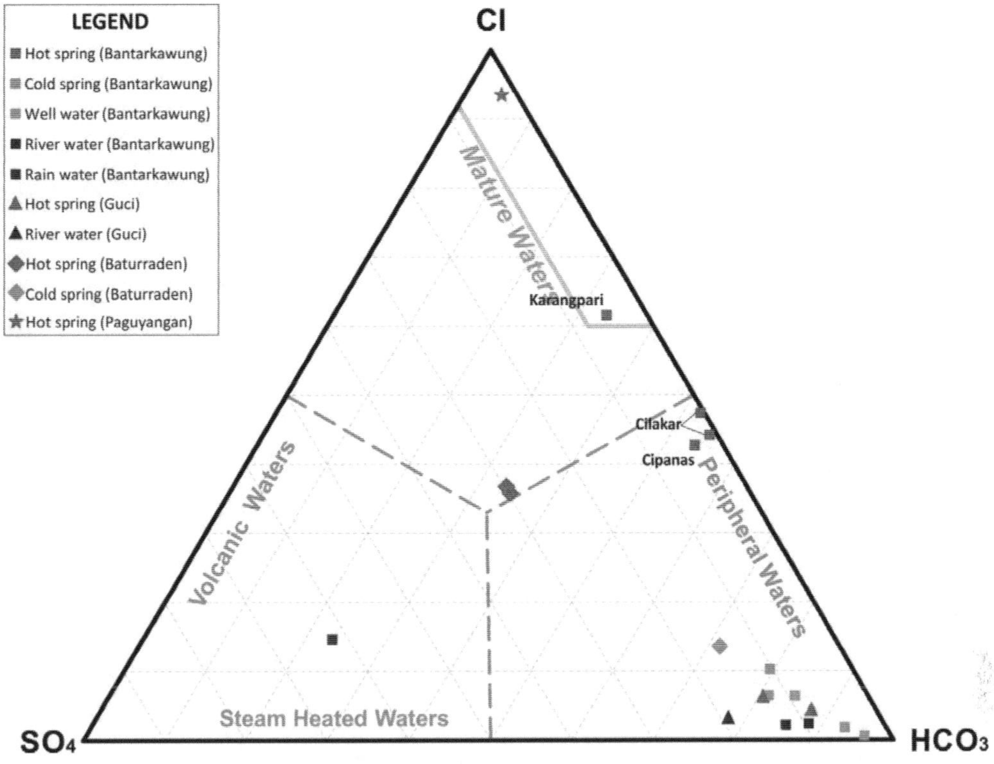

Figure 3. SO_4-Cl-HCO_3 ternary diagram plot for hot springs and meteoric water samples from research area

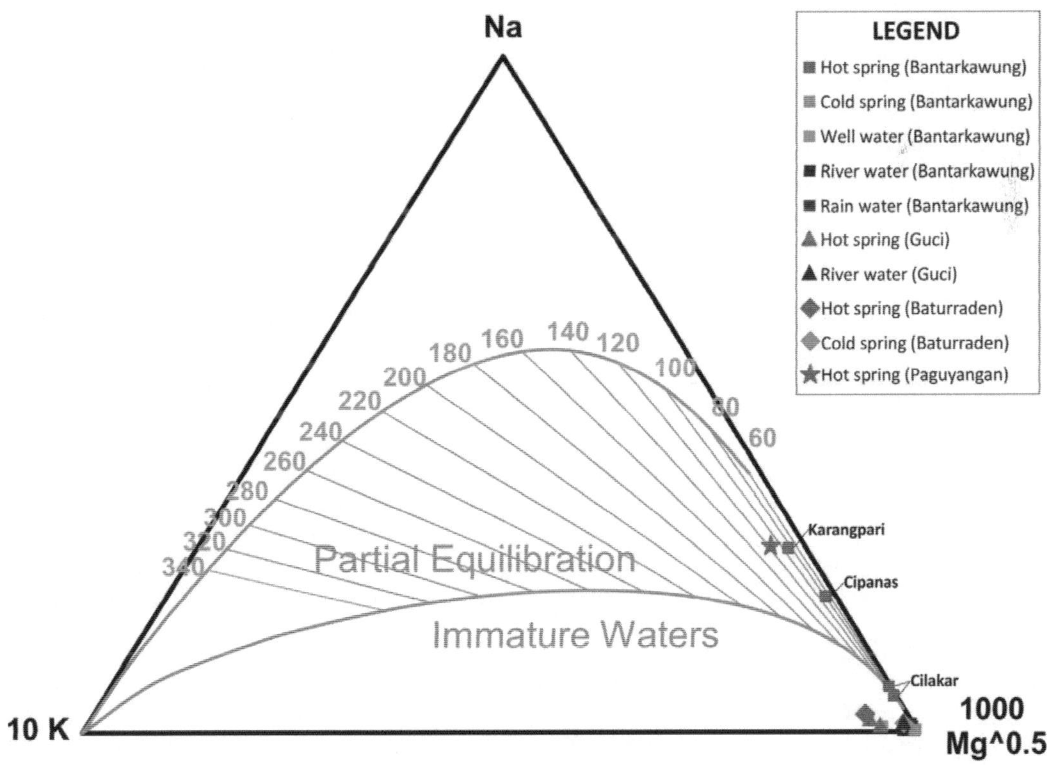

Figure 4. Na-K-Mg ternary diagram plot for hot springs and meteoric water samples from research area

The Paguyangan hot spring is exceptional among others hot springs as it is classified as chloride water. The Bantarkawung hot springs are chloride-bicarbonate waters. Though, the chloride concentration of the Paguyangan and Bantarkawung is low (414 ppm and <120 ppm). Moreover, this hot spring is included as partially equilibrium water based on concentration of Na, K and Mg (figure 4). Based on the graphical estimation of the Na-K geothermometer in Fig.4 the temperature of the reservoir is about 120°C. This temperature is too low for the volcanic hosted geothermal system. The Na/K ratio does not represent the Na/K ratio of the deep thermal water. Therefore it is hard to interpret that the Paguyangan hot spring is representative of deep geothermal water. Therefore the thermal water may mixed with high chloride fluid from other source beyond geothermal system.

Table 1. Geochemistry data of hot spring, cold spring, and other water samples from research area

Location	Source	T (°C)	pH	Cl^-	SO_4^{2-}	HCO_3^-	Li^+	Na^+	K^+	Mg^{2+}	SiO_2	B	Ionic Balance
Bantarkawung		27.4	6	5.07	28.6	197,64	0	11.56	0.92	6.90	17.85	0.34	-1,0
Bantarkawung	River	25.7	6	5.36	21.24	198,05	0	12.88	1.15	7.53	20.86	0.41	-2,5
Guci		22.4	7	3.31	20.72	84,38	0	10.02	4.30	7.05	54.59	0.48	-1,8
Bantarkawung	Rain	28.1	5	1.39	5.95	2,24	0	0.62	0.30	0.30	0.00	0.28	-12,4
Bantarkawung	Ground water	27.8	7	31.8	60.2	409,19	0	32.4	16.81	15.95	30.18	0.42	-2,4
Bantarkawung		28.4	8	28.3	40.6	374,54	0	31.6	3.90	14.13	30.44	0.40	-2,4
Bantarkawung		26.9	7	25.26	25.32	195,40	0	14	7.96	15.8	35.32	0.30	-1,6
Bantarkawung	Cold Spring	27.4	8	3.96	11.23	201,30	0	7.96	0.69	5.31	24.69	0.35	-2,5
Bantarkawung		26.2	7	1.2	6.16	185,85	0	27.52	0.26	9.11	93.57	0.33	-1,6
Baturaden		21	6	13.36	14.68	70,15	0.01	12.48	4.29	7.07	48.02	0.37	-2,3
Bantarkawung (Cipanas)		43	6	43.7	3.59	54,90	0.01	50.9	0.32	0.04	38.29	3.37	1,7
Bantarkawung (Cilakar)		42.6	7	113	2.39	140,30	0.01	89	0.56	2.48	42.38	1.48	-1,5
Bantarkawung (Cilakar)		43.1	7	120	1.20	132,17	0.01	94.9	0.59	1.86	44.15	1.64	-0,7
Bantarkawung (Karangpari)		56	8	105	8.48	56,93	0.01	86.2	0.69	0.05	51.70	7.40	-2,1
Baturaden (Pancuran 7)	Hot spring	50	7	754	609	687,27	0.67	389	76	185	169.14	4.40	-1,5
Baturaden (Pancuran 3)		46	6	724	600	695,40	0.58	377	76.7	185	163.75	3.97	-1,0
Guci (Pancuran 13)		40.5	8	17.3	32.5	345,67	0.02	57.3	24.35	29.8	121.33	2.84	0,6
Guci (Pengasihan		50.4	8	44.2	89	549,00	0.06	129	36.3	46.1	134.83	6.87	0,7
Paguyangan		72	7	414	8.51	20,33	0.16	193	2.92	0.23	57.67	3.72	-2,1

3.2. Hot springs relationship

The plot on the ternary diagram of Cl-B-Li indicates that hot springs in Baturaden are different from the hot springs in Guci (figure 5). The Guci hot springs have lower Cl/B ratio than that of Baturraden hot springs. Paguyangan hot spring has high Cl/B ratio and is plotted close to the Baturraden hot springs. Two of the Bantarkawung hot springs (Cilakar hot springs) are classified as high Cl/B ratio while the other two hot springs (Karangpari and Cipaas) are low Cl/B ratio. This plot revealed that geothermal system in Baturraden is different from the geothermal system in Guci. Paguyangan hot spring is probably the outflow of the Baturraden geothermal system.

Referring to the regional geology map, marine sedimentary rocks such as Rambatan Formation and Halang Formation crop out surrounding the Slamet Volcano. Those sedimentary formation may extend beneath Slamet Volcano and become the bed rocks of the volcano. The occurrence of oil seepages near Slamet volcano as reported by Tjahjono [10] and Tobing [11] indicate the petroleum system containing formation water. There may be influence of the marine sedimentary rocks and/or the formation water to the geothermal water composition surrounding Slamet volcano.

The binary diagram of Cl and B (figure 6) shows that all of the cold waters have similar B concentration, but their Cl concentration varies. The Cl/B ratio is represented by the gradient of the trend in that diagram. As the Cl/B ratio value is relatively constant therefore we may use the Cl/B ratio to differentiate the hot springs into two groups. The first is group of Guci trend. This trend includes hot springs in Guci, Cipanas and Karangpari (both are in Bantarkawung), and Cilimus cold spring. Those hot springs and cold spring are distributed at the northern part of Bantarkawung area. The second group is Baturraden group that includes Baturrden hot springs, Paguyangan hot spring, Cilakar hot springs, Baturraden cold spring, dug well water in Bantarkawung, and Warudoyong cold spring. The Baturraden group is distributed in the southern part of the Bantarkawung area.

Both the Guci and Baturraden groups have similar Cl/B ratio value. The Baturraden group has higher intercept on Cl concentration. Therefore it is interpreted that the thermal water that feed the Baturraden group have been mixed with high Cl fluid possibly formation water from the marine sedimentary rocks. It is estimated based on the extension of the trend line of Baturraden group that crosses the mixing line of condensate water and seawater (figure 6).

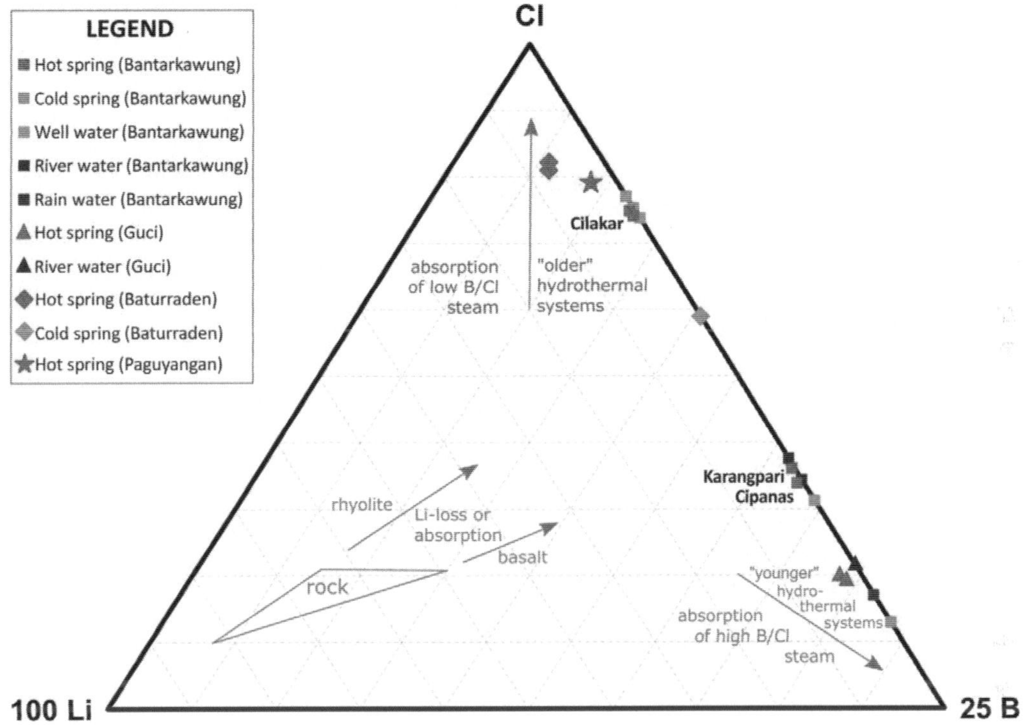

Figure 5. Cl-Li-B ternary diagram plot for hot springs and meteoric water samples from research area

4. Conclusions

Hot springs surrounding Slamet volcano are the out flow of the geothermal system beneath Slamet volcano. The Baturraden hot springs is a result of the mixing of volcanic water and surface water.

Cl/B ratio value can distinguish two geothermal system beneath Slamet. The first is the geothermal system related to Guci hot springs and is purely volcanic related geothermal system. The other geothermal system is feeding Baturraden hot springs that have contribution of the formation water. Both geothermal system flow laterally to the west until Bantarkawung area. The Baturraden system is situated in the southern area while the Guci system flows in the northern area.

Acknowledgement

The authors thank to Prof. Koichiro WATANABE, Prof. Ryuichi ITOI, Dr. Kotaro YONEZU and Rie YAMASHIRO san for their permission and help to do laboratories analysis. The authors also thank to JASSO for giving scholarship to Saefudin Juhri for his stay at Japan for short student exchange in Kyushu University.

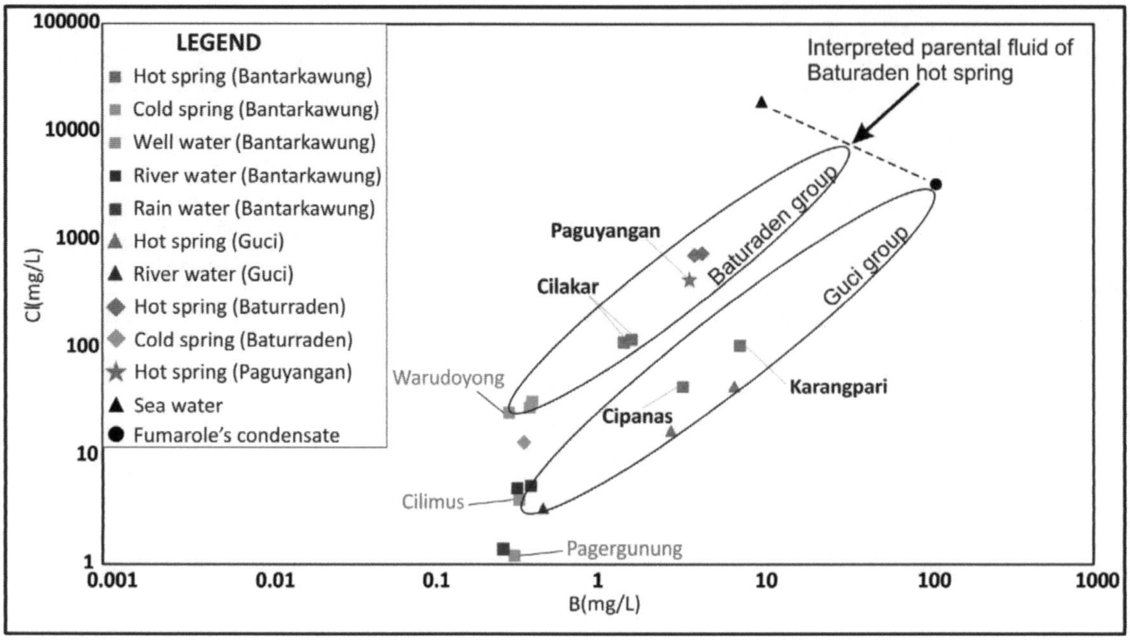

Figure 6. Cl vs B binary diagram plot for hot springs, meteoric water samples from research area, sea water [16] and condensate fumarole from Slamet crater [17].

References

[1] Ellis A J 1970 Quantitative interpretation of chemical characteristics of hydrothermal systems. *Geothermics 2* 516-527
[2] White D E 1970 Geochemistry applied to the discovery, evaluation, and exploitation of geothermal energy resources *Geothermics Special Issue 2* 58-80
[3] Founier R O 1977 Chemical geothermometers and mixing models for geothermal systems *Geothermics 5:* 41-50
[4] Founier R O 1979 Chemical and hydrological considerations of the use of enthalpy-chloride diagrams in the prediction of underground conditions in hot-spring systems *Journal of Volcanology and Geothermal Research 5* 1-16
[5] Arnórsson S 1985 The use of mixing models and chemical geothermometers for estimating underground temperatures in geothermal systems *Journal of Volcanology and Geothermal Research 23* 299-335
[6] Janik C J, Truesdell A H, Goff F, Shevenell L, Stallard M L, Trujillo P E, and Counce D 1991 A geochemical model for the Platanares geothermal systems, Honduras *Journal of Volcanology and Geothermal Research 45* 125-146
[7] Sutawidjaja I S, Aswin D, and Sitorus K 1985 *Geological Map of Slamet Volcano, Central Java, scale 1:100.000* (Bandung: Volcanological Survey of Indonesia)
[8] Djuri M, Samodra H, Amin T C, and Gafoer S 1996 *Geological Map of the Purwokerto and Tegal Quadrangle, Java, scale 1:100.000* (Bandung: Geological Research and Development Centre)
[9] Kastowo 1975 Geological Map of the Majenang Quadrangle, Java, scale 1:100.000 (Bandung: Geological Survey of Indonesia, Ministry of Mines)
[10] Tjahjono J A E 2002 Inventarisasi Bitumen Padat Daerah Banjarnegara, Kabupaten Banjarnegara Propinsi Jawa Tengah *Prosiding Koloqium Direktorat Inventarisasi Sumberdaya Mineral*
[11] Tobing S M 2002 Inventarisasi Endapan Bitumen Padat (Cannel Coal) Di Daerah Wangon Dan Sekitarnya, Kabupaten Banyumas Dan Kabupaten Cilacap, Propinsi Jawa Tengah *Prosiding Koloqium Direktorat Inventarisasi Sumberdaya Mineral*
[12] Nicholson K 1993 *Geothermal Fluids: Chemistry and Exploration Techniques* Springer-Verlag
[13] Dirjen EBTKE 2012 *Profil Potensi Panas Bumi Indonesia* (Jakarta: Kementerian Energi dan Sumber Daya Mineral)
[14] Giggenbach W F 1991 Chemical Techniques in Geothermal Exploration in D'Amore, F *Applications of Geochemistry in Geothermal Reservoir Development* (Rome: UNITAR/UNDP Centre on Small Energy Resources)

[15] Bogie I, Lovelock B, and Ussher G 2015 The Neutral High-Cl Thermal springs of Java, Indonesia. *Proceedings World Geothermal Congress 2015* Melbourne
[16] Purnomo B J and Pichler T 2014 *Geothermal systems on the Island of Java, Indonesia* Journal of Volcanlogy and Geothermal Research vol 285 p 47-59
[17] Sumarti S 1996 *Penyelidikan geokimia dan emisi gas SO2 di Gunung Slamet, Jawa Tengah.* Direktorat Vulkanologi Unpublished report

Geothermal system boundary at the northern edge of Patuha Geothermal Field based on integrated study of volcanostratigraphy, geological field mapping, and cool springs contamination by thermal fluids

Suryantini[1,2], C Rachmawati[1], and M Abdurrahman[1]
[1]Geology Engineering Study Program, Faculty of Earth Sciences and Technology, Institut Teknologi Bandung, Indonesia, Jl Ganesha 10, Bandung 40132
[2]Geothermal Technology Study Program, Faculty of Mining and Petroleum Engineering, Institut Teknologi Bandung, Indonesia, Jl Ganesha 10, Bandung 40132

e-mail: rcindytami@gmail.com

Abstract. Patuha Geothermal System is a volcanic hydrothermal system. In this type of system, the boundary of the system is often determined by low resistivity (10 ohm.m) anomaly from Magnetotelluric (MT) or DC-Resistivity survey. On the contrary, during geothermal exploration, the system boundary often need to be determined as early as possible even prior of resistivity data available. Thus, a method that use early stage survey data must be developed properly to reduce the uncertainty of the geothermal area extent delineation at the time the geophysical data unavailable. Geological field mapping, volcanostratigraphy analysis and fluid chemistry of thermal water and cold water are the data available at the early stage of exploration. This study integrates this data to delineate the geothermal system boundary. The geological mapping and volcanostratigraphy are constructed to limit the extent of thermal and cold springs. It results that springs in the study area are controlled hydrologically by topography of Patuha Volcanic Crown (complex) or so called PVC, the current geothermal field and Masigit Volcanic Crown (complex) or so called MVC, the dormant volcano not associated with active geothermal system. Some of the cold springs at PVC are contaminated by subsurface steam heated outflow while others are not contaminated. The contaminated cold springs have several characteristics such as higher water temperature than ambient temperature at the time it was measured, higher total disolved solid (TDS), and lower pH. The soluble elements analysis support the early contamination indication by showing higher cation and anion, and positive oxygen shifting of stable isotope of these cool springs. Where as the uncontaminated spring shows similar characteristic with cool springs occur at MVC. The boundary of the system is delineated by an arbitrary line drawn between distal thermal springs from the upflow or contaminated cool springs with the cool uncontaminated springs. This boundary is more or less in agreement with low resisitivity boundary derived from MT and DC resistivity survey. The area defined as part of geothermal area from this method is also validate with drilling data that give high temperature gradient. It suggests that the method use in this study is applicable and reliable.

1. Introduction

Study area is located in Patengan Village, Rancabali District, Bandung Regency, West Java Province, adjacent to Patuha Geothermal Field, with an area of 16 km2 (figure 1). Patuha Geothermal System is a volcanic hydrothermal system. Boundary of this system is characterized by a lateral outflow, characterized by the occurrence of acidic to neutral bicarbonate-chloride or chloride thermal springs manifestation.

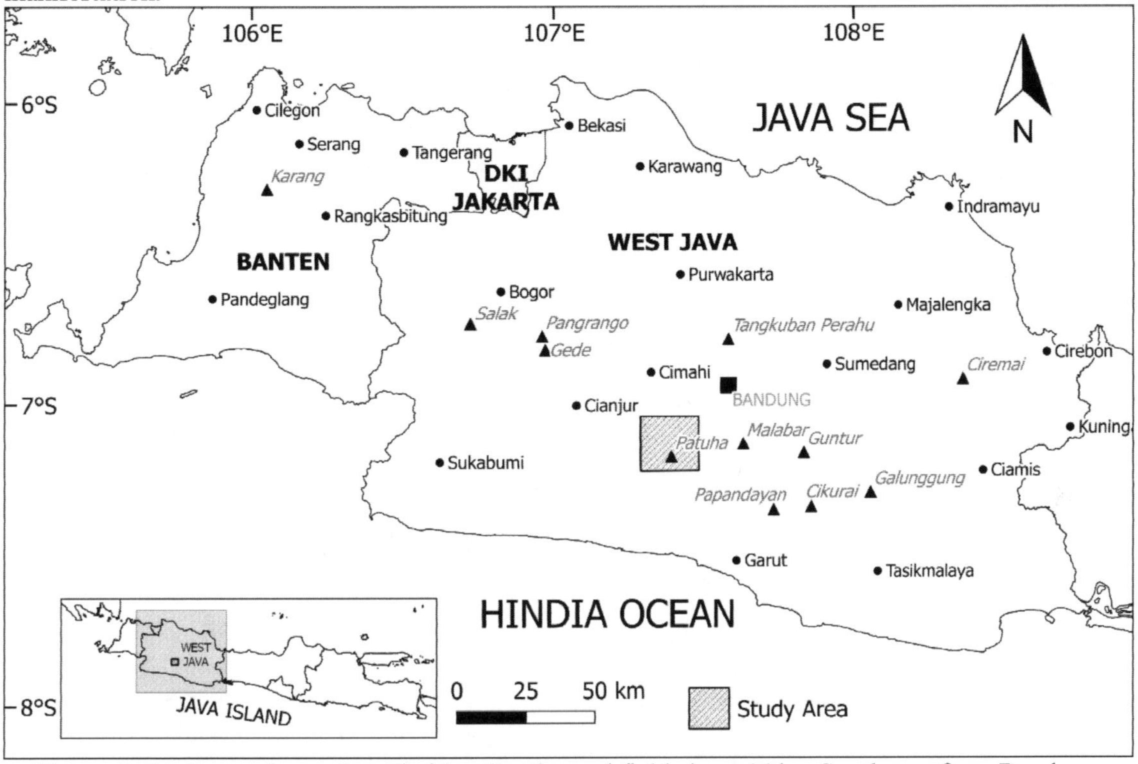

Figure 1. Study area located at Ciwidey Geothermal field about 46 km Southwest from Bandung.

The absence of this thermal spring will be a difficulty in determining the geothermal system boundary. In other side, its occurrence is predicted to contaminate shallow groundwater. Contaminated cold spring will be enriched with cations and anions, as well as the anomaly of pH and temperature from its common trend. The boundary of a geothermal system is expected to be found by examining cold water springs that are not contaminated by the steam heated thermal water.

The objective of this study is to determine the geothermal system boundary at the northern edge of Patuha geothermal field from integrating geological field mapping, volcanostratigraphy analysis and fluid chemistry of thermal water and cold water. The water samples are obtained from several springs in study area, in the early stage of geothermal exploration.

2. Materials and Method

Geological field mapping, volcanostratigraphy analysis and fluid chemistry of thermal water and cold water are the method used in the study. This study integrates these data to delineate the geothermal system boundary.

This study is conducted by classifying cold springs and thermal spring associated with geological analysis, including volcanostratigraphy, geological units, geological structure, from field data and literature. The geological mapping, including volcanostratigraphy are constructed to limit the extent of thermal and cool springs. It was done to determine their relation with the occurrence of cold springs and thermal springs in the research area and its association with Patuha geothermal system.

Chemical analysis, taken from one hot spring, six warm springs, and nine cold springs, is done to determine which cold water springs has been contaminated with thermal fluid. The location of springs samples is shown in Figure 2. This fluid chemistry analysis was completed by using temperature, pH, TDS, anions, cations, and isotopes data. Laboratory data that consists of anions, cations, and isotopes, are further evaluated by Schoeller Diagram and isotope graph. Geothermal system boundary is determined and validated low resisitivity boundary derived from MT and DC resistivity survey and also drilling data that give high temperature gradient.

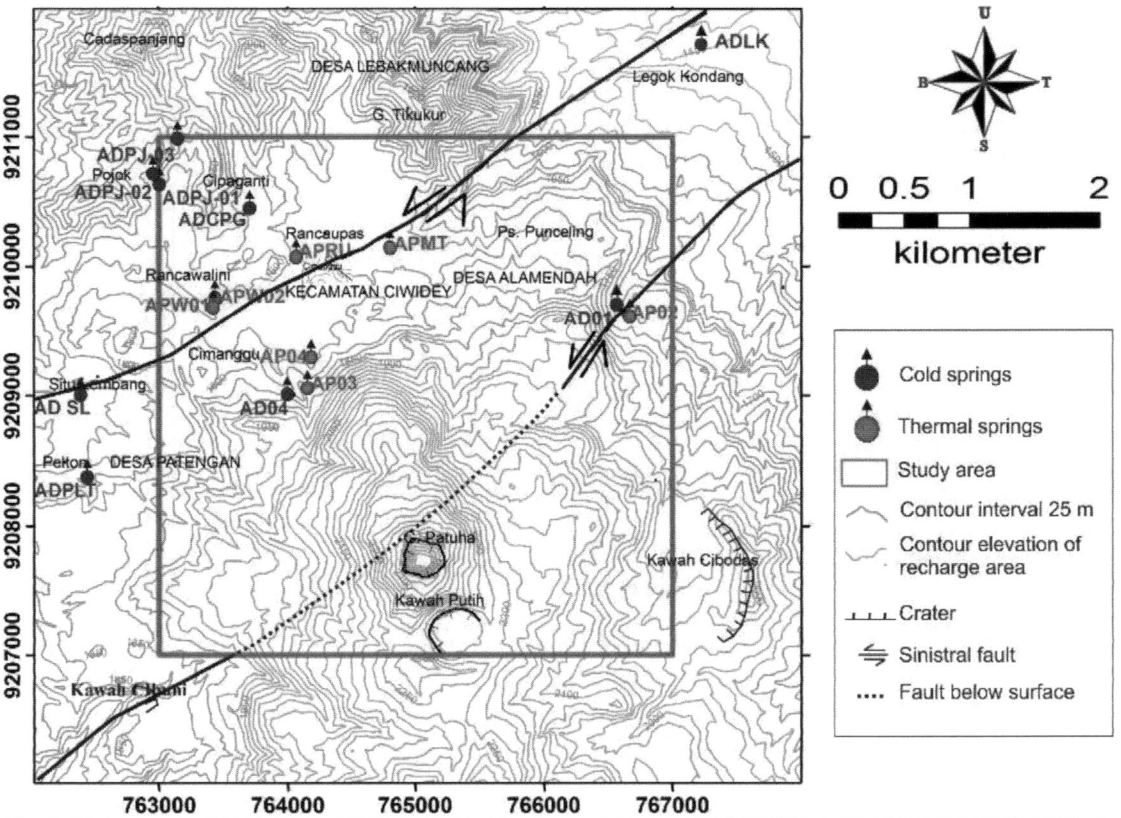

Figure 2. Location of springs sample and countour elevation of recharge area based on [1].

3. Geology

Geology in research area was based on identifying satellite imagery and topography map as well as observing field of study. The analysis includes volcanostratigraphy, geological unit and geological structure. Those analysis were then associated with the springs occurrence.

3.1. Volcanostratigraphy

Volcanostratigraphy analysis was constructed to limit the occurrence of thermal and cold springs. Identifying satelite imagery and topographic map was done to determine the ceruption center, classify the volcanic complex and define the relative age of the geological units. From satelite imagery (Figure 3), eruption center can be determined by identifying circular features that form a cone or depression with the distribution of its surrounding volcanic product, which build the volcanic body. Idenfying topographic map in the scale of 1:100.000 was done to classify study area into several volcanic crown (complex) or khuluk (Figure 4). It is a whole body of volcanic that was produced by one or more erution center [2].

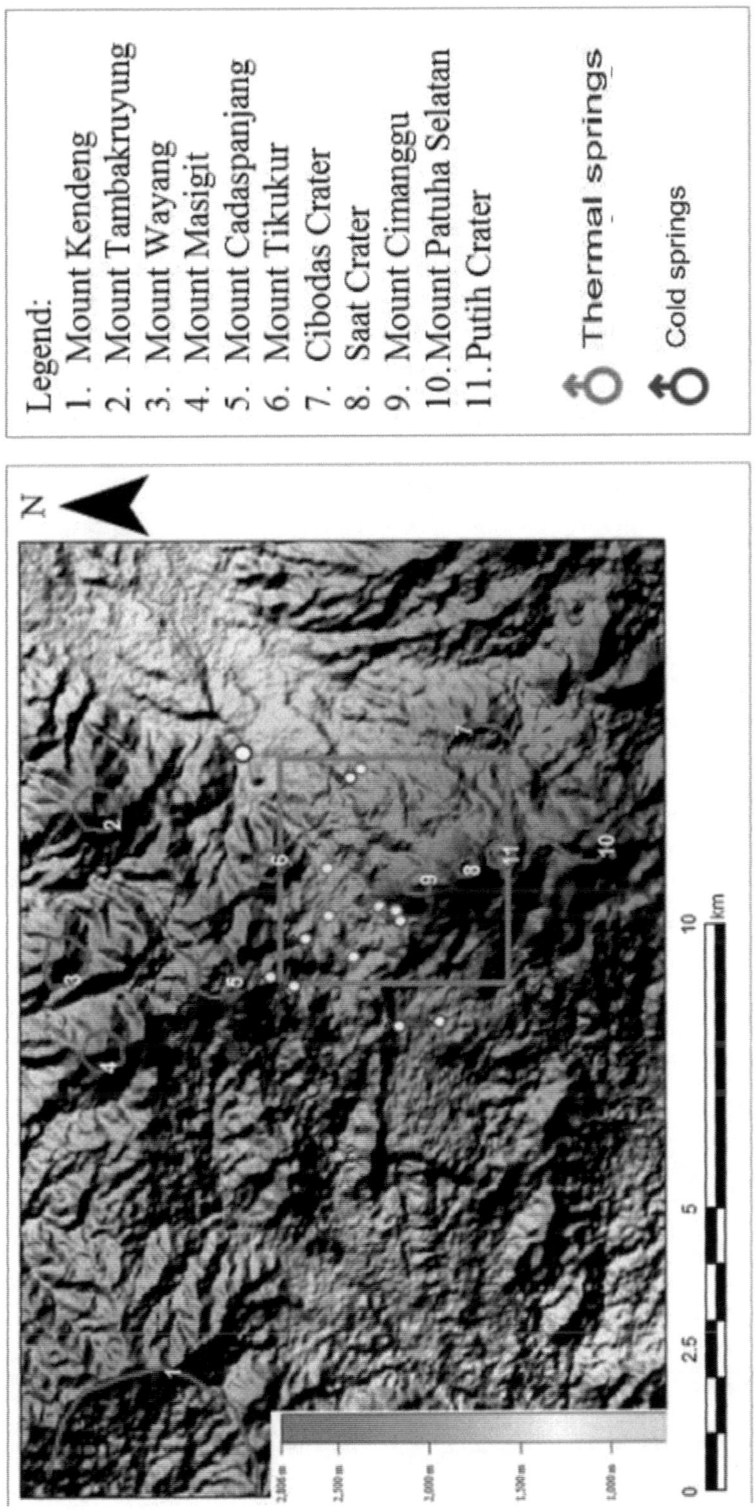

Figure 3. Eruption center from satelite imagery identification.

It results, as seen in the picture 4 that springs in the study area are controlled hydrologically by topography of Patuha Volcanic Crown (complex) or so called PVC, the current geothermal field and Masigit Volcanic Crown (complex) or so called MVC, the dormant volcano not associated with active geothermal system. The occurrence of the springs was associated with a closest volcanic summit that usually become its recharge area. Springs in Pojok, Cipaganti and Legok Kondang came from MVC recharge area, while the rest of them came from PVC recharge area.

Idenfying topographic map in the scale of 1:50.000 was done to classify volcanic complex in study area into more detailed volcanic body from each eruption center or gumuk. This classification will lead to further observation of geological units in geological field mapping.

3.2. Geological Units

Names of the unit were related to its eruption center, type of rocks, and relative age, as given in Sandi Stratigrafi Indonesia [2]. Study area consists of 11 geological, which are (from the oldest) Tikukur Tuf-Breccia Pyroclastics Flow (Ta), Cadaspanjang Tuf-Breccia Pyroclastics Flow (Ca), Patuha 1 Breccia Pyroclastics Flow (Pa1), Patuha 1 Andesitic Lava Flow (Pl1), Patuha 2 Andesitic Lava Flow (Pl2), Patuha 3 Andesitic Lava Flow (Pl3), Patuha 4 Andesitic Lava Flow (Pl4), Patuha Andesitic Lava – Tuf-Breccia Pyroclastics Flow (Pa2), Cimanggu Andesitic Lava Flow (Cil), Patuha Selatan Tuf-Breccia Pyroclastics Flow (Psa), dan Kawah Putih Breccia Pyroclastics Flow (Ka). Geological maps and volcanostratigraphy column is given below Figure 5 and Table 1.

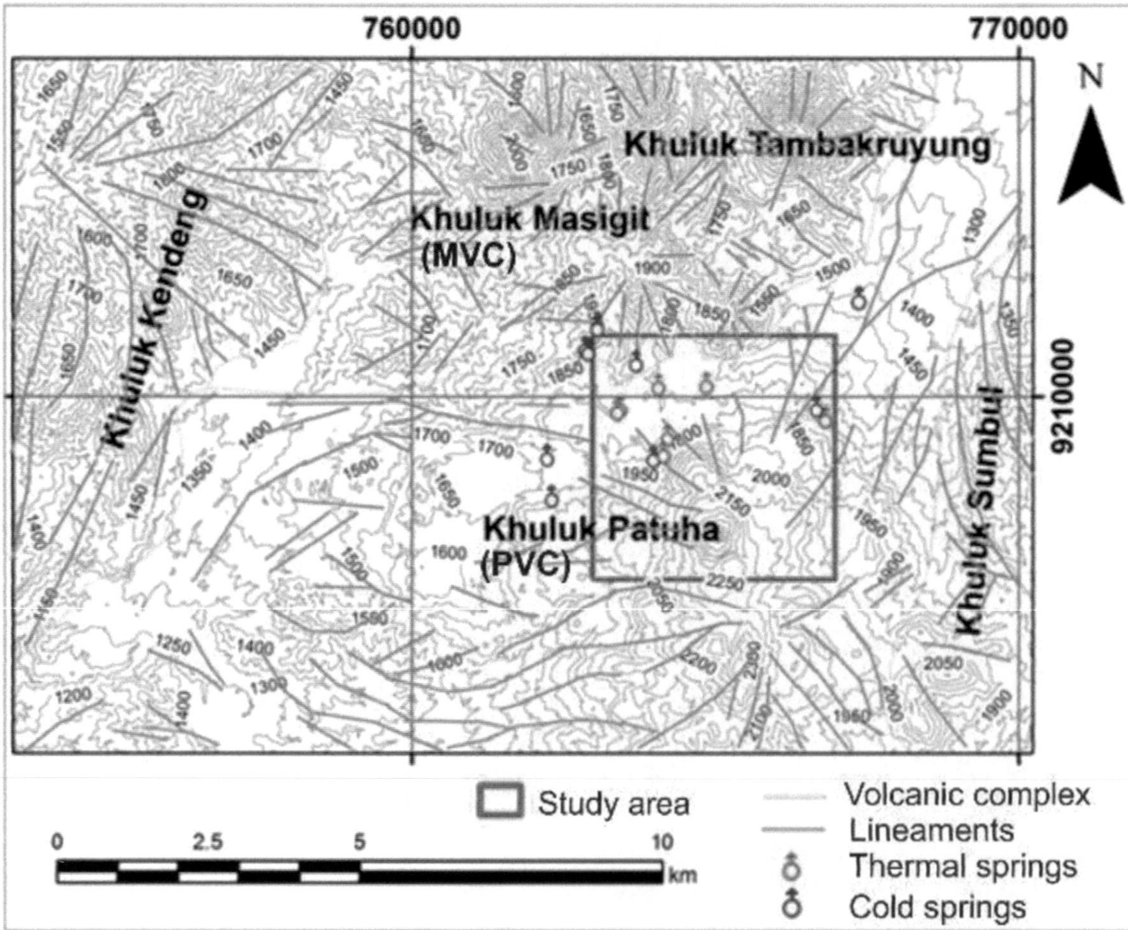

Figure 4. Volcanic complex classification

3.3. Geological Structure

Geological structure in study area consist of craters, which are Saat Crater and Putih Crater, and also faults, which are Punceling Sinistral Fault and Cimanggu Sinistral Fault. These faults were identified from lineaments, occurrence of thermal springs and altered outcrop (Figure 6). Punceling Sinistral Fault is proved by the manifestations occurred in Cibuni Crater and Punceling area. There are also some anomalies in air ground temperature data, ground pH, Hg and CO_2 near Punceling thermal spring and Cibuni Crater along the lineaments [3]. Cimanggu Sinistral Fault is also supported by the thermal springs occurred in Walini, Ranca Upas and EMTE area. There are also some anomalies in air ground temperature data near Walini thermal springs [3].

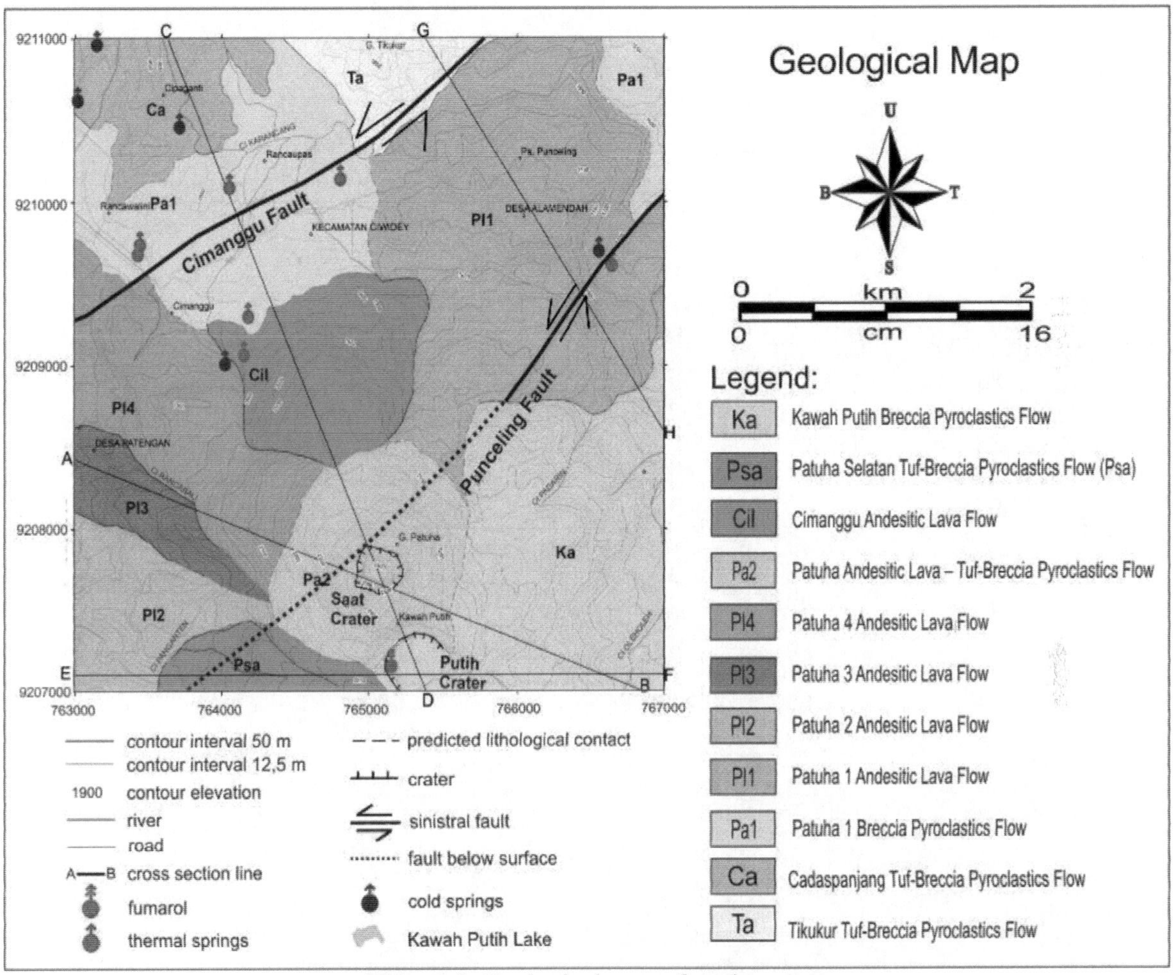

Figure 5. Geological map of study area.

Table 1. Volcanostratigraphy column of research area.

Period	Epoch	Volcanic complex	Eruption center	Primary volcanic deposits: Lava	Primary volcanic deposits: Pyroclastics flow
Quarternary	Holocene	Patuha	Kawah Putih		Ka
			Patuha Selatan		Psa
			Cimanggu	Cil	
			Patuha		Pa2
				Pl4	
				Pl3	
				Pl2	
				Pl1	
					Pa1
	Pleistocene	Masigit	Cadaspanjang		Ca
			Tikukur		Ta

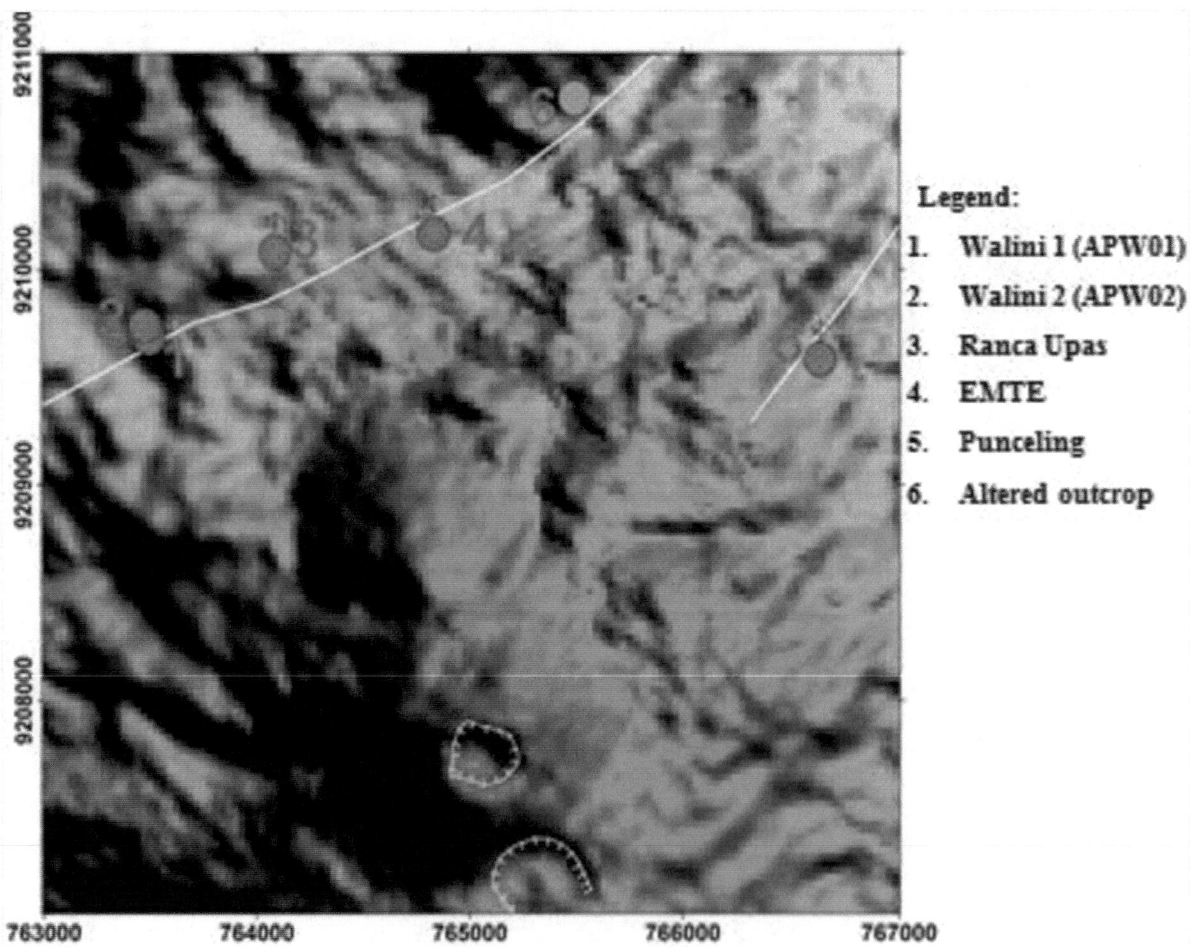

Figure 6. Structural linaments.

4. Fluid Chemistry

Water chemistry analysis was done to determine the contaminated and uncontaminated cold springs by thermal water, which is then integrated with the results of the springs classified from the geological analysis. Collecting of water data in study area started with the measurement of ambient or surrounding air temperature, water temperature, the amount of dissolved solids or Total Dissolve Solid (TDS), as well as pH, followed by water sampling anions, cations and isotope analysis in laboratory. Water samples were taken with the procedures described in [4], consisting of samples for analysis of dissolved elements and isotopes for analysis.

Data were taken from a hot springs (T> 50°C), six hot springs (T≤50°C), and nine cold springs as shown in Figure 2 and Table 2. Sampling was conducted in June - July 2016, with sunny to cloudy weather conditions.

4.1. Temperature, pH, and TDS Data Analysis

The purpose of this field data analysis is to find contaminated cold springs by the indication of higher water temperature than the surface temperature, lower value of pH than the average cold water pH in study area and high amount of dissolved solids (TDS> 100). These values were determined using several assumptions. The first assumption is that the temperature of cold springs uncontaminated with thermal water, equal to the outdoor air temperature when the measurements were taken. Second, the uncontaminated cold springs would have TDS <100 ppm [5]. Third, water with a very low pH value or below the average is assumed to be contaminated by lateral flow of geothermal fluid. Springs data in the form of temperature, pH, and TDS as the primary data is shown in Table 2.

Table 2. Temperature, pH, and TDS of springs samples

Location	Code	Coordinate		Elevation (m)	Temperatur			pH	TDS (ppm)
		mE	mN		Air	Water	ΔT		
Thermal springs									
Punceling	AP02	766647	9209614	1698,53	22.0	41.3	19.3	5.94	123
Cimanggu	AP03	764153	9209067	1837,50	22.7	72.9	50.2	6.33	438
	AP04	764185	9209301	1777,53	24.9	47.9	23.0	5.93	380
Walini	APW01	763428	9209697	1760,28	23.4	45.5	22.1	6.27	608
	APW02	763441	9209736	1756,53	22.1	39.6	17.5	6.57	540
EMTE	APMT	764810	9210143	1678,00	19.1	41.3	22.2	6.22	746
Ranca Upas	APRU	764057	9210088	1774,91	19.5	36.3	16.8	6.43	406
Cold springs									
Punceling	AD01	766560	9209700	1664,53	23.3	18.6	4.7	7.25	167
Pelton	ADPLT	762442	9208373	1759,53	19.5	17.9	1.6	7.87	66
Cipaganti	ADCPG	763710	9210465	1764,00	19.3	18.1	1.2	6.96	23
Pojok	ADPJ-01	763003	9210623	1768,53	18.9	20.1	-1.2	8.11	72
	ADPJ-02	762974	9210676	1771,53	21.8	17.8	4.0	6.04	17
	ADPJ-03	763150	9210965	1848,52	19.8	16.3	3.5	5.98	19
Cimanggu	AD04	764030	9209008	1818,53	19.2	19.6	-0.4	5.45	115
Legok Kondang	AD LK	767222	9211691	1712,00	23.0	20.9	2.1	6.32	156
Situ Lembang	AD SL	762390	9209011	1448,53	22.6	25.7	-3.1	6.42	252

4.1.1. Temperature

Springs can be classified by its temperature. Thermal springs, according to Bery et al [6], consists of hot springs with temperature above 50°C and warm springs with temperature less than or equal to 50°C. Cold springs naturally will follow the surrounding air temperature. The temperature of cold springs, in ideal conditions should not be higher than the surrounding air temperature, even should be lower. This can happen if the occurrence of these springs are covered by something, such as trees, houses, and others. The temperature of cold springs, if contaminated by thermal fluid, is expected to increase.

Hot springs manifestation was found only in Cimanggu with temperatures reaching 72,9oC. The temperature of the warm springs in the study area ranged between 36.3 – 47,9oC. Cold springs temperatures ranged between 16.3 - 25,7oC. In Table 2, it can be seen that there are three samples that have higher water temperature than surrounding air temperature, as in the sample ADPJ-01, AD04, and ADSL. In Pojok, only springs ADPJ-01 which has a temperature showed an anomaly value. Cold springs ADPJ-01 is located at an elevation of the lowest compared with the other cold springs in Pojok, and their water temperature increases as the elevation is lower (Table 2). The air temperature at ADPJ-01 was supposed to demonstrate the value that is higher than the air temperature at ADPJ-02 and ADPJ-03 which located at a higher elevation. However, existing data showed that air temperature of ADPJ-01 is lower, whereas the value of air temperature ADPJ-02 is much higher than the value of the temperature of cold springs ADPJ-01. Thus, the temperature anomaly at the point ADPJ-01 is estimated to be due to an error in measurement of air temperature data at the moment in ADPJ-01 location. Cold springs in Cimanggu (AD04) and Situ Lembang (ADSL) higher temperature are expected to be caused by contamination of the geothermal fluid lateral flow below the surface.

4.1.2. pH

Level of acidity of the water is indicated by pH value. The pH value can be influenced by the chemical constituents present in the water, so that the pH became one of the important indicators that could indicate changes in water chemistry. The groundwater will generally have pH values ranging from 6.5 to 8.5 [5], whereas the thermal water outflow zone will have a pH that is slightly acidic or almost near neutral [4]. Eyes cold water contaminated expected to have a pH lower than the pH value of the reference [5]. However, lower pH values can also occur due to the influence of agricultural fertilizers containing sulfate, thereby making the water more acidic and increasing pH.

Water samples in study area in general have pH value relatively close to neutral. The hot springs Cimanggu has a pH of 6.33. Warm springs have a pH ranging from 5.93 to 6.57. Meanwhile, cold springs have a slightly wider pH range, 5.45 to 8.11. Cold springs in Pojok (ADPJ-02 and ADPJ-03), Cimanggu (AD04), Legok Kondang (ADLK), and Situ Lembang (ADSL) has a relatively more acidic pH compared to the reference. Decreasing value of pH in the Pojok and Legok Kondang cold springs was thought to be influenced by agricultural fertilizer, as it is located on a plantation area resident. Whilst, decreasing value of pH in Cimanggu and Situ Lembang cold springs was expected due to contamination by lateral flow of geothermal fluid below the surface (Figure 7). The other cold water pH values showed no significant changes and therefore were not used as indicator to determine the conditions of uncontaminated cold springs.

4.1.3. Total Dissolved Solids (TDS)

TDS is a value that indicates the amount of solids or dissolved solids in the water. Most of the natural springs contain relatively small dissolved components (Piper, 1944). Fresh cold springs will have a value of TDS <100 ppm [5], so cold springs with TDS values> 100 ppm are expected to have been contaminated by thermal water.

The hot springs (AP03) in the study area has 438 ppm TDS. Warm springs in the study area has TDS value ranged from 123-746 ppm. The high value of TDS at warm springs in EMTE (APMT) was due to human activities, because this springs is used directly for pools. Cold springs in the study area has a TDS value is relatively low, ranging from 17 – 252 ppm. The lowest TDS value is located in

Pojok, Pelton and Cipaganti cold springs with a value of <100 ppm. Those cold springs is expected to be uncontaminated by thermal water, based on the analysis of the value of TDS. Cold springs in Punceling (AD01) and Cimanggu (AD04), which is very close to the manifestation of thermal springs, have a relatively high TDS values, which exceeds 100 ppm. Cold springs Situ Lembang (ADSL) and Legok Kondang (ADLK) also have relatively high TDS value. TDS value analysis showed that cold springs Punceling, Cimanggu, Situ Lembang, and Legok Kondang suspected to have been contaminated by a thermal fluid (Figure 7).

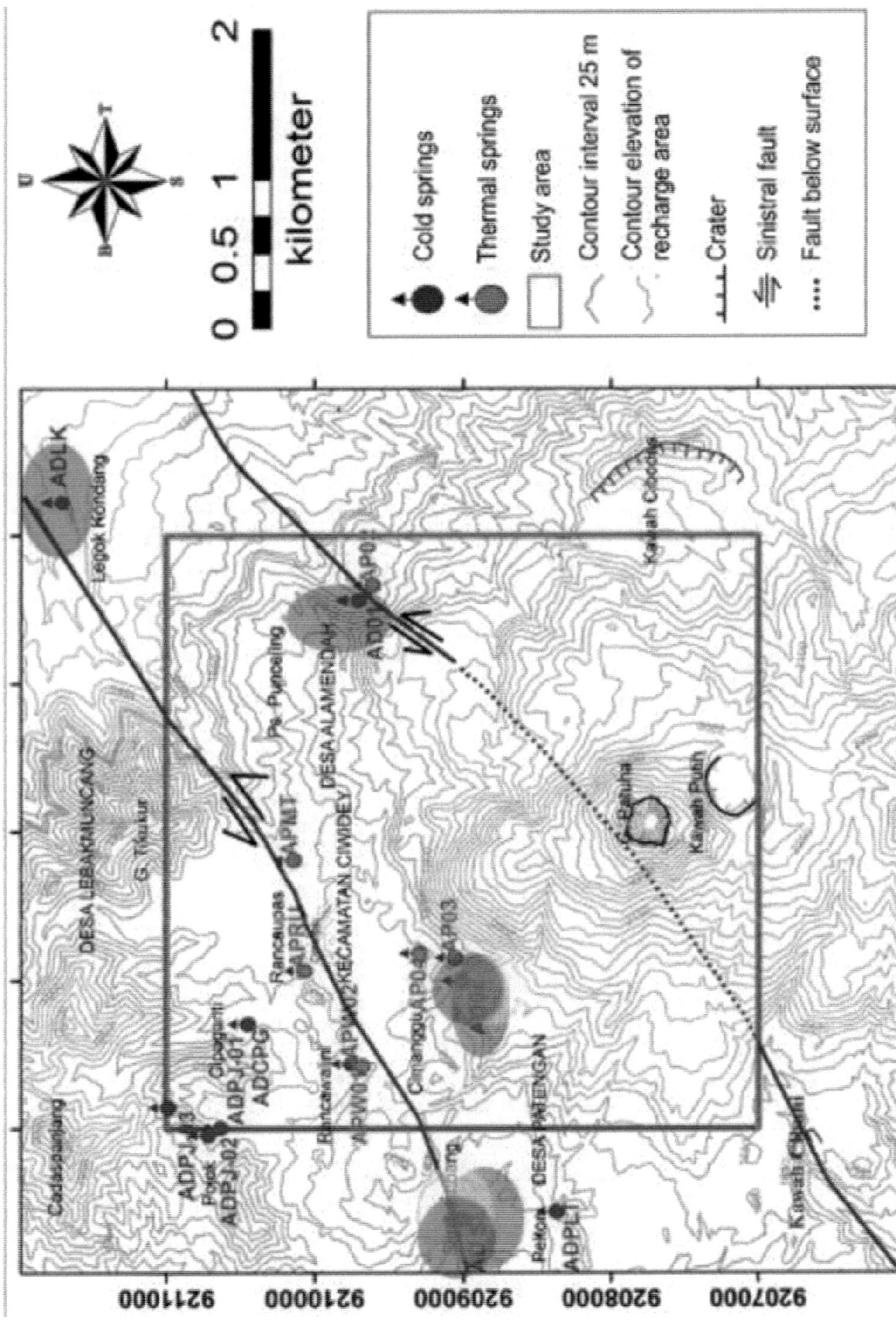

Figure 7. Contaminated cold springs based on temperature data (red circle), pH (yellow circle) and TDS (purple circle).

Based on the review of the water characteristics of above, cold springs that were expected to have been contaminated by thermal water are located in Punceling (AD01), Cimanggu (AD04), Situ Lembang (ADSL), and Legok Kondang (ADLK). Punceling cold springs and Legok Kondang have TDS values greater than 100 ppm. Cimanggu cold springs and Situ Lembang has a higher temperature than the air temperature, the pH is relatively low, and the value of TDS is more than 100 ppm.

4.2. Laboratory Data Analysis

The purpose of analyzing dissolved chemical elements are to identify cold springs that have a relatively high value of cations and anions compared to those of vicinity cold springs. Moreover, this analysis was also done to identify cold springs that have isotope 18O shifting values from the local meteoric line. Some assumptions are used in this analysis. First, cold springs that located quite far away from the geothermal system and have no anomalies in the data of temperature, pH, and TDS, are assumed as uncontaminated cold springs as well as a baseline reference. Second, uncontaminated cold springs from the same volcanic complex or khuluk that associated with geothermal systems will contain relatively similar cations and anions to those in baseline reference. Third, the enrichment of isotope 18O value in cold springs was caused by contamination of the thermal fluid.

Ions data from chemistry laboratory test results are presented in Table 3, while the isotope primary data from study area is shown in Table 4.

4.2.1. Schoeller Diagram

Examination of the Schoeller diagram that shows similar slope of lines connecting solute concentrations is an indication of groundwater from a similar source [7]. If the waters have different concentrations, they will appear on the graph one above the other, thus indicating relative movement of the groundwater between the points of origin of these analyses; the flow being from a lower concentration to a higher concentration [7]. Based on the increasing composition of meteoric water, it can be concluded that the total dissolved solids and dissolved ions concentration can be an indication of contaminated water by thermal springs below the surface. However, this must be supported with other evidence, especially the location of the springs.

Table 3. Ions concentrations of springs sample.

Location	Code	Fe	Ca	Mg	Na	K	NH_4	Cl	SO_4	HCO_3	Σ Cations (meq/L)	Σ Anions (meq/L)
Thermal springs		mg/L										
Punceling	AP02	2.18	250.60	50.58	136.10	37.80	1.00	401.76	480.47	294.36	23.71	26.17
Cimanggu	AP03	0.65	43.07	16.66	84.66	18.12	0.83	50.29	59.34	306.19	7.74	7.68
	AP04	2.54	49.74	15.38	59.34	15.39	0.73	32.43	49.61	292.44	6.86	6.75
Walini	APW01	0.08	70.84	45.69	60.09	20.33	0.08	114.96	172.00	239.24	10.45	10.75
Ranca Upas	APRU	0.08	46.89	30.70	49.29	16.85	0.41	87.42	109.80	161.41	7.48	7.41
Cold springs		mg/L										
Punceling	AD01	0.08	27.71	15.06	13.37	4.42	0.49	63.13	46.28	62.41	3.35	3.77
Pelton	ADPLT	0.24	10.40	2.31	6.24	2.15	1.09	3.47	16.00	37.42	1.10	1.04
Cipaganti	ADCPG	0.64	3.00	1.00	1.56	1.25	2.00	2.00	6.50	14.88	0.47	0.44
Pojok	ADPJ-02	0.18	2.00	0.71	2.00	1.02	1.50	0.61	4.40	14.00	0.36	0.34
	ADPJ-03	0.26	2.38	0.94	1.70	1.25	2.00	0.44	4.20	14.91	0.42	0.34
Cimanggu	AD04	0.13	29.06	3.95	10.06	3.68	0.73	0.92	13.96	117.65	2.36	2.24
Legok Kondang	AD LK	0.02	27.67	10.91	13.94	6.10	0.29	51.44	40.06	50.93	3.06	3.14
Situ Lembang	AD SL	0.29	39.35	22.97	26.26	8.47	0.33	68.59	64.55	122.31	5.24	5.30

Plot results on Schoeller Diagram shows the difference of ions concentration in general between cold and thermal springs (Figure 8). Thermal springs have relatively higher ions concentration than cold springs. Cold springs with the lowest ions concentration is located in Pojok (ADPJ-02 and ADPJ-03) and Cipaganti (ADCPG). Those springs have a relatively similar trend and close location, and therefore expected to come from the same source, recharge area in MVC. Relatively low ion concentration is also an indication that the springs is not contaminated by the fluid geothermal systems.

Other cold springs contain relatively higher ions concentration. This enrichment of the dissolved ions with a similar trend line in Schoeller Diagram is an indication of contamination occurred by the geothermal fluid. Cold springs in PVC are also from the recharge area in PVC. The only cold springs in the study area that are located in PVC, but from recharge area in MVC is Legok Kondang cold springs (ADLK), which is marked by a green circle in Figure 9. Legok Kondang springs are cold springs are not contaminated.

4.2.2. Isotope

Rain water isotope data [1] was used to determine the local meteoric water line. The line is used as a reference in determining the value of isotope shifting in the cold springs. Isotope values of springs in the study area are shown in Table 4. Plot of oxygen and deuterium isotope values as well as the meteoric line is shown in Figure 10. Cold springs that originated from meteoric water that is not contaminated by thermal water will have a value of isotopes that are still along the local meteoric water line. Plot results of hot and warm springs showed an ^{18}O isotope shifting due to evaporation or steam heated.

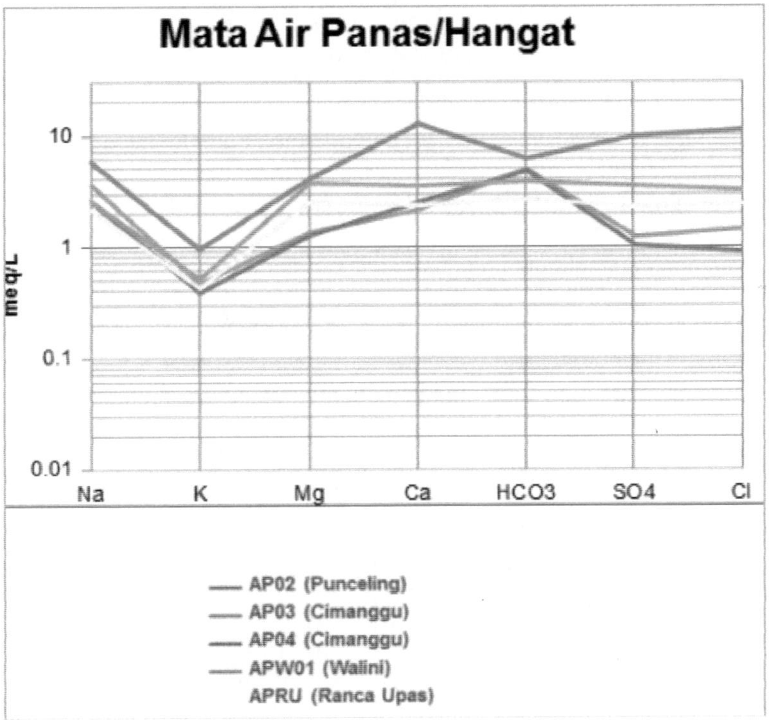

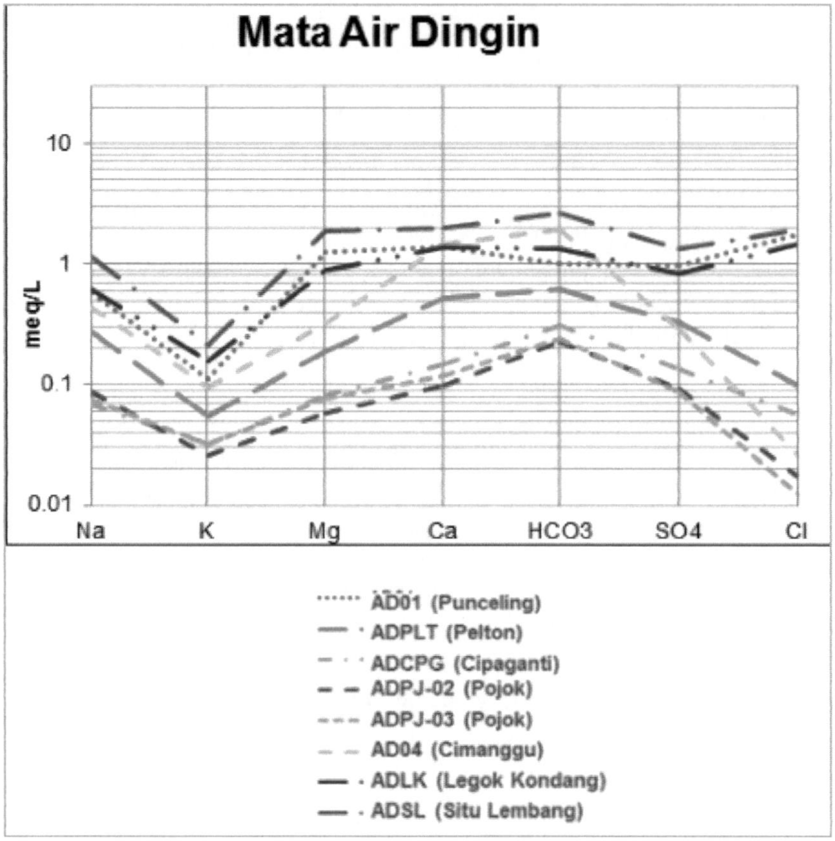

Figure 8. Ions concentrations plot results in Schoeller Diagram

Table 4. ^{18}O dan ^{2}H isotope data.

Location	Code	$d^{18}O$	δD
Thermal springs		°/₀₀	
Punceling	AP02	-8.50	-55.44
Cimanggu	AP03	-8.96	-56.32
	AP04	-8.95	-57.08
Walini	APW01	-8.69	-55.42
Ranca Upas	APRU	-8.76	-55.65
Cold springs			
Punceling	AD01	-8.66	-54.94
Pelton	ADPLT	-9.25	-58.04
Cipaganti	ADCPG	-9.07	-55.44
Pojok	ADPJ-02	-9.00	-55.62
	ADPJ-03	-9.23	-56.75
Cimanggu	AD04	-9.47	-58.21
Legok Kondang	AD LK	-8.56	-52.02
Situ Lembang	AD SL	-8.89	-54.68

In Figure 10, we can see that cold springs in Legok Kondang (ADLK), Pojok (PJ02 and PJ03), Cipaganti (ADCPG) and Cimanggu (AD04) are plotted along the local meteoric water line. Legok

Kondang, Cipaganti and Pojok area are not located adjacent to manifestations of geothermal system, so the cold springs in these areas are not expected to be contaminated by the fluid thermal. The presence of spring water that is not contaminated is indicated by green circle in Figure 8 Cimanggu cold springs (AD04) is located along the local meteoric water line, but the location of cold springs are located very close to the manifestation of Cimanggu thermal springs, the AP03 and AP04, therefore this cold springs is expected to be contminated by the thermal springs. Cimanggu thermal springs AP03 and AP04 have a value of 18O isotope shifting about 0.5 – 0.6 0/00 from local meteoric water line. Unshifting value of δ18O isotope in AD04 can be due to the short residence time of meteoric water below the surface. Therefore, the relatively short local ground water flow will result in no significant change of isotope values when cold springs occur at the surface.

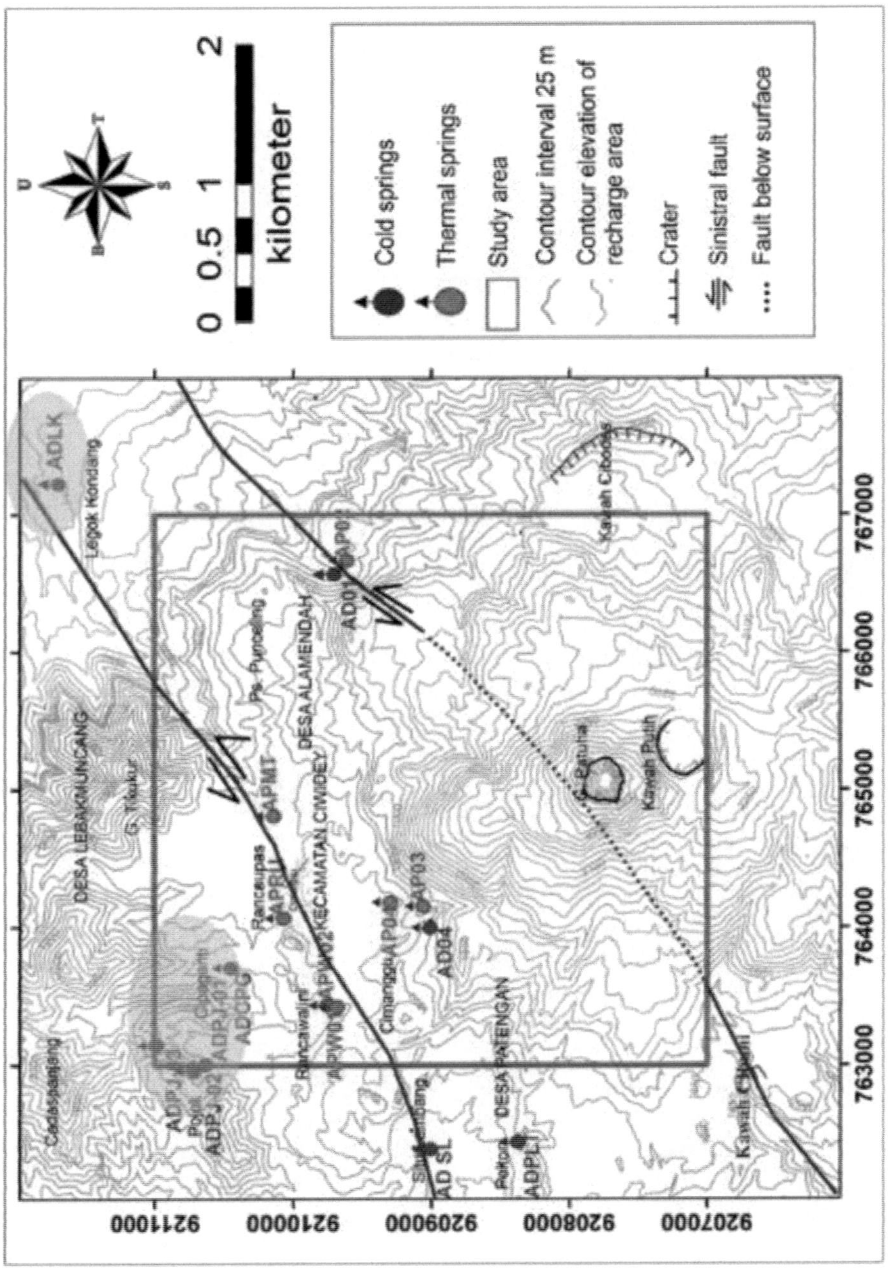

Figure 9. Uncontaminated cold springs based on Schoeller Diagram and isotope analysis (green circle).

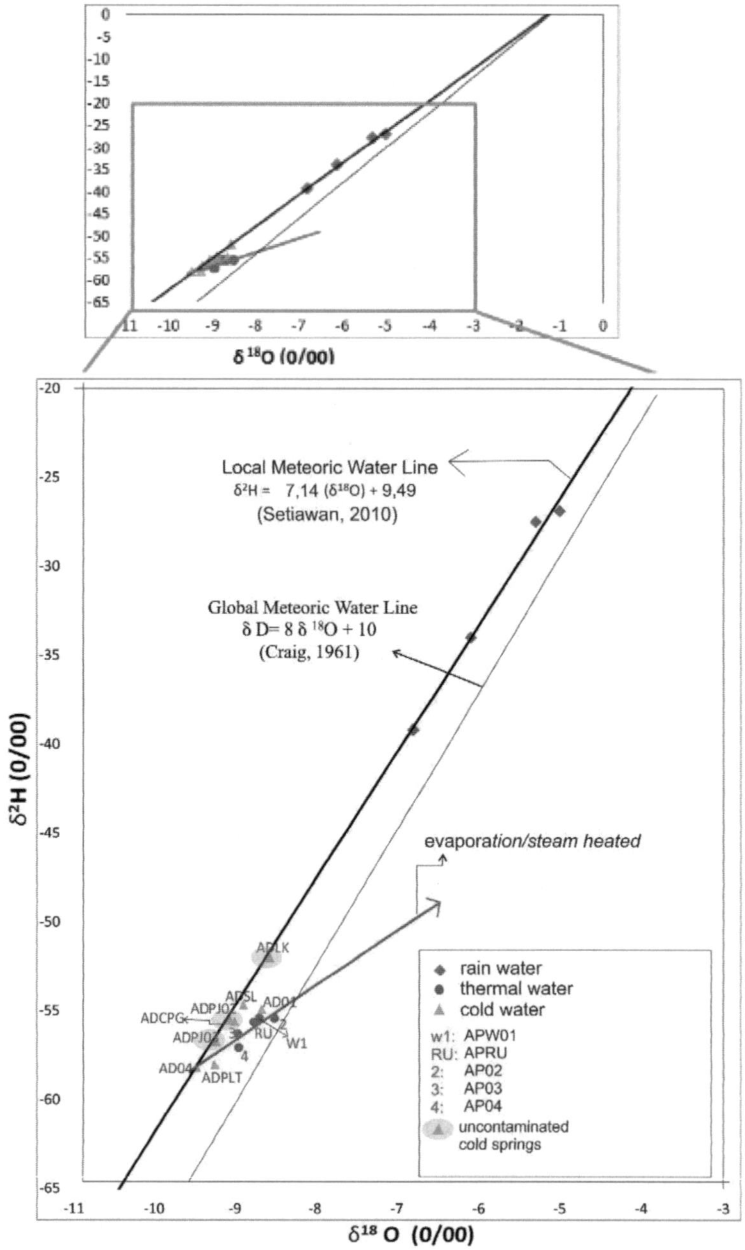

Figure 10. ^{18}O and ^{2}H Stable Isotope Plot

The other cold springs in Pelton (ADPLT), Situ Lembang (ADSL) and Punceling (AD01) showed a shifting value towards the local meteoric line approaching the line of evaporation / steam heated. Shifting to more positive value can occur due to isotope exchange between water with rocks at a higher temperature [4]. Cold springs in the area Punceling, Pelton and Situ Lembang are expected to be contaminated by a thermal fluid. This is also supported by their location which are quite closed with the surrounding thermal springs manifestation in PVC.

5. Boundary Results

Based on the geological analysis, springs in the study area are grouped into springs PVC, which are springs in Cimanggu, Punceling, Situ Lembang, Pelton and Legok Kondang, and springs in MVC, which are springs in Pojok and Cipaganti. The occurrence of these springs are also controlled by permeable zones in Cimanggu Fault and Punceling Fault, as well as the depression zone in the form of intersection spring soil to the slopes.

Fluid chemistry analysis indicates that cold springs Punceling (AD01), Pelton (ADPLT), Cimangggu (AD04), and Situ Lembang (ADSL) located in PVC have increased in TDS, higher temperature than the surrounding air or ambient temperature, decreased in pH, increased in ions concentration, and isotope ^{18}O shifting due to contamination of the Patuha geothermal system. Those contaminated springs are also located relatively close to the thermal spring manifestations. On the other hand, springs in Legok Kondang (ADLK), Cipaganti (ADCPG), and Pojok (ADPJ-02 and ADPJ-03) are not contaminated by the thermal fluid. The boundary of the system is delineated by an arbitrary line drawn between distal thermal springs from the contaminated cold springs with the uncontaminated cold springs (Figure 11).

This boundary is validated afterward with low resistivity boundary derived from DC and MT resistivity survey (Figure 12 and Figure 13, respectively). Low resistivity values (<20 ohm.m) represents the cap rocks in the system. Reservoir of geothermal systems is expected to correlate with those cap rocks. This boundary is also validating with drilling data that give high temperature gradient. Figure 12 shows temperature gradient that ranges between 100 – 200°C/km. In the normal area, a temperature gradient is typically ± 25°C [8]. It shows that the temperature anomaly beneath the surface has a high value and could be an indication of an active geothermal systems. Those validations suggest that the method use in this study is applicable and reliable.

6. Conclusions

Based on geological studies, springs in the study area are controlled hydrologically by topography of Patuha Volcanic Crown (complex) or so called PVC, which consist of springs in Cimanggu, Punceling, Situ Lembang, Pelton, and Legok Kondang, and of Masigit Volcanic Crown (complex) or so called MVC, which consist of springs in Pojok and Cipaganti. Thermal springs manifestations in study area are present as an outflow zone in Patuha geothermal systems.

Some of the cold springs in PVC are contaminated by subsurface steam heated outflow while others are not contaminated. The contaminated cold springs have several characteristics such as higher water temperature that ambient temperature at the time it was measured, higher total dissolved solid (TDS), and lower pH. The soluble elements analysis support the early contamination indication by showing higher cation and anion, and positive oxygen shifting of stable isotope of these cool springs. Whereas the uncontaminated spring shows similar characteristic with cool springs occur at MVC.

The geothermal system boundary location is estimated to be in between uncontaminated cold springs and the farthest thermal springs from upflow zone or the closest contaminated cold springs to outflow zone. This boundary is validated by using low resistivity value (≤20 ohm.m) and temperature gradient data that shows similarity in the boundary location, and it can be concluded that the method used is valid

IOP Conf. Series: Earth and Environmental Science **103** (2017) 012016 doi:10.1088/1755-1315/103/1/012016

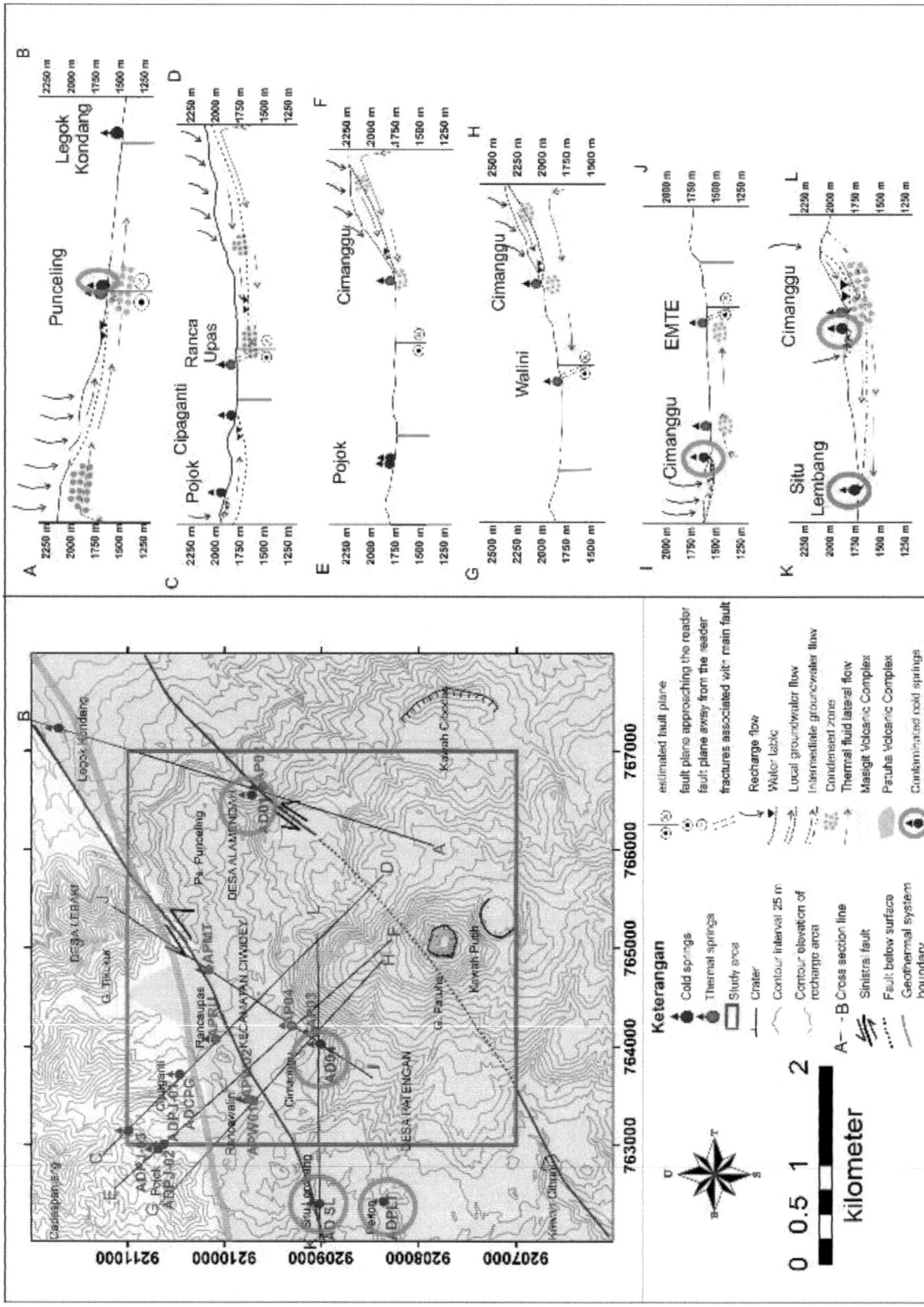

Figure 11. Groundwater flow and thermal water flow model in study area.

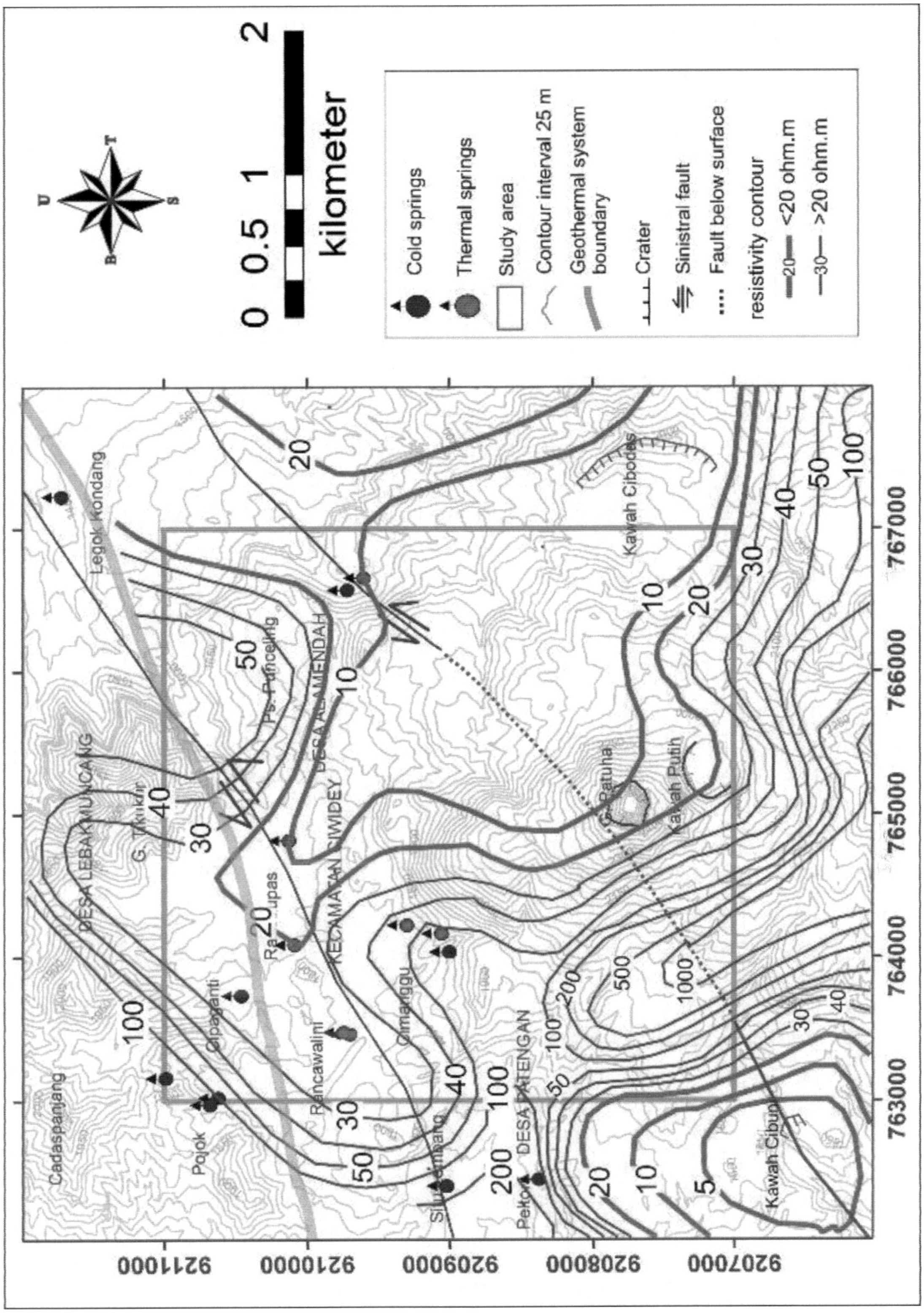

Figure 12. Geothermal system boundary validation with DC-Resistivity, Schlumberger, AB/2 = 500 meter [9].

IOP Conf. Series: Earth and Environmental Science **103** (2017) 012016 doi:10.1088/1755-1315/103/1/012016

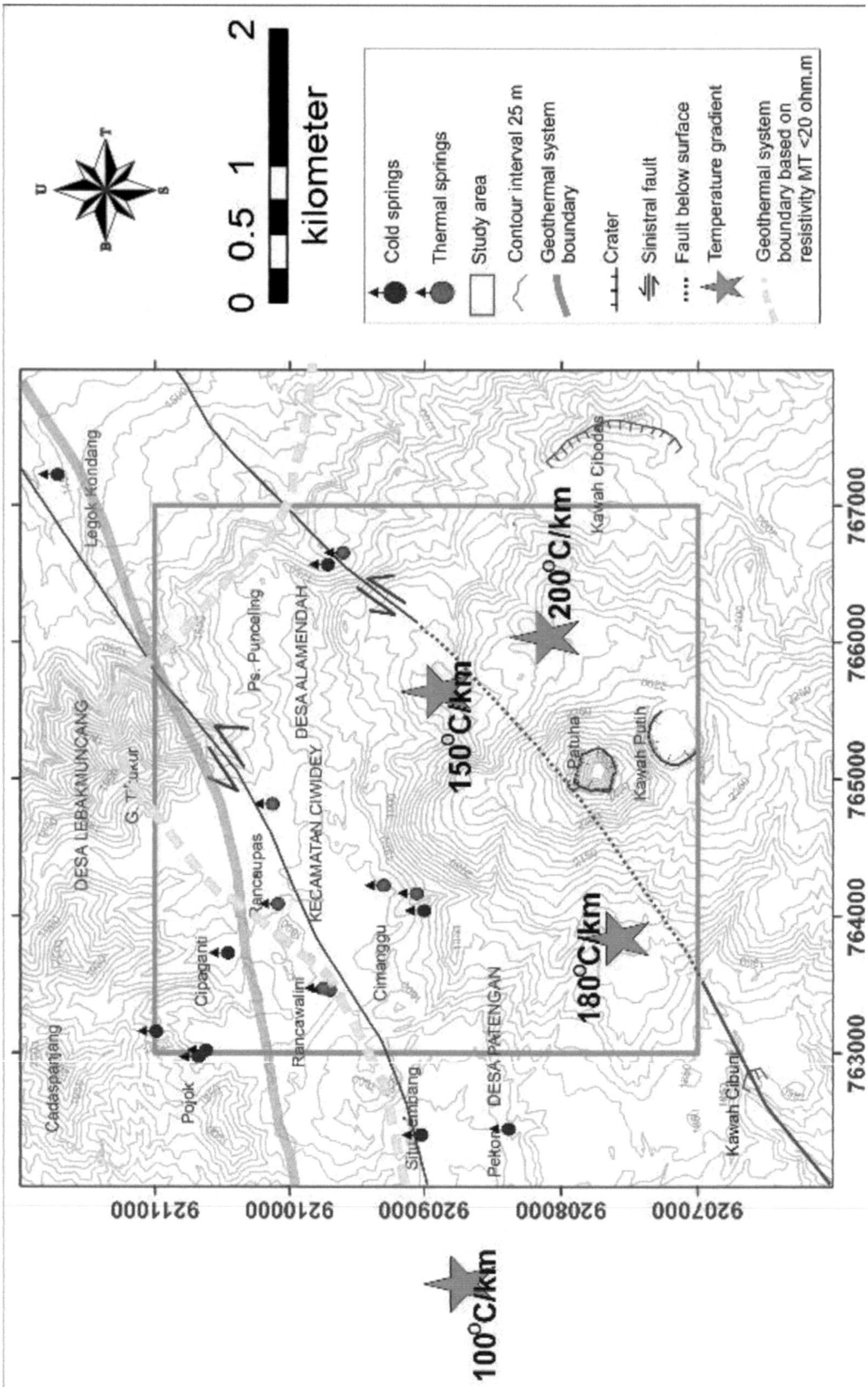

Figure 13. Geothermal system boundary validation with resistivity MT and gradient temperature [10].

References

[1] Setiawan Iwan F 2010 *Penentuan Daerah Resapan Lapangan Panasbumi Patuha, Kabupaten Bandung, Jawa Barat* Master Theses Teknik Geologi Institut Teknologi Bandung.
[2] Martodjojo S and Djuhaeni 1996 *Sandi Stratigrafi Indonesia* Jakarta Ikatan Ahli Geologi Indonesia.
[3] Pradipta R A 2016 *Penentuan Zona Permeabilitas Menggunakan Citra ALOS Palsar, Pemetaan Geologi, dan Geokimia Tanah dan Udara Tanah (Hg-CO2), Studi Kasus Lapangan Panasbumi Ciwidey, Kabupaten Bandung* Master Theses Institut Teknologi Bandung.
[4] Nicholson K 1993 *Geothermal Fluids: Chemistry and Exploration Techniques* Berlin: Springer-Verlag.
[5] WHO 2011 *Guidelines for Drinking-Water Quality* Fourth Edition WHO Press
[6] Berry G W, Grim P J and Ikelman J A 1980 Thermal Spring List for The United States *National Oceanic and Atmospheric Administration Key to Geophysical Records Documentation* 12.
[7] Piper Arthur M 1944 A graphic procedure in the geochemical interpretation of water-analyses *Transactions American Geophysical Union* **25** p 914-928.
[8] Wolfson R 2012 *Energy from Earth and Moon in Energy, Environment, and Climate, 2nd ed* W.W. Norton & Company New York.
[9] Alhamid I 1989 The resource potential of The G. Patuha Geothermal Area, West Java *IPA 18th Annual Convention Proceeding* **2** p 1-19.
[10] JBIC 2007 *Feasibility Study for Patuha Geothermal Power Development. Final Feasibility Report* West Japan Engineering Consultants, Inc.

IOP Conf. Series: Earth and Environmental Science **103** (2017) 012017 doi:10.1088/1755-1315/103/1/012017

Preliminary determination of geothermal working area based on Thermal Infrared and Synthetic Aperture Radar (SAR) remote sensing

Indra Agoes Nugroho[1], Beta Kurniawahidayati[1], Reza Syahputra Mulyana[1], Asep Saepuloh[1,2]

[1]Geothermal Engineering Master Program, Faculty of Mining and Petroleum Engineering Institut Teknologi Bandung (ITB), Jl. Ganesha No. 10, Bandung, West Java, 40132, Indonesia.
[2]Geology Engineering Study Program, Faculty of Earth Sciences and Technology, Institut Teknologi Bandung (ITB), Jl. Ganesha No. 10, Bandung, West Java, 40132, Indonesia.

E-mail: indraanugroho@live.com

Abstract. Remote sensing is one of the methods for geothermal exploration. This method can be used to map the geological structures, manifestations, and predict the geothermal potential area. The results from remote sensing were used as guidance for the next step exploration. Analysis of target in remote sensing is an efficient method to delineate geothermal surface manifestation without direct contact to the object. The study took a place in District Merangin, Jambi Province, Indonesia. The area was selected due to existing of Merangin volcanic complex composed by Mounts Sumbing and Hulunilo with surface geothermal manifestations presented by hot springs and hot pools. The location of surface manifestations could be related with local and regional structures of Great Sumatra Fault. The methods used in this study were included identification of volcanic products, lineament extraction, and lineament density quantification. The objective of this study is to delineate the potential zones for sitting the geothermal working site based on Thermal Infrared and Synthetic Aperture Radar (SAR) sensors. The lineament-related to geological structures, was aimed for high lineament density, is using ALOS - PALSAR (Advanced Land Observing Satellite - The Phased Array type L-band Synthetic Aperture Radar) level 1.1. The Normalized Difference Vegetation Index (NDVI) analysis was used to predict the vegetation condition using Landsat 8 OLI-TIRS (The Operational Land Imager – Thermal Infrared Sensor). The brightness temperature was extracted from TIR band to estimate the surface temperature. Geothermal working area identified based on index overlay method from extracted parameter of remote sensing data was located at the western part of study area (Graho Nyabu area). This location was identified because of the existence of high surface temperature about 30°C, high lineament density about 4 - 4.5 km/km2 and low NDVI values less than 0.3.

1. Introduction

Remote sensing is a common exploration technique for preliminary survey and aiding the complicated methods such as geology and geophysical surveys. Remote Sensing is the science and art of obtaining information about an object, area, and phenomenon through the analysis of data acquired by a device that is not contact with the object, area, and phenomenon under investigation [1]. In geothermal studies, this method is not only used for exploration, but also in exploitation to monitor geothermal working area [2][3]. Remote sensing was used satellite imageries including optic and Synthetic Aperture Radar (SAR), to obtain reflectivity, emissivity, and scattering information from the target at ground surface. The results may be used as a basis for geological, geophysical and geochemical surveys [4].

This method is used to interpret surface condition based on color, texture, and structure from corrected imageries. The color difference from image can be interpreted as different physical properties of object or material at the ground surface [5]. Texture in the image might show the geomorphic stage of volcano, for example the smooth surface is relatively young stage due to less erosion process regardless rock type. Linear, circular, and radial structures of the ground surface could be identified based on Geomorphologic and Structural Features (GSF) in the imageries [6]. Based on GSF, major structures could be identified by lineament distribution and the high permeability could be interpreted by lineament density analysis [7][8]. In this study, the remote sensing data were used to detect the volcanic products, vegetation cover, and surface temperature.

Merangin area, in Jambi, was selected as study area because of the existence of surface manifestations with the remaining geomorphic of the eruption centers (figure 1). Therefore, the area is expected to have a geothermal system beneath the ground surface. However, there is not any exploration activity yet. The government has already made an auction for beginning exploration at the end of 2016 [9]. Thus, this study may hopefully usable for the exploration activity.

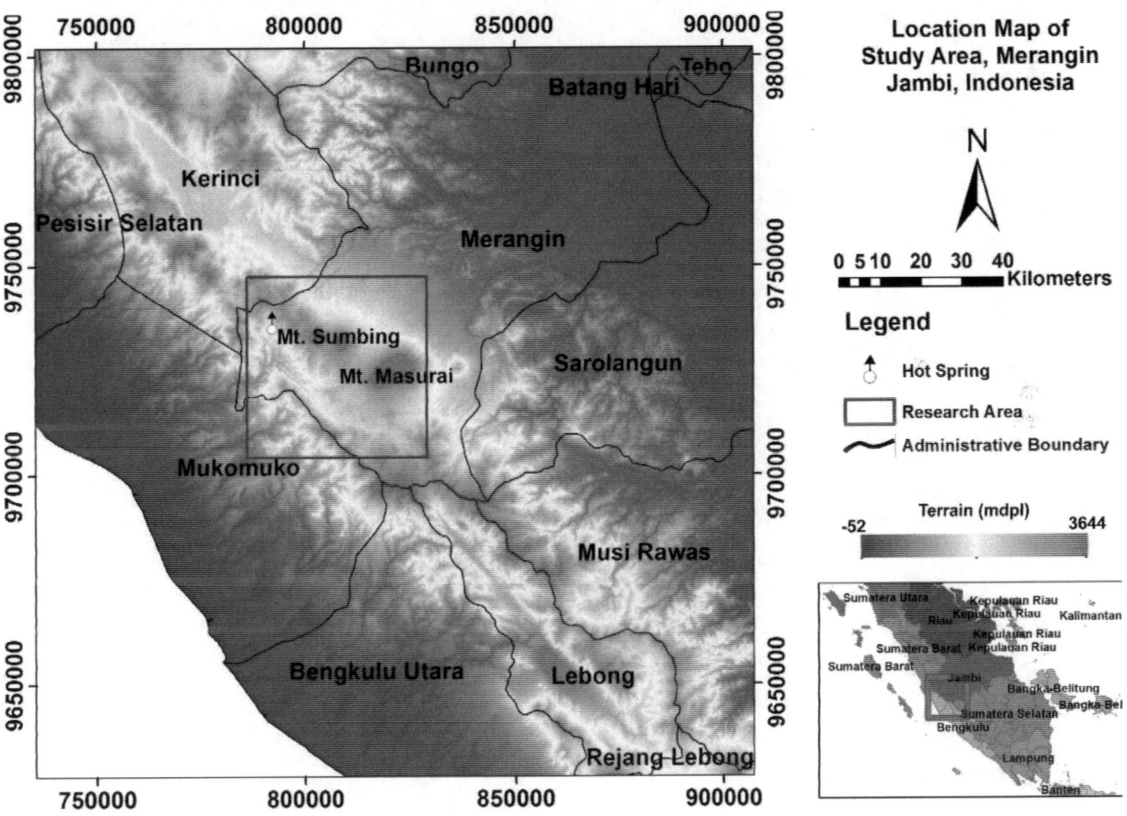

Figure 1. Location of study area showed by blue rectangle of the DEM SRTM 1 arcs and the subset image is Sumatra Island

The study area was composed by Hulu Simpang Formation, as the oldest formation, Granodiolit Langkup, Volcanic Rhyo-Andesit Unit and Volcanic Breccia-Tuff Unit as the youngest formation. Quartenary Volcanic Lithology (Volcanic Breccia-Tuff Unit) formed the Mount Tua, Sumbing, Hulunilo and Masurai (figure 2). The major structure near the study area is the Great Sumatra Fault from Northwest – Southeast that control the surrounding minor faults [10].

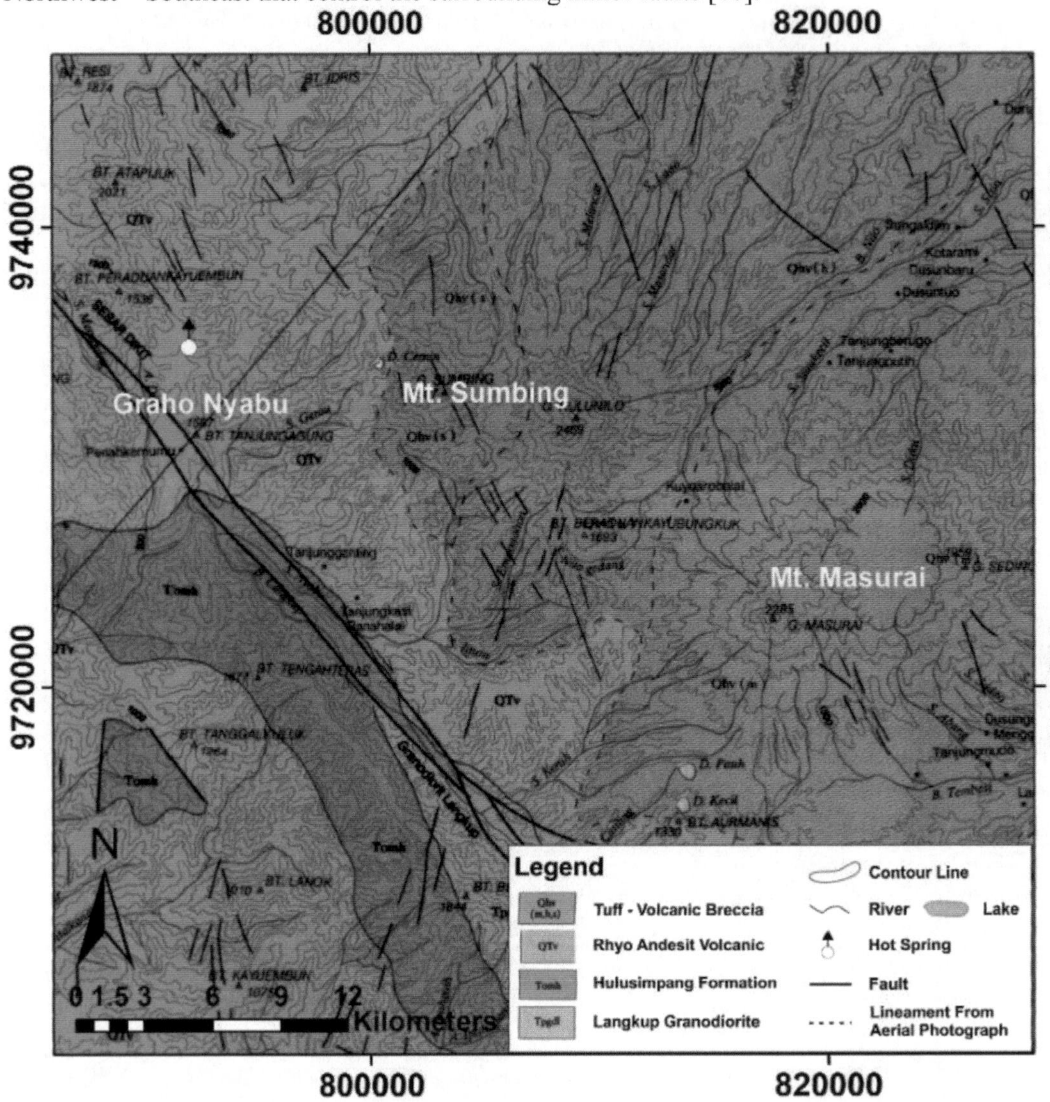

Figure 2. Regional geological map of study area from [11] shows volcanic lithology distribution and great Sumatra fault from northwest to southwest.

2. Image Processing

2.1. Optical Image

Optical image used for this study is Landsat 8 OLI-TIRS containing eleven bands, from visible to thermal infrared region. The acquisition date on July 23rd, 2014. Eruption center, volcano base, geological structure on the surface, and vegetation cover were identified from Landsat 8 image.

The image processing was performed using ENVI 5.2 including radiometric calibration, atmospheric, and geometric corrections (figure 3). Radiometric calibration was used to convert digital number to radiance so that the data can be processed for atmospheric correction. The Fast Line of sight

Atmospheric Analysis of Spectral Hypercubes (FLAASH) embedded in ENVI was used for atmospheric correction. The reflectance produced after atmospheric correction is crucial to reduce the water vapor, oxygen, carbon dioxide, methane, ozone, molecule distribution, and aerosol in atmosphere. Following the atmospheric correction, geometric correction was applied to correct the geometric distortion. We used a Digital Elevation Model (DEM) from Shuttle Radar Topography mission (SRTM) data for georeferencing, registration, and rectification.

After corrections were applied, the color composite of RGB were used as basis of visual interpretation. RGB for band 4, 3, 2 showed the natural surface color, while RGB 5, 6, 7 showed false color. Both composites were used for volcanic products and lineament analyses. The Normalized Difference Vegetation Index (NDVI) was also used to identify the vegetation covers for estimating the leaf condition related to geothermal system beneath the surface [12]. The NDVI was calculated by incorporating the division of near infrared (NIR) and visible red bands. The NDVI indicated the distribution of rare and dense vegetation cover that the one might serve as an indicator of geothermal surface manifestation [12]. Visual observation based on natural color image is needed to confirm that low NDVI probably originated from surface manifestation, rather than clouds and infrastructures.

Landsat 8 OLI-TIRS provided thermal infrared bands for estimating brightness surface temperature in a specific location. Band 10 and 11 from Landsat 8 OLI-TIRS were used as input for calculating brightness surface temperature (figure 3). Radiometric calibration to generate radiance value from image digital number is presented by [13]:

$$L_\lambda = \left(\frac{L_{max,\lambda} - L_{min,\lambda}}{Q_{calmax} - Q_{calmin}}\right)(DN - Q_{calmin}) + L_{min,\lambda} \qquad (1)$$

Where L_λ is sensor's top atmosphere radiances, *DN* is digital number, L_{max} and L_{min} are TOA Radiance scaled to Q_{calmax} and Q_{calmin}, and λ is band wavelength.

The brightness temperature was calculated using following expression [14]:

$$T = \frac{K_2}{ln\left(\frac{K_1}{L_\lambda} + 1\right)} \qquad (2)$$

T is sensor brightness temperature (K), K_2 and K_1 are calibration constant, depends on sensor and L_λ is spectral radiance. Brightness temperature can represent the surface temperature [14].

2.2. Synthetic Aperture Radar (SAR) Image

Synthetic Aperture Radar satellite is an active satellite that produce its own electromagnetic wave to earth surface, and obtain the backscattering wave from earth surface. The backscattering depends on roughness and dielectric property of the ground surface, thus the data are effective to produce a great quality image for geomorphologic and structural features identification [6]. SAR image has two acquisition methods, ascending, the acquisition from south to north, and descending in vice versa. SAR image used for this study is from ALOS PALSAR satellite level 1.1 data in ascending orbit. The acquisition time is November 9th, 2010.

We used SNAP 4.0 for SAR image processing. The processing contains radiometric calibration, multi-looking, polarimetric matrix generation, polarimetric speckle filtering, and geometric correction (figure 3). Radiometric calibration was used to convert the digital number to backscattering. Multi-looking process to increase the image focus without decreasing the resolution significantly.

In the SNAP SAR processing, the digital number must be in covariance and coherence matrices because it preserves the vector of the electromagnetic radiations and therefore, is preferred for complete analysis of the backscatter [15]. Therefore, the backscattering data were converted to covariance and coherence using polarimetric matrix generation [15]. The HH and HV polarisation band from ALOS PALSAR 1.1 were converted to C11, C12_real, C12_imag, and C22 using matrix generation C2. Matrix generation C2 consists of 2×2 matrix dimension with C11, C12, C12_image, and C22 are the scalar vector in the matrix [16]. Speckle filtering was applied to reduce the speckle noise in image using Refined Lee with window size 7×7 pixels.

After all calibration and polarimetric processing, the image needs a geometric correction. There were two geometric corrections applied: ALOS de-skewing and terrain correction. The ALOS de-skewing was used to convert inverse geometric to Doppler geometric. Whereas the terrain correction was used to evaluate the distortion from SAR geometry into map geometry using a Digital Elevation Model (DEM) from the Shuttle Radar Topography Mission (SRTM). The terrain correction is very important to reduce the foreshortening, layover, and shadow effect in the SAR image.

The local structures presented by lineament were extracted visually from corrected SAR images. Then, lineament density was calculated from the total of lineament in a specific area with a certain grid [8]. In this study, we used 100 × 100 m grid size, in accordance with the map scale, to represent the density of lineament in the study area more detail than 1×1 km grid size. The area with high density of lineament was interpreted as intensive structural zone.

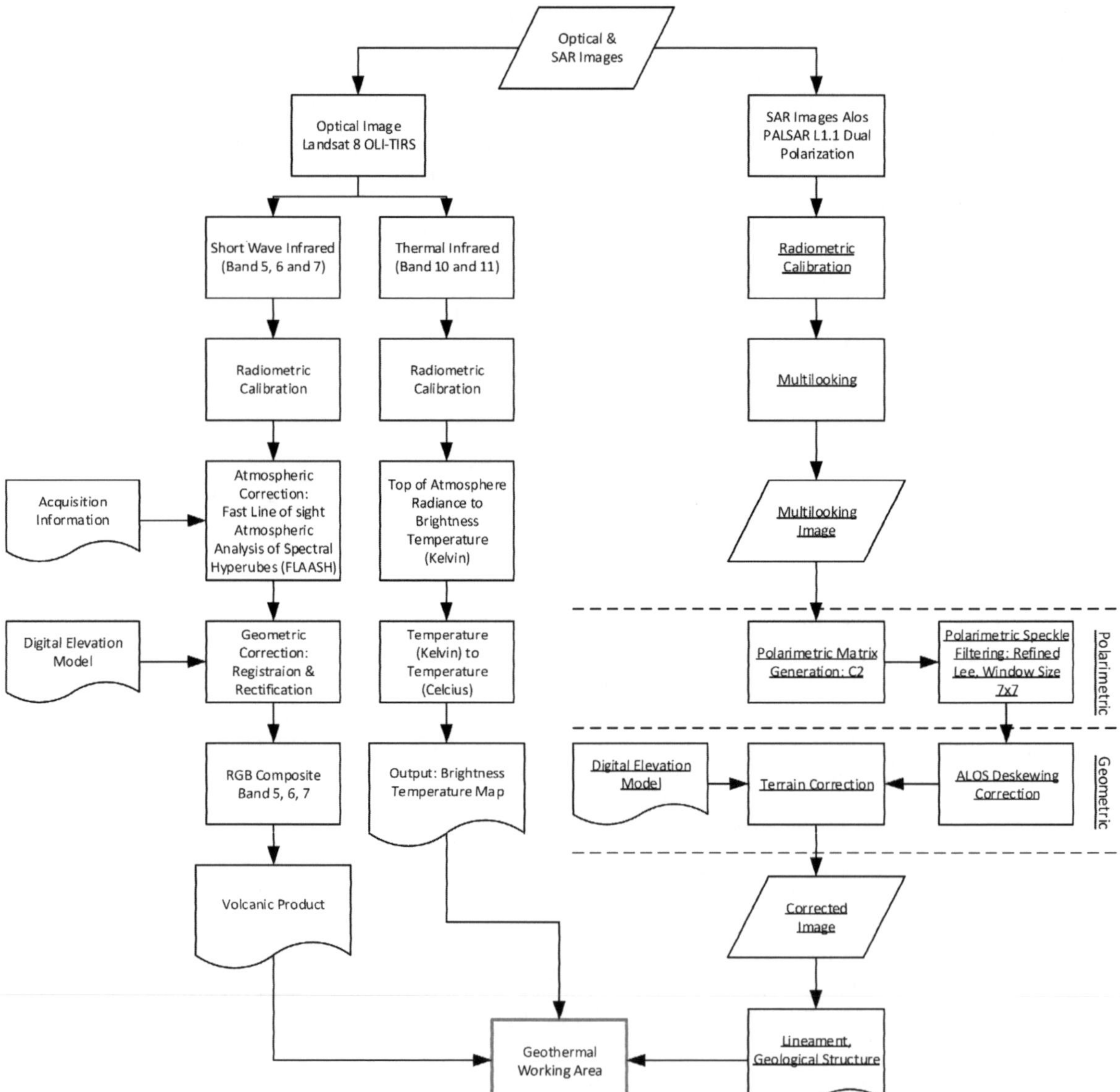

Figure 3. Flowchart of data processing showed how to generate geothermal working area using optical and SAR image.

2.3. Delineating Geothermal Working Area

Geothermal working area was delineated visually based on index overlay method [17]. Index overlay method used three parameters which were processed, indexed and calculated for each detected target parameter including volcanic products, brightness surface temperature, and lineament density. Prospected geothermal area should have at least the heat source potential with high permeability reservoir. Since the most of geothermal heat source in Indonesia are volcanoes [18], the heat source possible location could be estimated at surface by delineating the volcanic formation including crater rim and volcanic products using shortwave infrared band [5]. The possible high surface permeability was estimated using lineament density (LD) from SAR image. We interpreted that high LD are associated with dense geological structures such as faults, fractures, and damage zones [8]. The existences of geothermal surface manifestation and high surface temperature indicated that there is path of hydrothermal fluid from the sub-surface to the ground [12][19]. Geothermal surface manifestation can be indicated by rare vegetation that is not caused by human activities [12]. NDVI analysis can be used to delineate the geothermal surface manifestation [12]. To analyse NDVI, true color composite image was needed to compare the rare vegetation and good vegetation index distribution. Brightness surface temperature was calculated using thermal infrared band from Landsat 8 [13, 14]. Using equation (1) and (2) surface brightness temperature can be generated. Combined heat source location, good permeability zone, and high surface brightness temperature can be used to delineate the geothermal working area. Detailed working flows was depicted by figure 3.

3. Results and Discussions

3.1. Volcanic products detection

Following Indonesia stratigraphic code [20], the volcanostratigraphic of study area are consisted of five hummocks which are divided into two crowns. Hummocks were formed by some rocks or deposits from one or more eruptions which form a volcano, and crowns are volcano stratigraphic unit which consists of eruption rocks or deposits from two or more hummocks. The first is Masurai Crown which have three hummocks, in order from the oldest to the youngest are Masurai Hummock, Kumbang Hummock and Mabuk Hummock. Masurai Hummock have a rougher texture than Kumbang Hummock, whereas Mabuk Hummock have no rough texture from erosion.

The second is Sumbing Crown. Sumbing Crown located in the west of study area. This crown contains Sumbing Hummock and Hulunilo Hummock. According to their conical shape and texture, these hummocks is quartenary and Hulunilo Hummock is relatively older than Sumbing Hummock. The maturity of hummocks were predicted by the image which show Hulunilo Hummock already erode more intensive than Sumbing Hummock. Sumbing Crown was selected as a potential geothermal working area because it is younger than Masurai Crown interpretatively. According to geological map from [11], both crowns was formed by Quaternary volcanic breccia with tuff matrix. Figure 4 shows volcano geology map with Sumbing Crown was delineate with yellow boundary and Masurai Crown with blue boundary.

3.2. Geothermal Surface Manifestation

NDVI analysis shows that there are low vegetation covers with index less than 0.3 (figure 5). The NDVI values were divided into five classes: red for very rare vegetation, yellow for poor vegetation, light blue for medium vegetation, dark blue for dense vegetation, and green for very good vegetation. NDVI using minimal and maximum value to get the comparison value for rare and dense vegetation, therefore the NDVI value is an index value, not an absolute vegetation covers.

IOP Conf. Series: Earth and Environmental Science **103** (2017) 012017 doi:10.1088/1755-1315/103/1/012017

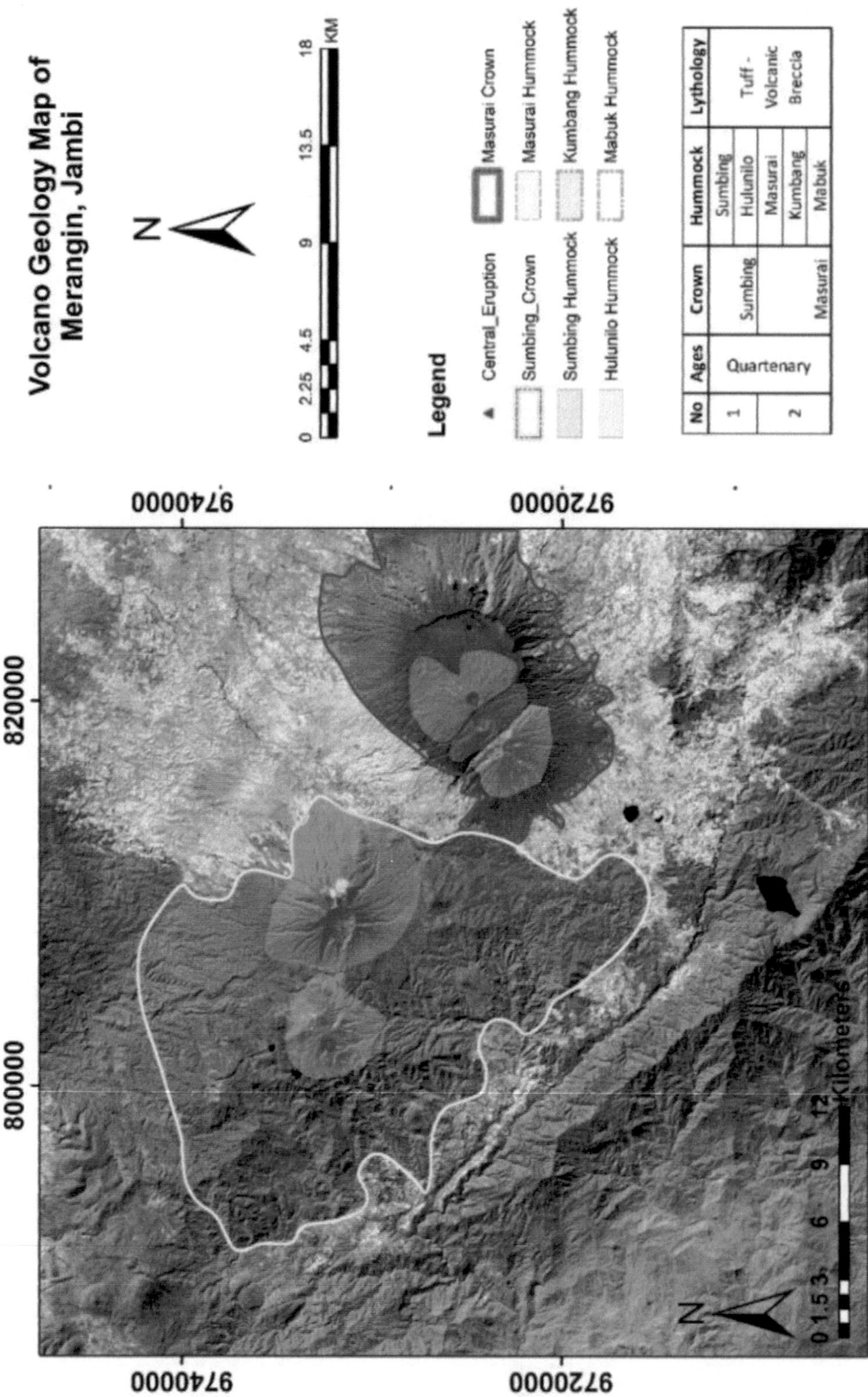

Figure 4. Interpreted volcano geology map of study area using color composite RGB 5, 6, 7 for Landsat-8.

The rare vegetation are located in Masurai and Sumbing Crater and hot springs in Graho Nyabu. The NDVI value is lower than surrounding area about 0.3. Visual analysis using natural color Landsat-8 image shows that this area is not a manmade. It may imply that rare vegetation is probably caused by the effect from the ground such as hydrothermal fluids. This area is a potential for groundtruthing in the next steps.

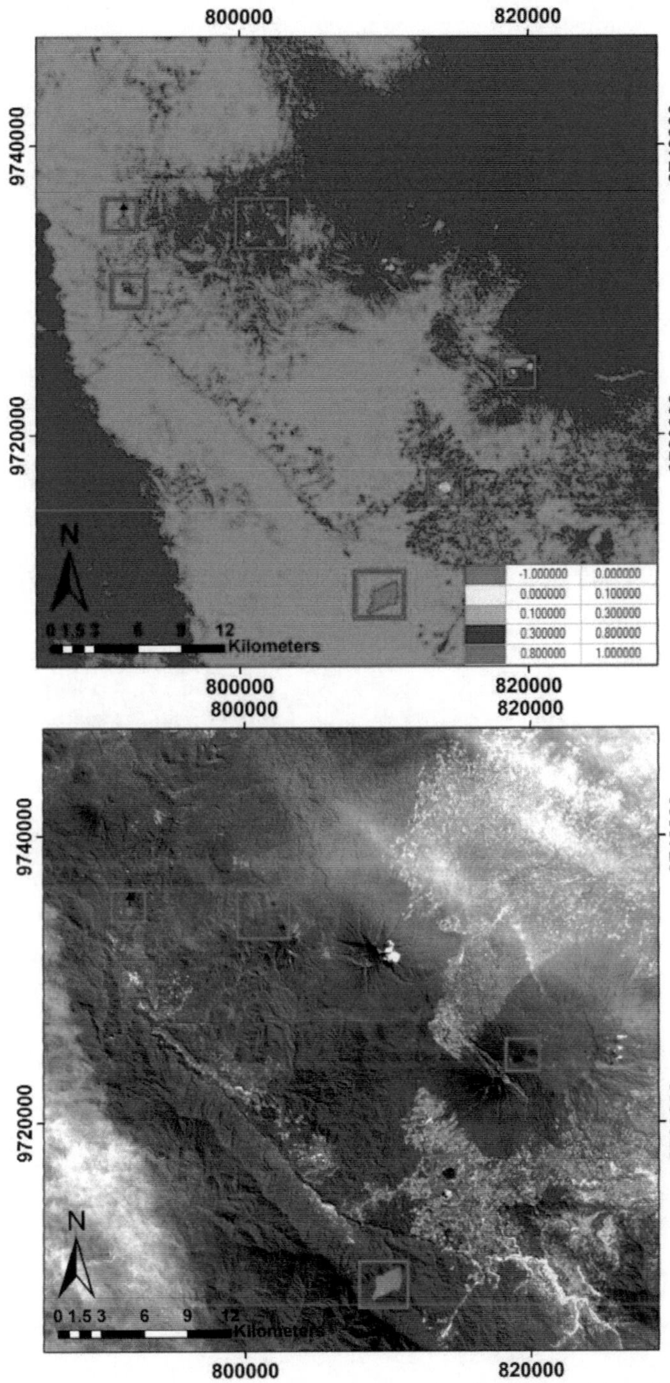

Figure 5. NDVI Map (up) and natural color image (down). Low vegetation index marked with red squares.

IOP Conf. Series: Earth and Environmental Science **103** (2017) 012017 doi:10.1088/1755-1315/103/1/012017

3.3. Brightness Temperature

Brightness surface temperature shows that there are several areas with temperature above 30 °C (figure 6). Comparing to NDVI map, we classified the target area into two area: with and without human activities. According to NDVI and brightness temperature maps, rare vegetation in the southeastern part of study area, rare vegetation are more likely to be originated from housing and manmade (red square in figure 6).

Following the NDVI and brightness temperature analyses, the target area is predicted at the eastern part of study area which has high density of vegetation (green square in figure 6). The surface temperature at the target area is about 30 °C.

We interpreted that that the emitted heat comes from the existences of geothermal surface manifestation. This interpretation was confirmed by the existence of a hot spring presented by white dotted arrow in figure 6.

Dense vegetation with lower brightness temperature about 28 °C were detected near the central eruption (black square at figure 6). It can be indicated that the volcanoes are not active anymore but still have heat source of geothermal system beneath the surface [1]. Nevertheless, this interpretation still needs direct groundcheck in the area.

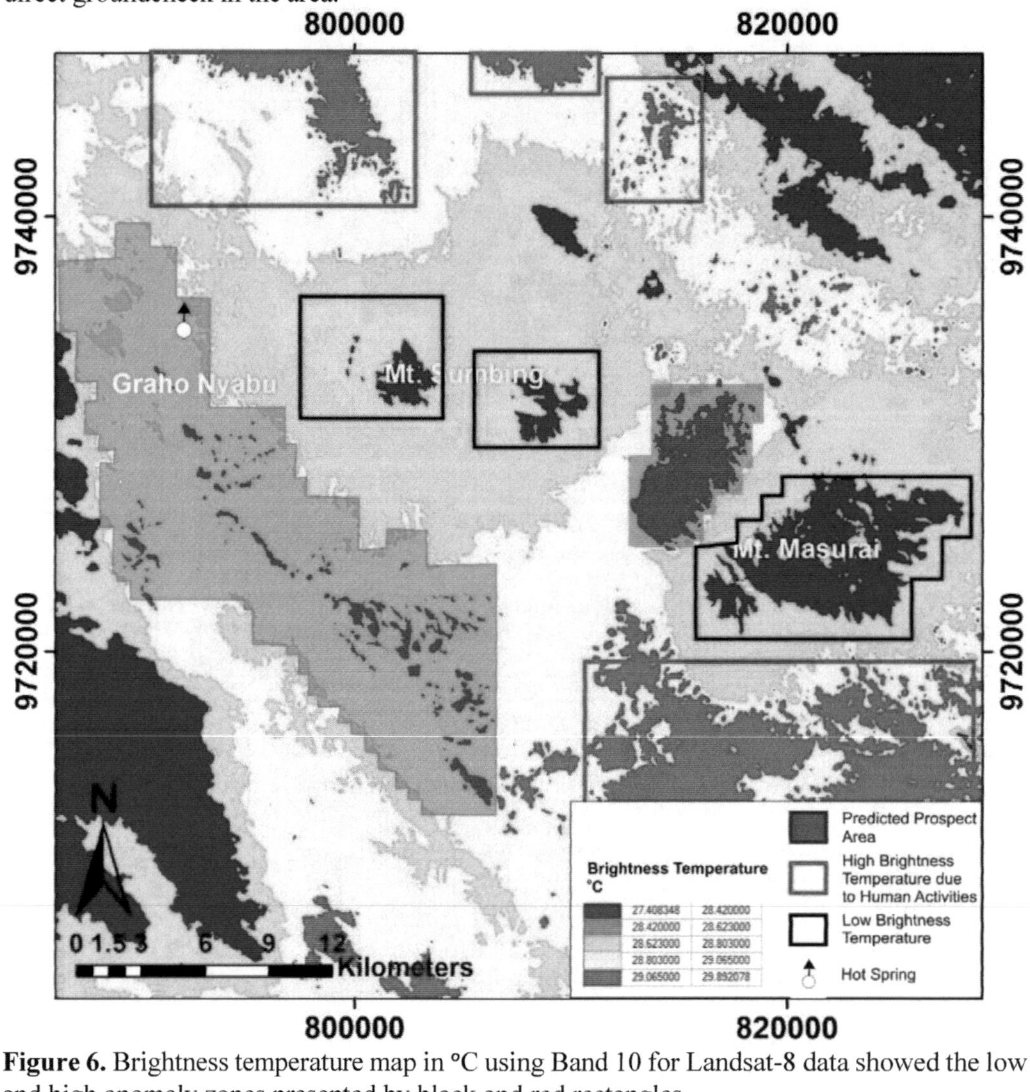

Figure 6. Brightness temperature map in °C using Band 10 for Landsat-8 data showed the low and high anomaly zones presented by black and red rectangles.

IOP Conf. Series: Earth and Environmental Science **103** (2017) 012017 doi:10.1088/1755-1315/103/1/012017

3.4. Lineament Density

Lineament density was analysed using SAR image because it shows the structure clearer than optical image [21]. The small and local structure could be detected extensively.

Geothermal prospect area is associated with high lineament density. Geothermal surface manifestation presented by hot springs were detected in the area with high lineament density, about 4.5 km/km^2, on Graho Nyabu (figure 8). The lower lineament density is about 2 km/km^2 and no surface manifestation found in the east of Graho Nyabu.

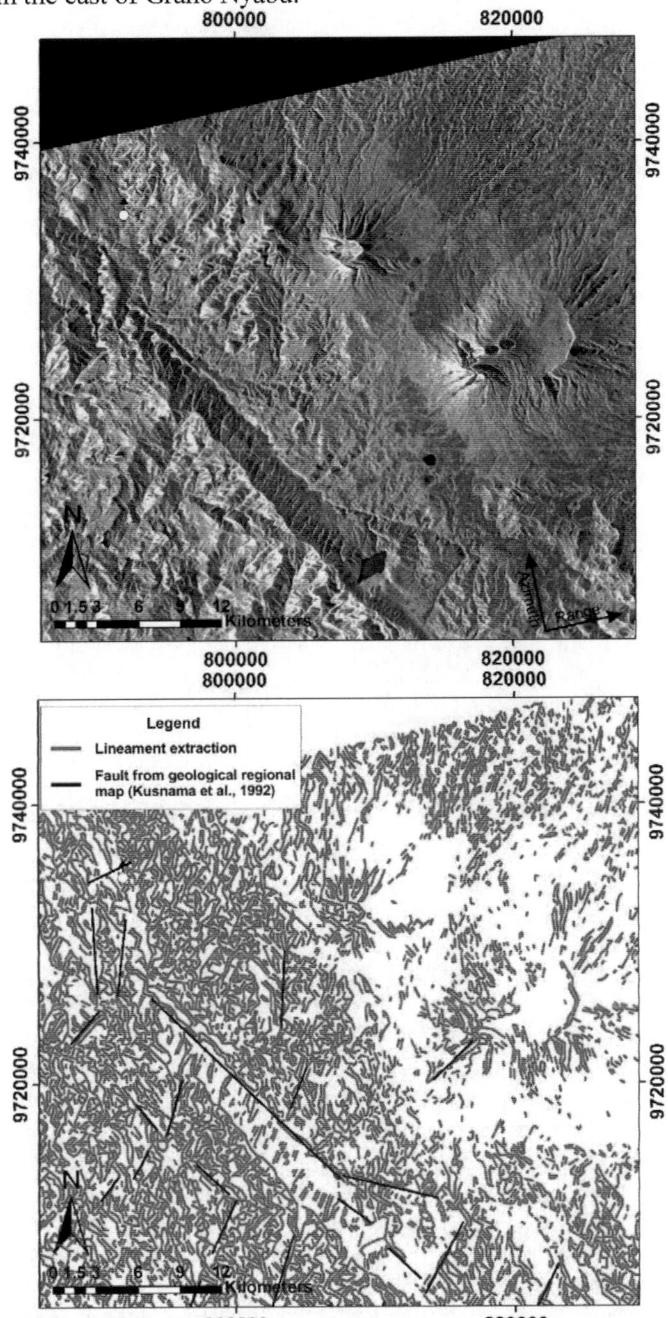

Figure 7. Processed SAR image RGB C11, C22, C11/C22 (up) and lineament extraction from the left image (down).

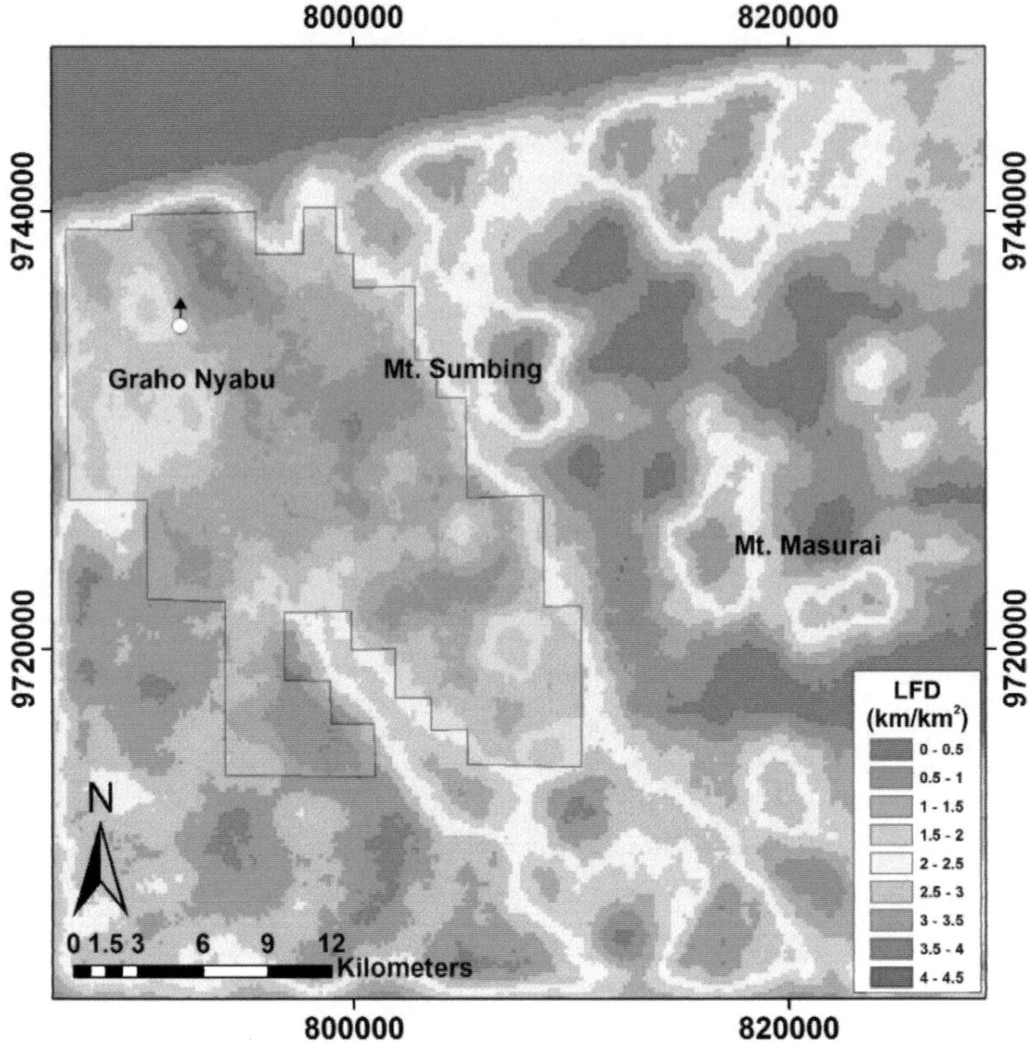

Figure 8. Lineament density image with target area presented by grey rectangle.

4. Suggested Working Area

Index overlay method used to delineated geothermal working area using processed, indexed and calculated parameters include volcanic products, geothermal surface manifestation, surface temperature, and lineament density. Volcanic product distribution can be delineated using volcano geology analysis with target area at Sumbing Crown (figure 4). Surface temperature can be estimated using Thermal Infrared Sensor and the result showed that the area with high brightness temperature is located in the west study area (figure 6). The area with high lineament density was located in Graho Nyabu, west of study area, with Great Sumatra Fault serve as a permeable structure (figure 8). The existence of hot springs indicated that there is geothermal surface manifestations. Other manifestation which predicted using NDVI and brightness temperature analyses was detected in south area of Graho Nyabu, but still need to be evaluated with groundcheck in the next steps (figure 5). Local structure that surrounding Graho Nyabu can be the system boundary, because area outside boundary structure have a low lineament density. Geothermal working area was predicted based on overlay of all these maps as depicted by figure 9.

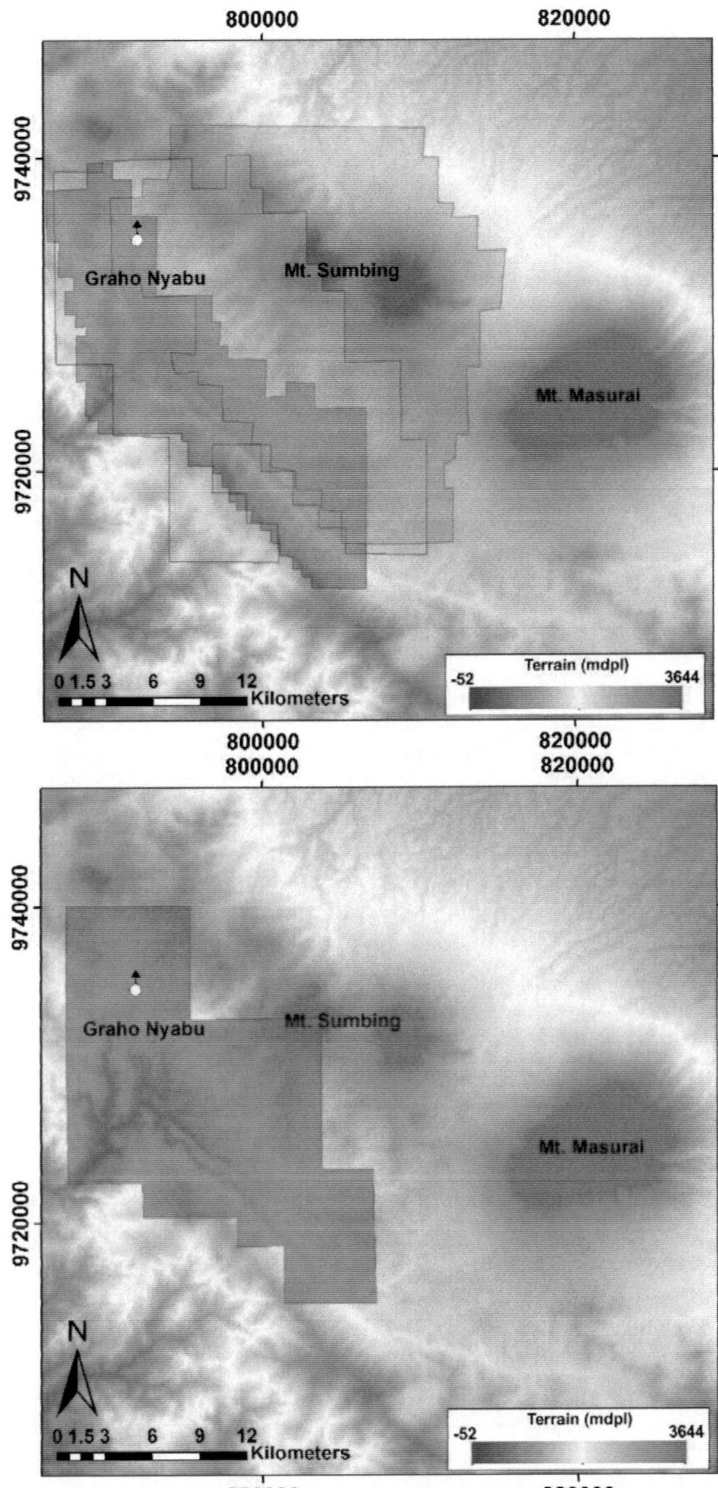

Figure 9. Index overlay map of the three parameters: brightness surface temperature anomaly in green rectangle, volcanic product in blue rectangle, and lineament density anomaly in pink rectangle (left) were used to estimate the geothermal working area (right).

5. Conclusions

Remote sensing is an effective method in an early stage of geothermal preliminary exploration. From this method, the brightness temperature, volcanic product, and lineament density could be extracted to interpret the possibility of geothermal working area based on index overlay method. The identified area was located at the west of study area (Graho Nyabu area) because of the existence of high brightness temperature about 30°C, high lineament density about 4 - 4.5 km/km2, and low NDVI value less than 0.3. Addition parameters are necessary to be included to obtain a comprehensive working area such as local government boundaries as well as forest conservation. Ground-truthing are aimed to the next step for validating the result so that delineated working potential could be used for guiding the geological, geophysical, and geochemical surveys.

References

[1] Lillesand T M Kiefer R W and Chipman J W 2015 *Remote Sensing and Image Interpretation*, 7th ed (New Jersey: John Wiley & Sons)
[2] Saepuloh A 2013 *Potential Use of Synthetic Aperture Radar (SAR) Data for Geothermal Exploration* (Yogyakarta: Proceedings of International Conference on Geological Engineering)
[3] Bromley C Ashraf S Seward A and Reeves R 2015 *Monitoring and Quantifying Heat Loss from Significant Geothermal Areas Via Remote Sensing* (Taupo: Proceedings of 37th New Zealand Geothermal Workshop)
[4] Sumintadireja P and Saepuloh A 2013 *Significant Role of Remote Sensing Technology in Geology Exploration* (Denpasar: Proceeding of the 34th Asian Conference on Remote Sensing 2013)
[5] Saepuloh A Koike K Sumintadireja P and Nugraha A D 2008 *Digital Geological Mapping using ASTER Level-1B in Relation with Heat Source of Geothermal System in an Active Volcano* (Jakarta: ISTECS Journal) **X**
[6] Saepuloh A Koike K Urai M and Sri S J T 2015 *Identifying Surface Materials on an Active Volcano by Deriving Dielectric Permittivity from Polarimetric SAR Data* IEEE Geoscience and Remote Sensing Letters (GRSL) (New Jersey: IEEE) **12** 1620-1624.
[7] Akbari D and Saepuloh A 2016 *Identification of Surface Manifestation at Geothermal Field Using SAR Dual Orbit Data* IOP Conf. Series: Earth and Environmental Science **42**
[8] Iswahyudi S Saepuloh A and Widagdo A 2014 *Delineating Outflow Zones Using Linear Features Density (LFD) Derived from Landsat Imagery at Paguyangan, Brebes, Central Java* (Bandung: Proceedings of 3rd International ITB Geothermal Workshop 2014)
[9] *Investment Catalogue 2017* 2017 Ministry of Energy and Mineral Resources Republic of Indonesia
[10] Barber A J Crow M J and Milsom J S 2005 *Sumatra: Geology, Resources and Tectonic Evolution* Geological Society (London: Memoirs)
[11] Kusnama R Pardede S, Mangga A, Sidarto 1992 *Geological Map of the Sungaipenuh and Ketaun Quadrangle, Sumatra* (Indonesia: Geological Research and Development Centre)
[12] Suryantini Wibowo H Rahman K R Woldai T 2013 *Application of Normalized Different Vegetation Index (NDVI) Method to Identify Thermal Anomaly Area from Remote Sensing* (Jakarta: Proceeding 13th Indonesia International Geothermal Convention & Exhibition 2013)
[13] Barsi J A Barker J L and Schott J R 2003 *An Atmospheric Correction Parameter Calculator for a Single Thermal Band Earth-Sensing Instrument* (New Jersey: IEEE)
[14] Chander G Markam B L and Helder D L 2009 *Summary of Current Radiometric Calibration Coefficients for Landsat MSS, TM, ETM+, and EO-1 ALI Sensors* Journal Remote Sensing of Environment. (Amsterdam: North-Holland/American Elsevier) **113** 893 – 903
[15] Verma V 2012 *Polarimetric Decomposition Based on General of Scattering from Urban Areas and Multiple Component Scattering Model* Thesis of Faculty of Geo-Information Science and Earth Observation of The University of Twente (Twente)
[16] ESA SNAP 4.0 Module 2016 Scientific Exploitation of Operational Missions

[17] Suminar W Saepuloh A and Meilano I 2016 *Identifying Hazard Parameter to Develop Quantitative and Dynamic Hazard Map of an Active Volcano in Indonesia* AIP Conference Proceeding (New York: AIP Publishing)
[18] Hochstein M P and Sudarman S 2015 *Indonesian Volcanic Geothermal System* Proceedings of World Geothermal Congress
[19] Darajat F I Maris E E P Akib A A Guswinanda H and Saepuloh A 2016 *Analyses of Landsat 8 Imageries for Preliminary Assessment to Determine Geothermal Potential Area under Torrid Zones* (Strasbourg: European Geothermal Congress 2016)
[20] Sandi Stratigrafi Indonesia Ed 1996 2010 Ikatan Ahli Geologi Indonesia
[21] Joshi N Baumann M, Ehammer A Fensholt R Grogan K Hostert P Jepsen M R Kuemmerle T Meyfroidt P Mitchard E T A Reiche J Ryan C M and Waske B 2016 *A Review of the Application of Optical and Radar Remote Sensing Data Fusion to Land Use Mapping and Monitoring* Remote Sens

Topographic map analysis to determine Arjuno-Welirang volcanostratigraphy and implication for geothermal exploration

Lestari Apriani[1], Joshua Satriana[1], Citra Aulian Chalik[1], Reza Syahputra Mulyana[1], Muhammad Hafidz[1], Suryantini[1]

[1]Magister Program of Geothermal Engineering, Faculty of Mining and Petroleum Engineering, Bandung Institute of Technology, Indonesia

Email: aprianilestari04@gmail.com

Abstract. Volcanostratigraphy study is used for supporting geothermal exploration on preliminary survey. This study is important to identify volcanic eruption center which shows potential area of geothermal heat source. The purpose of volcanostratigraphy study in research area is going to distinguish the characteristics of volcanic eruption product that construct the volcanic body. The analysis of Arjuno-Welirang volcanostratigraphy identification are based on topographic maps of Malang sheet with 1:100.000 scale, 1:50.000 scale, and a geological map. Regarding to the delineation of ridge and river, we determine five crowns, three hummocks, one brigade and one super brigade. The crowns consist of Ringgit, Welirang, Arjuno, Kawi, and Penanggungan, the hummocks comprise of Kembar III, Kembar II, and Kembar I, the brigade is Arjuno-Welirang, and the super brigade is Tengger. Based on topographic map interpretation and geothermal prospect evaluation method analysis, shows that Arjuno-Welirang prospect area have good geothermal resource potential.

1. Introduction

Geographic position of the research area is in longitude 112°29'12"– 112°37'39" and latitude 7°37'56"– 7°49'51" or UTM coordinates is in 665500 – 679250 mE and 9139000 – 9158200 mN, and under the Malang district, Mojokerto district, and Pasuruan district, East Java Province. Volcano-stratigraphy is stratigraphy related to volcanism and its products. The volcano-stratigraphy layering system are based on volcanic source, deposit type, and sequence of time relative. The volcano-stratigraphy interpretation consist in relation among topographic map, geologic regional and the Arjuno-Welirang volcano-stratigraphy analysis are using the topographic map of Malang sheet, scale 1:100.000 and scale 1:50.000. The name of extinct volcano in the west of Arjuno-Welirang is Anjasmoro. Malang Plain is considered as maar volcano[1].

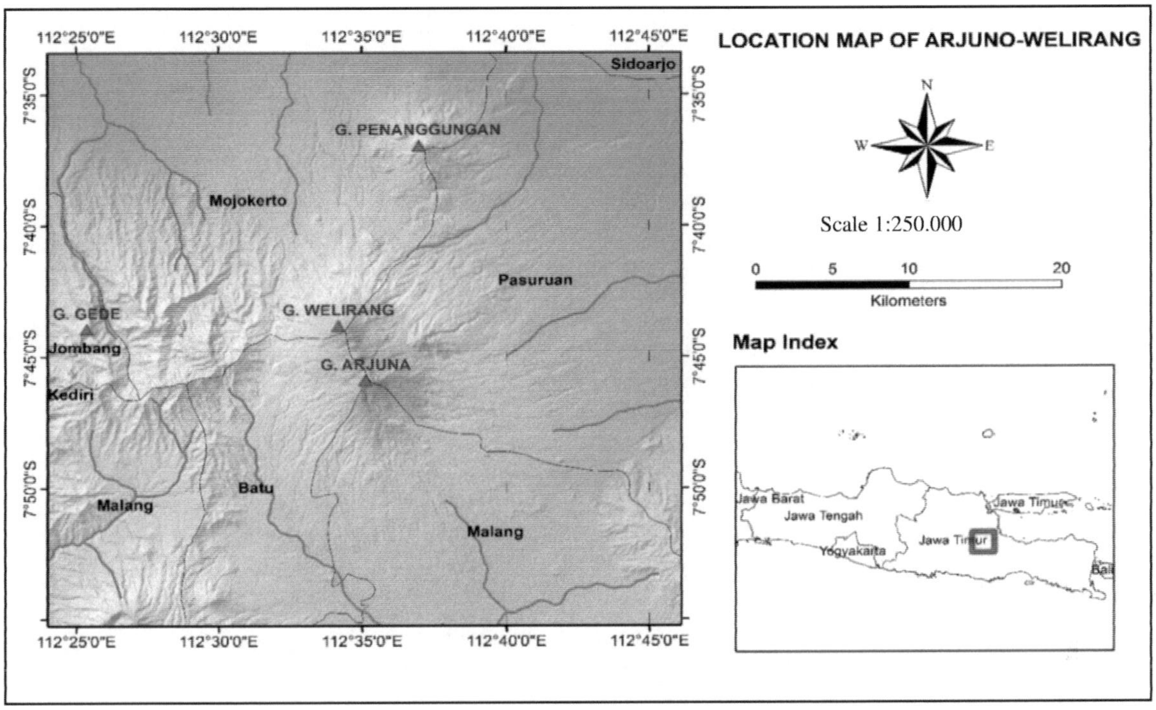

Figure 1. Location Map of Research Area.

2. Study Objective

This study objective is to identify eruption center of volcano in order to delineate volcanostratigraphy units such as Hummock (*Gumuk*), Crown (*Khuluk*), Brigade (*Bregada*), Super Brigade (*Manggala*), and Arc (*Busur*) [2]. After delineating, it is integrated with geological data, DEM data, and manifestation occurrence in the research area. By doing all of the steps, it can predict geothermal potential area for further exploration survey.

3. Method

Before applying volcanostratigraphy, geologists need to understand basic volcanology, e.g. volcanic geology concepts, genetic volcanism, volcanic landscapes, variation of volcanic rocks, volcanic facies, paleovolcanoes, and super-imposed volcanism. On the basis of a principal in geology, i.e. the present is the key to the past, a volcanic geology concept emphasizes that there is a continuous processes from magmatism through volcanism to sedimentation [3]. The research started with identify the top of mountain in a topographic map, then marking the rivers flow system, and the mountain ridges alignment. The Interpretation process is based on the topographic map and the geologic data of the research area by determining the relative ages of all volcanoes in the area. To determine the relative ages of the volcanoes, we interpreted the texture of the volcano itself and looked at the geology data.
All these information will lead us to the geothermal resource potential of this research area. Other interpretation methods is divided by each volcanic eruption center into some smaller volcano stratigraphy units. A volcanic eruption center can be devided into a hummock, several of hummock create a crown, a group crown become a brigade, some brigades and super brigades form an arc. The interpretation result is used as a reference to estimate the geothermal resource potential from a geothermal area, based on to the Geothermal Exploration of Volcano Composition flow chart [4].

IOP Conf. Series: Earth and Environmental Science **103** (2017) 012018 doi:10.1088/1755-1315/103/1/012018

A geothermal field potential can be calculated by knowing the volume of the volcano using this formula:

$$V_{volcano} = \frac{1}{3} x\, \pi\, x\, r^2\, x\, t \tag{1}$$

Which :

$V_{volcano}$: Volume of volcano (km^3)

r : Radius of volcanic cone (km)

t : Height of volcano (km)

Furthermore, it is critical to know the volcano's magma characteristic, such as the degree of partial melting for determining the magma types is produced, whether it is basaltic magma, andesitic magma, or rhyolitic magma. Knowing the volcano age, its structure, and the tectonic features are also important. If the measurement of all factors result are a valueable potential, it can be continued with a detail survey.

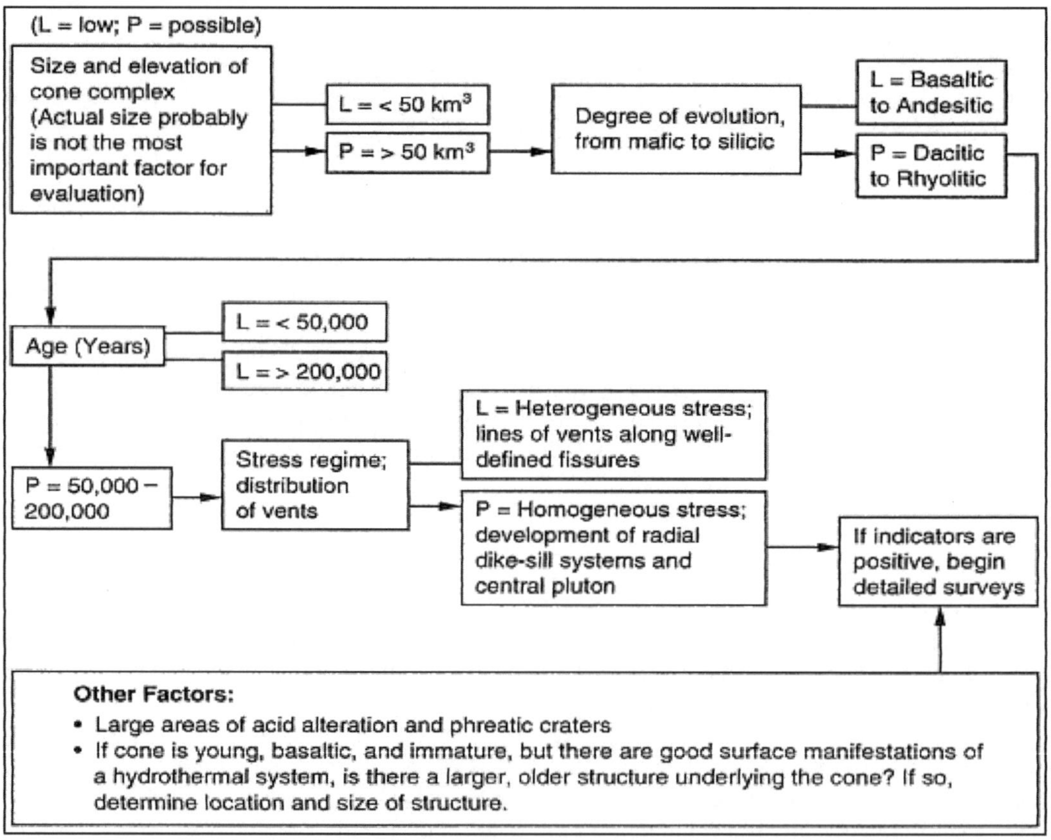

Figure 2. Geothermal Potential Evaluation on Stratovolcano [4].

4. General Geology

Based on the geological region data and East Java tectonic system, the research area is part of the Kendeng zone which is an anticlinorium formed by igneous rocks and sedimentary rocks [5].

In general stratigraphy of the research area from the oldest to the youngest are Linting followed by Arjuno and finally Welirang [6].

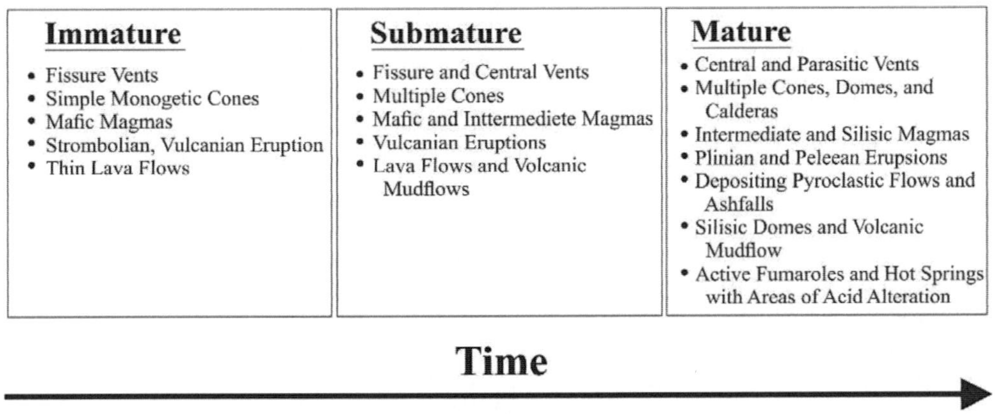

Figure 3. Geothermal Potential Based on Volcano Maturity [4].

There are 5 (five) eruption centers i.e. Mt.Welirang, Mt. Arjuno, Mt. Kembar I, Mt. Kembar II, Mt. Kembar III, Mt. Bakal and side vent eruption in Mt. Balak. Each of the volcano products are classified as magmatic eruption that produced lava and pyroclastic material.
A strong tectonic deformation affected the Arjuno-Welirang volcanic activity and caused a difficulty to find the main crater of Linting. Then, there was a big eruption ejected the old Arjuno-Welirang volcanic material and created an abyss in the old Arjuno-Welirang products, which is shown by the present of ring fracture and an collapse zone.
Furthermore the abyss would facilitate the coming of new volcanic material, the young Welirang-Arjuno, after it was formed. The volcanic process continued with the forming of Mt.Kembar I, followed by Mt.Kembar II and Mt.Bakal, Southeast trending. The last eruption that produced magmatic material was came from Mt. Kembar II located between Mt. Welirang and Mt. Arjuno.

5. Geological Structure

The geology structures in study area not only influnce by tectonic but also volcanic activity. The SE-NW trend of volcanic centers (Arjuno, Kembar I, Kembar II & Welirang) and hot spring of Padusan at Pacet Mojokerto District (SW slope of Welirang) may be controlled by subsurface fault.

a. North-South Trending Fault
   The north-south trending fault was represented by The Cangar Fault, The Puncung Fault , and The Claket Fault, which was an alignment manifestation, with the present of fault scarp, waterfall, and the topography difference, that is moderately steep.
b. North-West Trending Fault
   This fault was predicted as the antithetic of the major fault, trending southwest-northeast, paralel with the Meratus fault system. This fault was represented by The Kemiri Fault, and The Bakal Fault. The Kemiri Fault and The Claket Fault was assumed to form a graben, where the downward area was filled by the pyroclastic flow from Welirang. This faults became the trigger of thermal manifestation around Padusan area.
c. Southwest-Northeast Trending Fault
   This fault was the major fault that causing the forming of Arjuno-Welirang volcano area. And it was associated to the forming of Mt.Penanggungan, trending towards the Sidoarjo Mud. This fault was represented by The Welirang Fault, The Kembar Fault and The Balak Fault.
d. West-East Trending Fault
   This fault was represented by The Ledug Fault and The Ringit Fault.
e. The ring fault of Anjasmoro Caldera

This fault type is a normal fault, that formed a steep and circular scarp. This fault appearance can be seen clearly on the way to reach Cangar. This fault was thought to be formed from the remaining of an old caldera that appeared by volcano tectonic activity in Anjasmoro area.

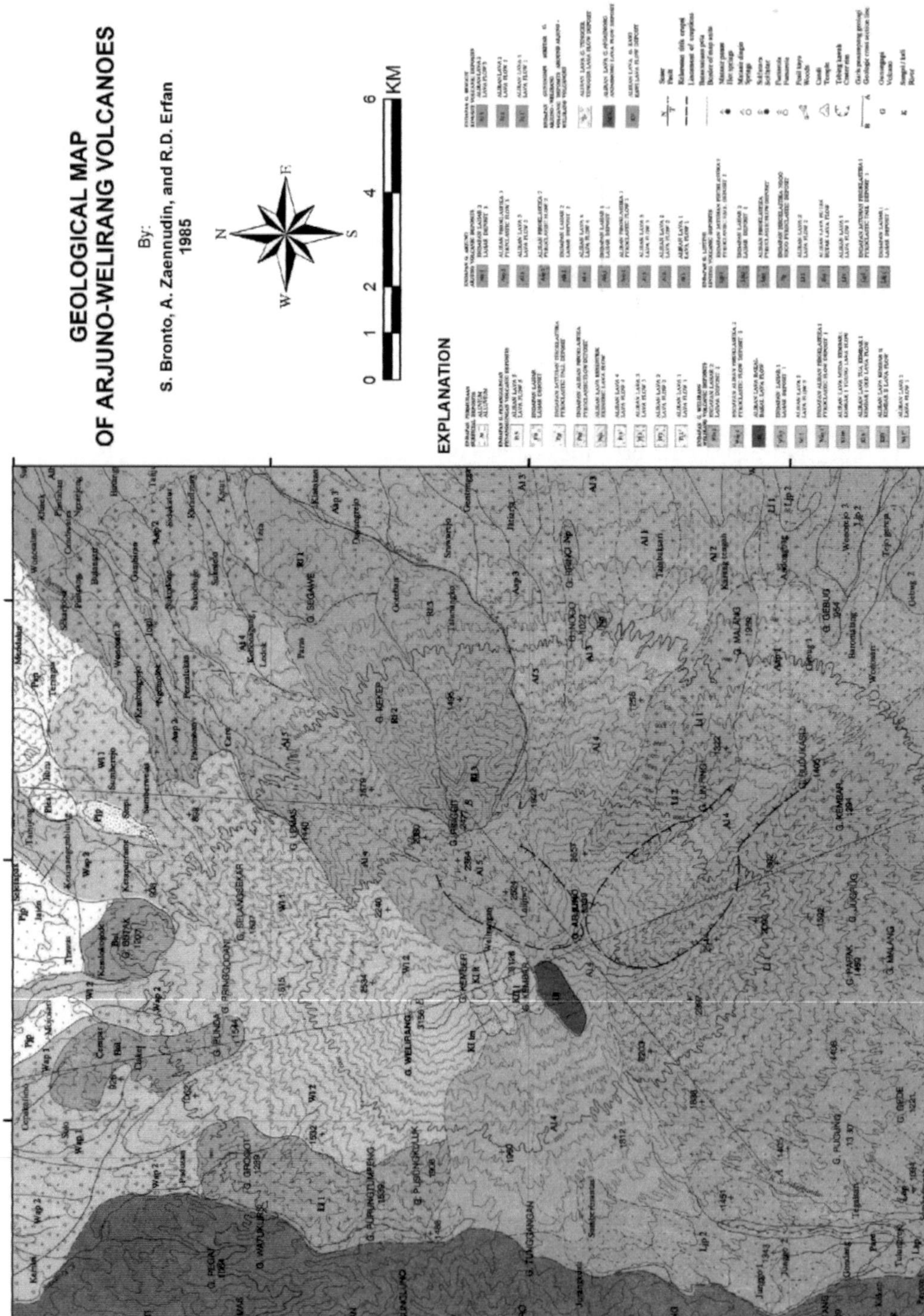

Figure 4. Geological Map of Arjuno-Welirang Volcanoes [6].

IOP Conf. Series: Earth and Environmental Science **103** (2017) 012018 doi:10.1088/1755-1315/103/1/012018

6. Result and Discussion

The topographic map analysis scale 1:100.000, identified that there are five crowns i.e. Welirang, Arjuno, and Kawi crowns, three hummocks, one brigade and one super brigade (Figure 6).The Tengger Super Brigade that is located in the Southeast is identified as the oldest unit, and followed by Kawi Crown, Arjuno Crown, Welirang Crown, and Penanggungan Crown as the youngest located in the north of Arjuno-Welirang Brigade.Based on the topographic map of Malang sheet with scale 1:50.000, focus on Arjuno-Welirang Brigade, identified there are five crowns, they are Ringgit, Arjuno, and Welirang Crown and several hummocks (Figure 7). In Arjuno-Welirang Brigade, there are three hummocks i.e. Kembar III, Kembar II, and Kembar I. Based on integrated geological data, volcano-stratigraphy and manifestation occurrence, it is predicted that Welirang Crown has geothermal potential resources.

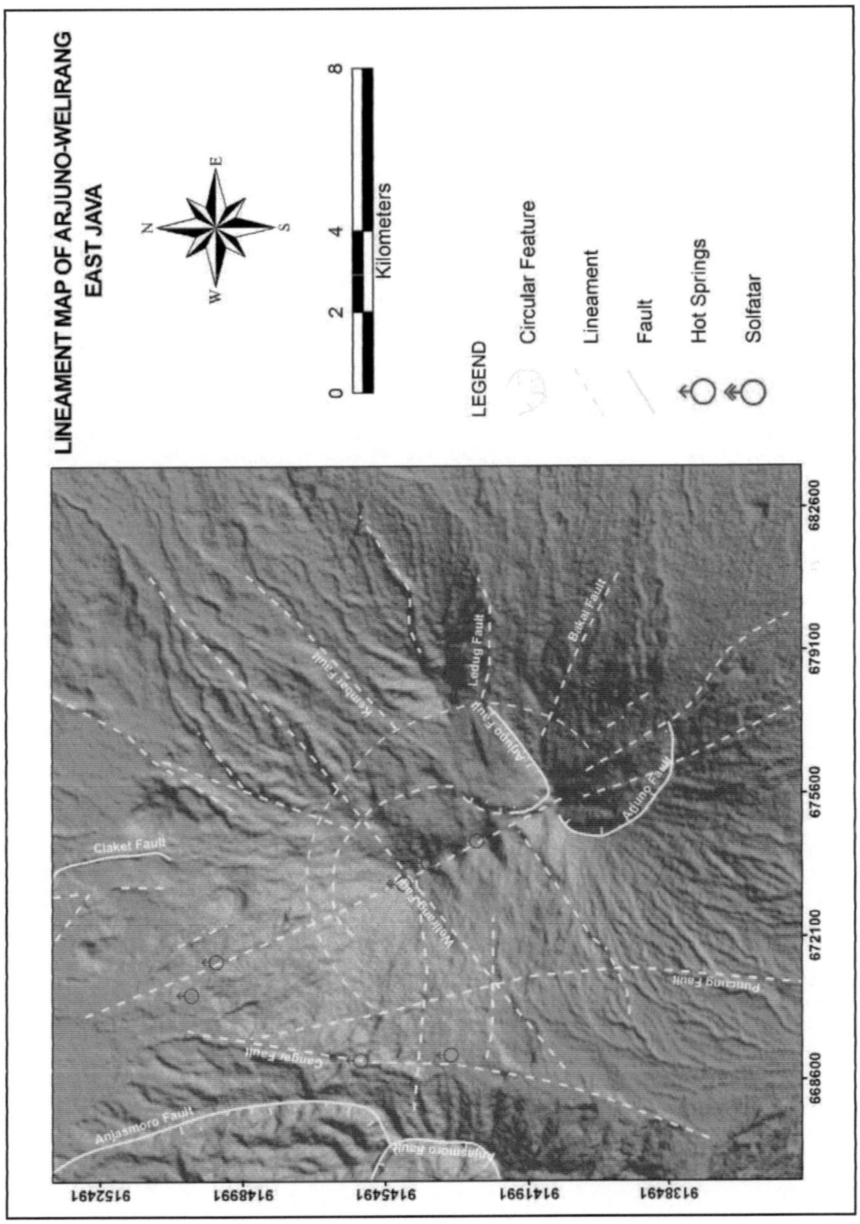

Figure 5. Lineament Map of Mt.Arjuno-Welirang [5].

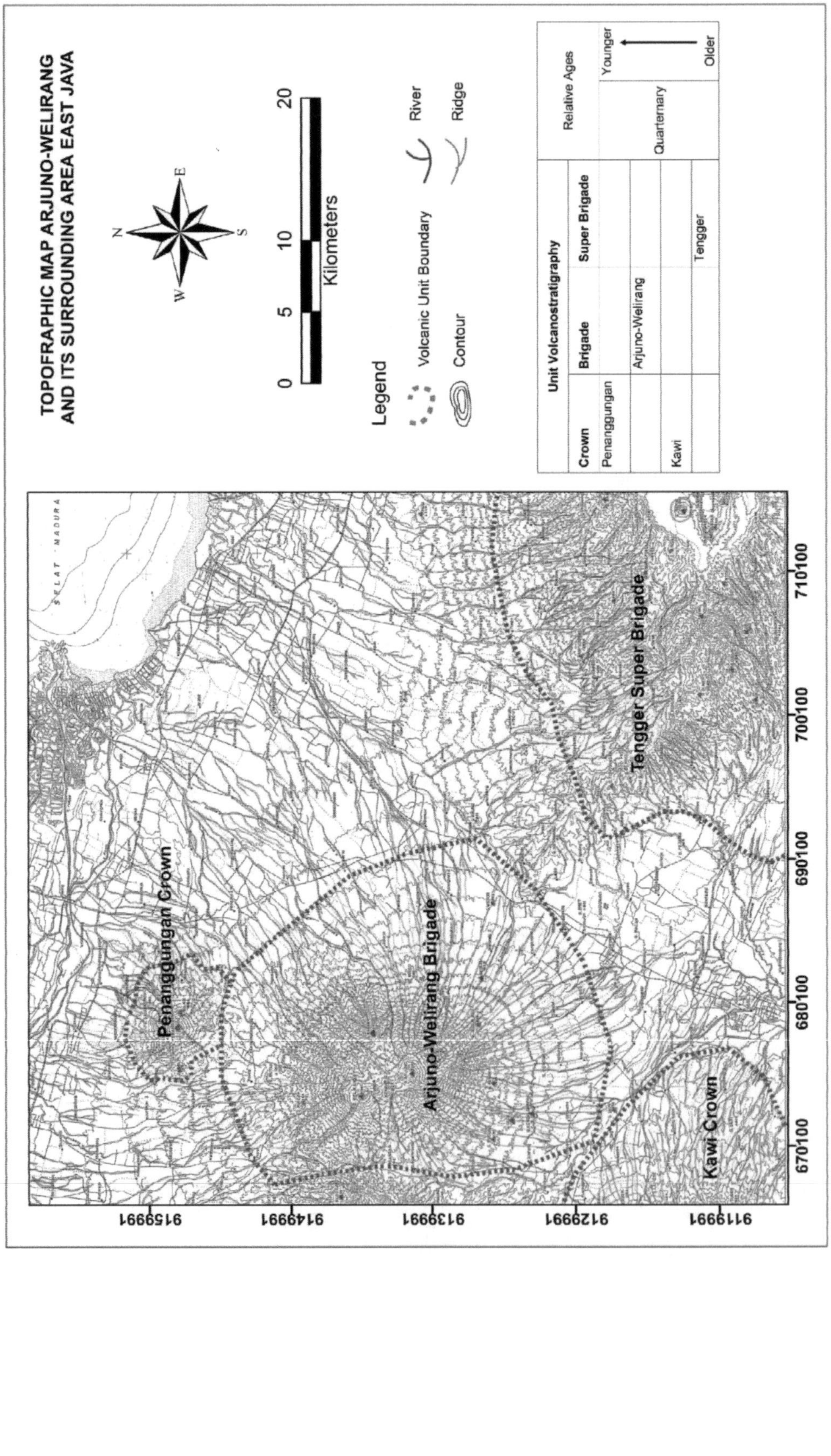

Figure 6. Topographic Map of Arjuno-Welirang and Its Surrounding Area (map scale 1:100000).

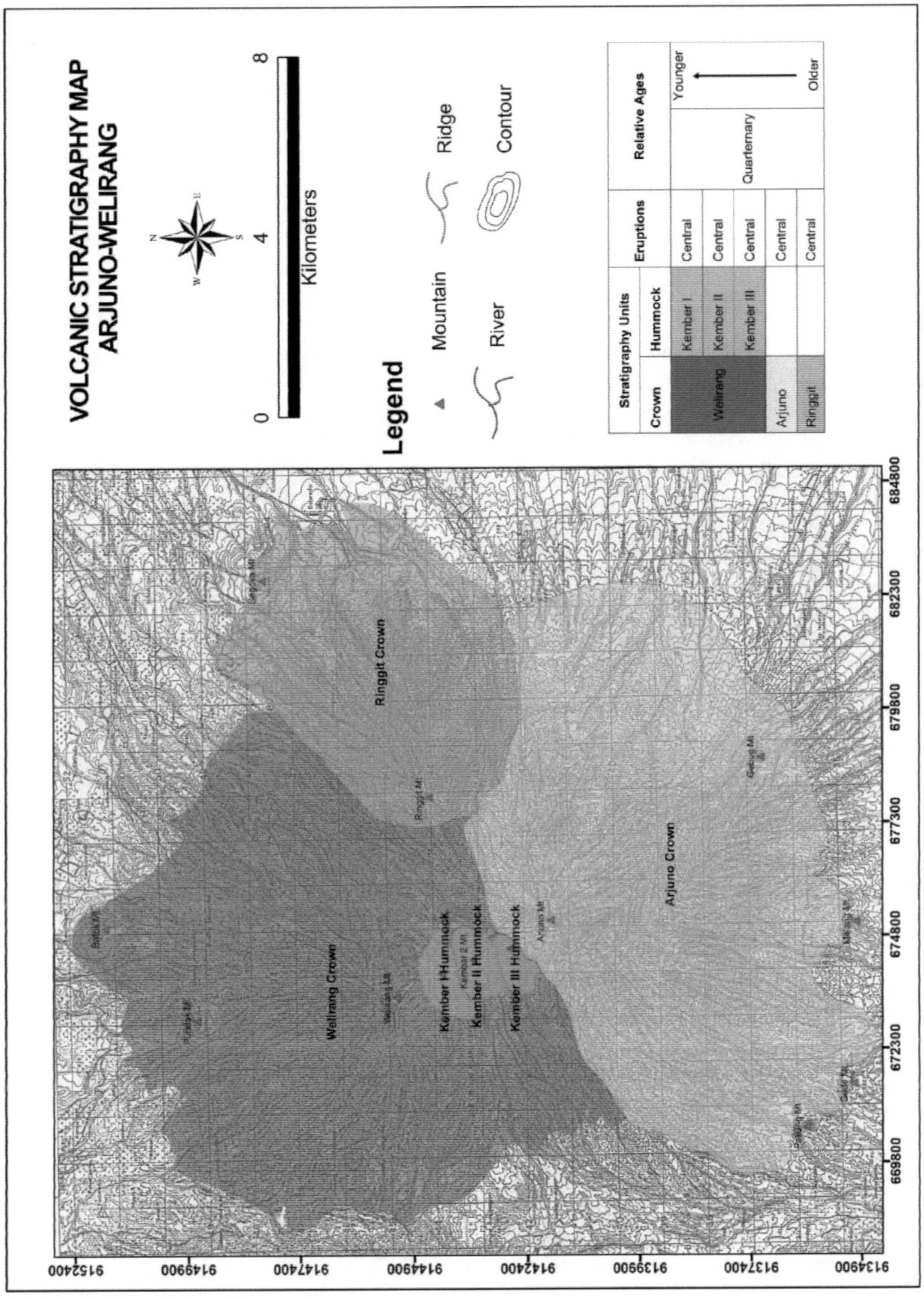

Figure 7. Topographic Map of Arjuno-Welirang and Its Surrounding Area (map scale 1:50000).

5.1 Welirang Geothermal System

Welirang has a central vent eruption that indicated the present of heat resource below the eruption center. There are several geothermal surface manifestations on the top of Mt.Welirang such as solfatara and fumarole (94-137°C) which acts as the upflow zone, and Padusan hot spring at

Kecamatan Pacet, Mojokerto District and another hot spring is called Cangar hot spring, located in the west slope of Kembar I (55°C) with neutral pH acts as the outflow zone in the northwest side of Mt.Welirang [7].The volcano stratigraphy analysis on a scale map 1:50.000 identified on Welirang crown. This crown data is used to calculate the geothermal potential using the geothermal potential evaluation on stratovolcano method [4].

The volume measurement of Welirang crown shows the diameter value is approximately 10.4 kilometers with the highest elevation is 3339 meters, and the volcano volume is estimated about 102 km^3. This volume indicates that Welirang has a big geothermal resources potential. The Welirang volcanic rocks consist of volcanic breccia, lava, and tuf, which age is Middle Quartenary, around 200.000 years old [4].

6. Conclusion

Volcano-stratigraphy units in Arjuno-Welirang Brigade has three hummocks (Kembar III, Kembar II, and Kembar I), five crowns (Ringgit, Welirang, Arjuno, Kawi, and Penanggungan), one brigade (Arjuno-Welirang), and one super brigade (Tengger). Welirang is a stratovolcano that classified into a high terrain geothermal system. The geothermal potential evaluation on stratovolcano method on Mt.Arjuno-Welirang indicate that the volcano has geothermal prospect to be developed by doing further exploration.

7. Recommendation

A geothermal exploration takes a complex information also some sufficient and qualified data for doing further exploration like geological survey for the purpose of creating a detailed geological map and the geothermal manifestations mapping, thus the interpretation process will be more accurate.

References

[1] Siebert L, Simkin, T, Kimberly P 2011 *Volcanoes of the World* ISBN: 9780520268777 University of California Press

[2] Martodjojo S 1975 *Stratigraphic Code of Indonesia* The Commission for the Stratigraphic Code of Indonesia IAGI Jakarta 19

[3] Bronto S, Sianipar J Y, and Pratopo A K 2016 Volcanostratigraphy for supporting geothermal exploration *IOP Conference Series: Science and Environmental* vol 42

[4] Wohletz K and Heiken G 1992 *Volcanology and Geothermal Energy* Berkeley University of California Press ISBN 0-520-07914-0

[5] Hadi M.N, Dedi K, Rezky Y 2010 *Buku Bidang Energi: Penyelidikan Terpadu Geologi, dan Geokimia Daerah Panas Bumi Arjuno – Welirang* (Bandung: Pusat Sumber Daya Geologi)

[6] Bronto S, Zaenudin A, Erfan RD 1985 *Geologic map of Arjuno-Welirang Volcanoes, East Java, scale 1: 70,000* (Bandung: Volcanology Survey of Indonesia)

[7] Erfan R D, Setiawan Tj Nanlohy F, Rab ES 1983 *Laporan geologi daerah panasbumi Cangar-Padusan Komplek Arjuno-Welirang, Jawa Timur* Unpublished Report VSI

Preliminary study of near surface detections at geothermal field using optic and SAR imageries

Beta Kurniawahidayati[1], Indra Agoes Nugroho[1], Reza Syahputra Mulyana[1], Asep Saepuloh[1,2]

[1]Geothermal Engineering Master Program, Faculty of Mining and Petroleum Engineering, Bandung Institute of Technology (ITB)
[2]Faculty of Earth Sciences and Technology, Bandung Institute of Technology (ITB)
Jl. Ganesha No.10, Bandung, West Java, Indonesia. 40132

Email: betakurnia@gmail.com

Abstract. Current remote sensing technologies shows that surface manifestation of geothermal system could be detected with optical and SAR remote sensing, but to assess target beneath near the surface layer with the surficial method needs a further study. This study conducts a preliminary result using Optic and SAR remote sensing imagery to detect near surface geothermal manifestation at and around Mt. Papandayan, West Java, Indonesia. The data used in this study were Landsat-8 OLI/TIRS for delineating geothermal manifestation prospect area and an Advanced Land Observing Satellite(ALOS) Phased Array type L-band Synthetic Aperture Radar (PALSAR) level 1.1 for extracting lineaments and their density. An assumption was raised that the lineaments correlated with near surface structures due to long L-band wavelength about 23.6 cm. Near surface manifestation prospect area are delineated using visual comparison between Landsat 8 RGB True Colour Composite band 4,3,2 (TCC), False Colour Composite band 5,6,7 (FCC), and lineament density map of ALOS PALSAR. Visual properties of ground object were distinguished from interaction of the electromagnetic radiation and object whether it reflect, scatter, absorb, or and emit electromagnetic radiation based on characteristic of their molecular composition and their macroscopic scale and geometry. TCC and FCC composite bands produced 6 and 7 surface manifestation zones according to its visual classification, respectively. Classified images were then compared to a Normalized Different Vegetation Index (NDVI) to obtain the influence of vegetation at the ground surface to the image. Geothermal area were classified based on vegetation index from NDVI. TCC image is more sensitive to the vegetation than FCC image. The later composite produced a better result for identifying visually geothermal manifestation showed by detail-detected zones. According to lineament density analysis high density area located on the peak of Papandayan overlaid with zone 1 and 2 of FCC. Comparing to the extracted lineament density, we interpreted that the near surface manifestation is located at zone 1 and 2 of FCC image.

1. Introduction

Preliminary survey is the earliest part in geothermal exploration. In this stage of exploration, the geologists map the prospect area to determine it's geological setting as early recognition of the geothermal system. The results of this survey are geological and surface manifestation map that can be applied to determine the focus area for next exploration and early prediction of the geothermal system that occurred in that area.

Geological mapping in geothermal exploration usually begins with remote sensing study to reduce the cost of field investigation [1] This study used optical and Synthetic Aperture Radar (SAR) images. Optical image is a remote sensing imagery that use electromagnetic radiation from optical wavelength (visible, near infra-red, short infra-red, and thermal infra-red). On the other hand, SAR image use microwave wavelength. Each image type has its advantageous and disadvantageous in the exploration. Optical image can be used to detect alterations related to hydrothermal activity even though the result will be limited by its low grade, area, erosion and scattered distribution of exposed surface manifestation due to vegetation cover, even though data from vegetation backscattering can provide alteration information with particular signal [2]. On the contrary, the SAR back-scattering intensity data represents the power return to the satellite as function of several surface parameters such as roughness, gradient slope, magnetic, and dielectric property of the ground regardless atmospheric condition [2]. This study conducts remote sensing analysis using both image to produce a low cost, easily understood preliminary study of near surface manifestation in Papandayan geothermal field, Garut, West Java, Indonesia.

Papandayan geothermal field located in Garut District, West Java, Indonesia has been selected as study area because there is some existence of geothermal manifestations from volcanic system (figure 1) in the form of extensive alteration, high temperature fumarole, and hot springs on the surface [3]. Papandayan is an "A" type volcano in Indonesian classification, which means this volcano has eruption history after 1600 [4][5]. The latest eruption noted in 2002 produced phreatic - phreatomagmatic eruption [3]. Those volcanic activities affirm that there is possibility of geothermal system below the surface of Papandayan area. There is a possibility that not all thermal manifestation can proceed to the surface due to the lithology, structural, or vegetation coverage boundary. Therefore, this research will give a preliminary study of geothermal near surface manifestation that will be useful to asse geothermal potential development in Papandayan.

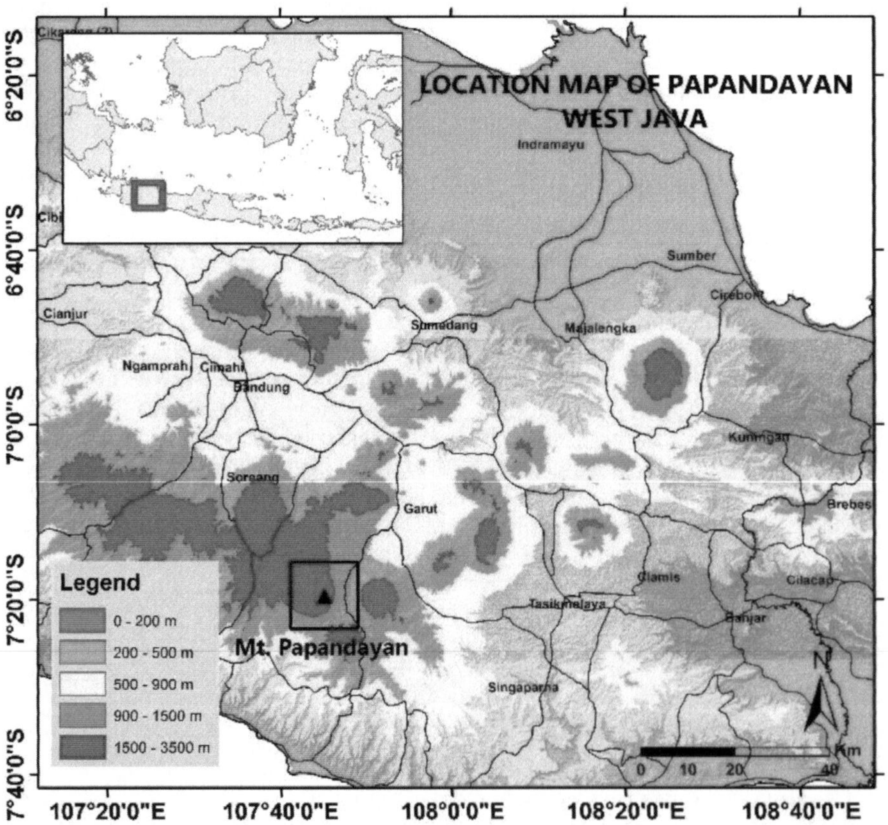

Figure 1 Location map of study area at Mt. Papandayan, West Java, Indonesia presented by black rectangle overlaid on DEM SRTM 30 m.

2. Data Collection and Processing

We used Landsat-8 OLI TIRS and SAR images in this study. The acquisition date of Landsat-8 OLI TIRS is 6 September 2016. The radiometric calibration, atmospheric correction, and geometric correction were applied before visual interpretation using ENVI 5.2 software.

Radiometric calibration is a process to improve image quality by converting digital number to radiance for further processing. Then the process was followed by an atmospheric correction to reduce atmosphere effect and convert the radiance to the reflectance. We used the *Fast Line of sight Atmospheric Analysis of Spectral Hyperubes* (FLAASH) to reduce atmospheric effect in the image. The FLAASH is a module in ENVI software to correct the effect of water vapor, oxygen, carbon dioxide, methane, ozone, molecule, and aerosol distribution in the atmosphere [6] as cited [7]. The geometric correction was also applied to locate pixel image into referenced coordinate system. Corrected and calibrated optic image was used to recognize the geothermal surface manifestation. The geothermal surface manifestation recognition was performed based on visual interpretation to the texture, tonal, and colour in the True Colour Composite (TCC) RGB for band 4,3,2 and False Colour Composite (FCC) RGB for band 5,6,7 (figure 2A and 2B). The identified geothermal surface manifestations were then compared to Normalized Different Vegetation Index (NDVI) image to obtain the vegetation condition at the ground surface. Geothermal potential area was interpreted to the area with low vegetation cover tendency as a result of heat from below the surface [8].

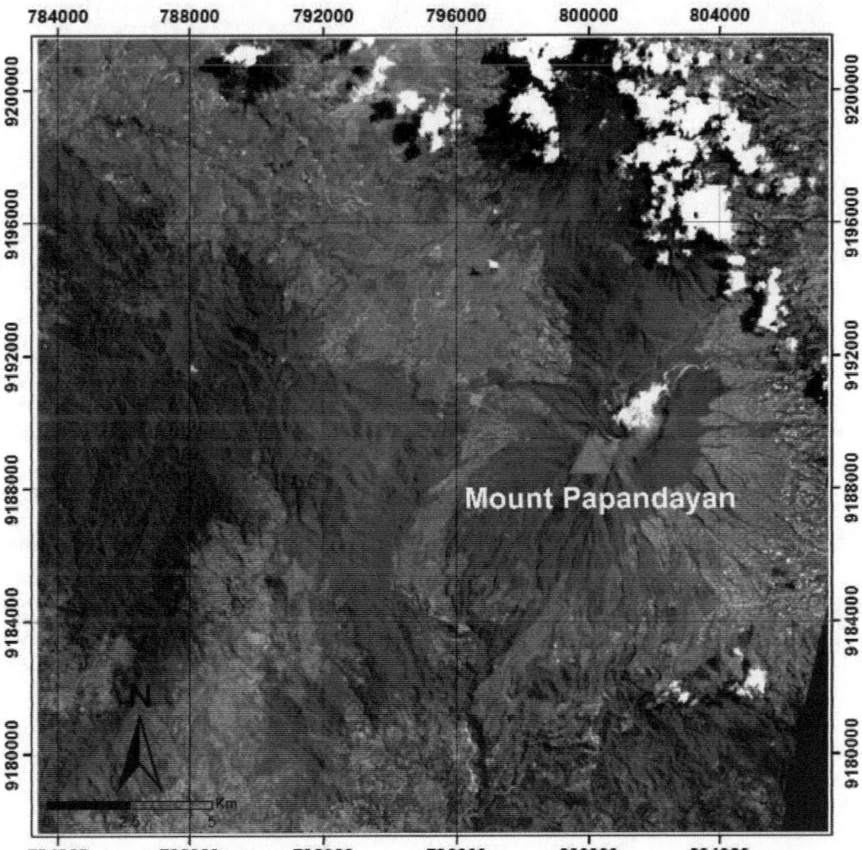

a

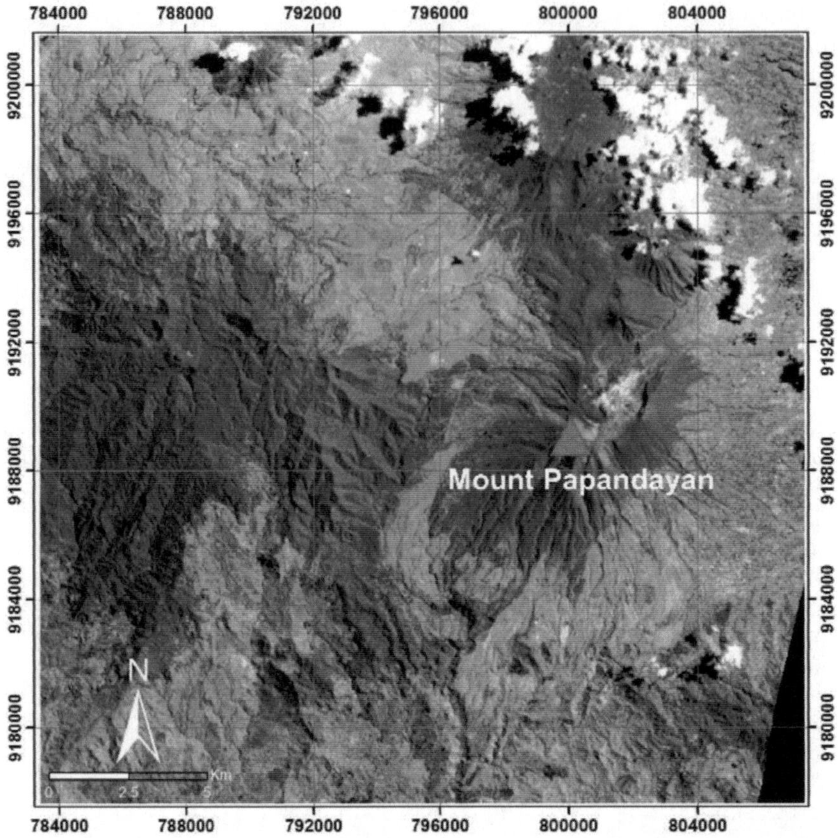

b

Figure 2 True Color Composite (TCC) RGB for band 4,3,2 at Mt. Papandayan and surrounding area (a) and False Color Composite (FCC) RGB for band 5,6,7 (b) showed that TCC image represents actual color compared FCC image have better visualization of surface condition with better contrast and clearer texture.

We also used the SAR image for analysing near surface structures in this study. Advanced Land Observing Satellite(ALOS) Phased Array L-type Synthetic Aperture Radar(PALSAR) images level 1.1 were selected due to long wavelength (=23.6 cm) and capability to reduce the canopy vegetation effect [9] with detail information of the data is listed in table 1.

Table 1 Detailed of ALOS PALSAR data used in this research

No	Data Type	Orbit	Date
1	Dual Polarization	Descending	10/6/07
2	Dual Polarization	Descending	10/6/07
3	Dual Polarization	Ascending	19/10/08
4	Dual Polarization	Ascending	19/10/08

ALOS PALSAR image processing including radiometric calibration, multi-looking, polarimetric, and geometric correction were applied using Sentinel Application Platform (SNAP) software.

Radiometric calibration function is to correct SAR image so that the produced pixels values are represent the real backscatter from earth reflection [10]. Multi-looking is calibration process to improve the image focus. This step is necessary to enhance the image quality and transformed into Single Look Complex (SLC) data [1].

Geometric corrections in SNAP software are including ALOS deskewing to transform reversed geometry of ALOS 1.1 image into Doppler geometry and terrain correction for geocode the image by correcting SAR geometric distortion with digital elevation model (DEM) and create map projection. Corrected ALOS PALSAR image was used to analyse geological structure and lineament density in ascending and descending orbits of ALOS PALSAR Images as depicted by figure 3 [1].

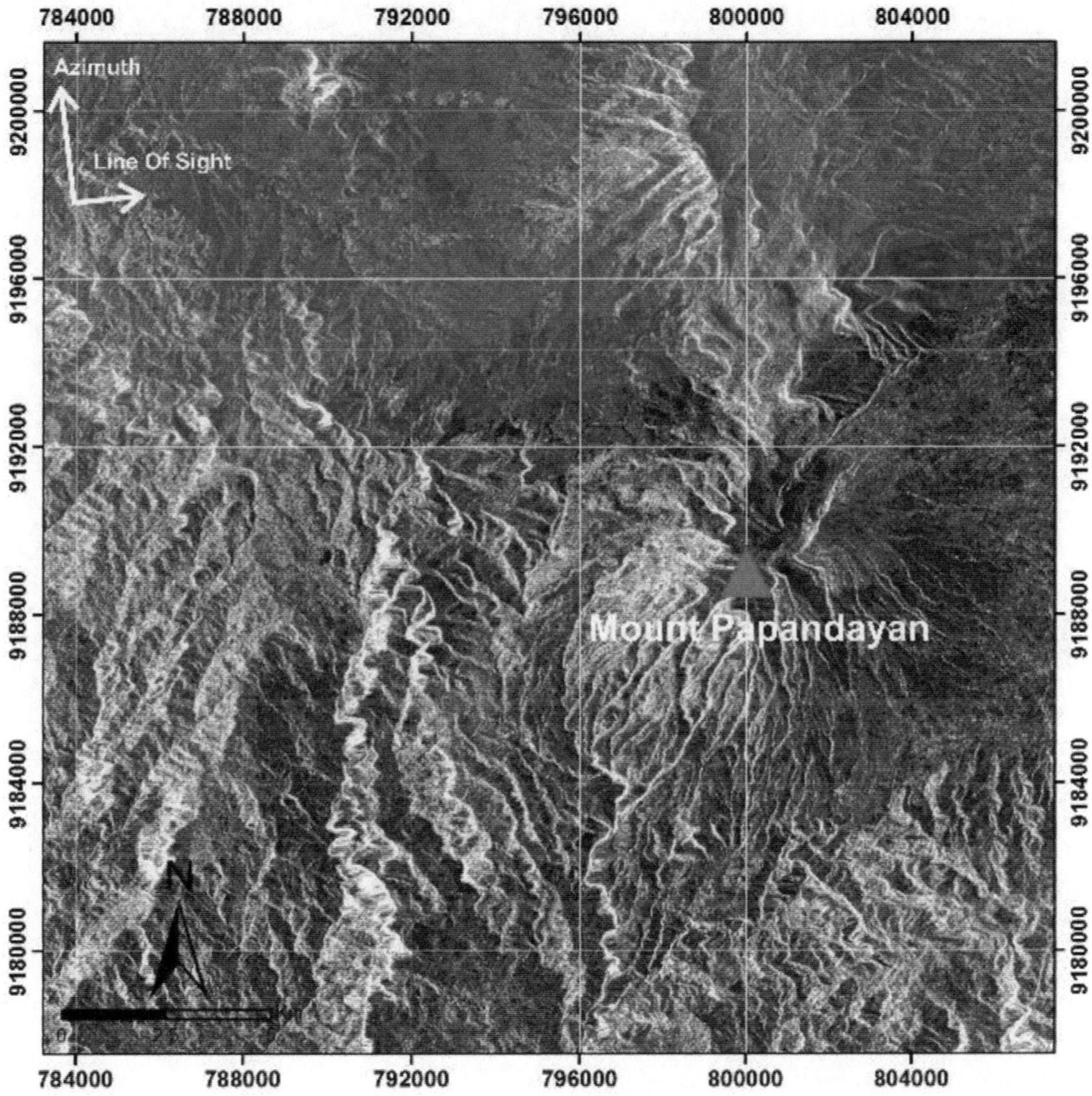

a

Figure 3 Images of ALOS-PALSAR ascending (a) and descending (b) orbits show surface feature in opposite Line of Sight (LOS).

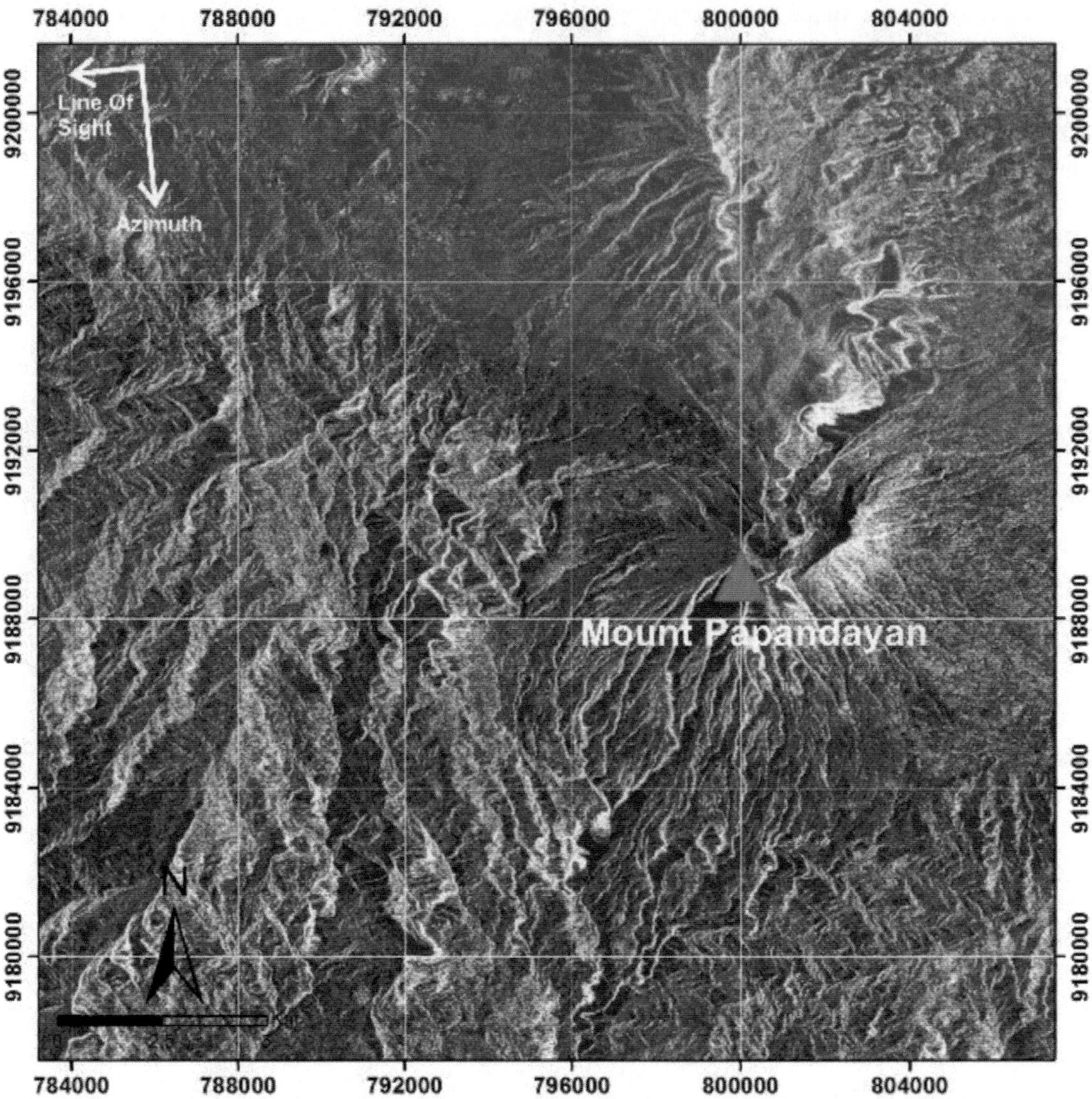

b

Figure 3 Images of ALOS-PALSAR ascending (a) and descending (b) orbits show surface feature in opposite Line of Sight (LOS).

3. Detected Near Surface Manifestation

3.1 Geothermal Surface Feature on Landsat-8 OLI TIRS

Remote sensing imagery has various bands with different wavelength. Theoretically, the longer a band's wavelength it makes a better penetration into the earth [11]. Detection of surface manifestation using TCC (figure 2a) of Landsat 8 data shows 6 zonation according to its RGB color (figure 5) with reflectance value detail in table 2. According to the delineation could be seen that TCC image can differentiate surface feature quite well by its visual, but still can't differentiate the manifestation and cloud that shows same color.

Table 2 Reflectance value of TCC for the classified geothermal surface feature zones

Zone	Reflectance					
	R (Band 4)		G (Band 3)		B (Band 3)	
	Min	Max	Min	Max	Min	Max
1	0.003698	0.016398	0.005432	0.017202	0.007629	0.019887
2	0.004721	0.025978	0.006080	0.024120	0.008372	0.023749
3	0.004332	0.011643	0.005819	0.012078	0.008096	0.011458
4	0.003722	0.041292	0.005568	0.038431	0.007912	0.036169
5	0.003550	0.014269	0.005228	0.014023	0.007535	0.014548
6	0.003642	0.018908	0.005388	0.015928	0.008084	0.014602

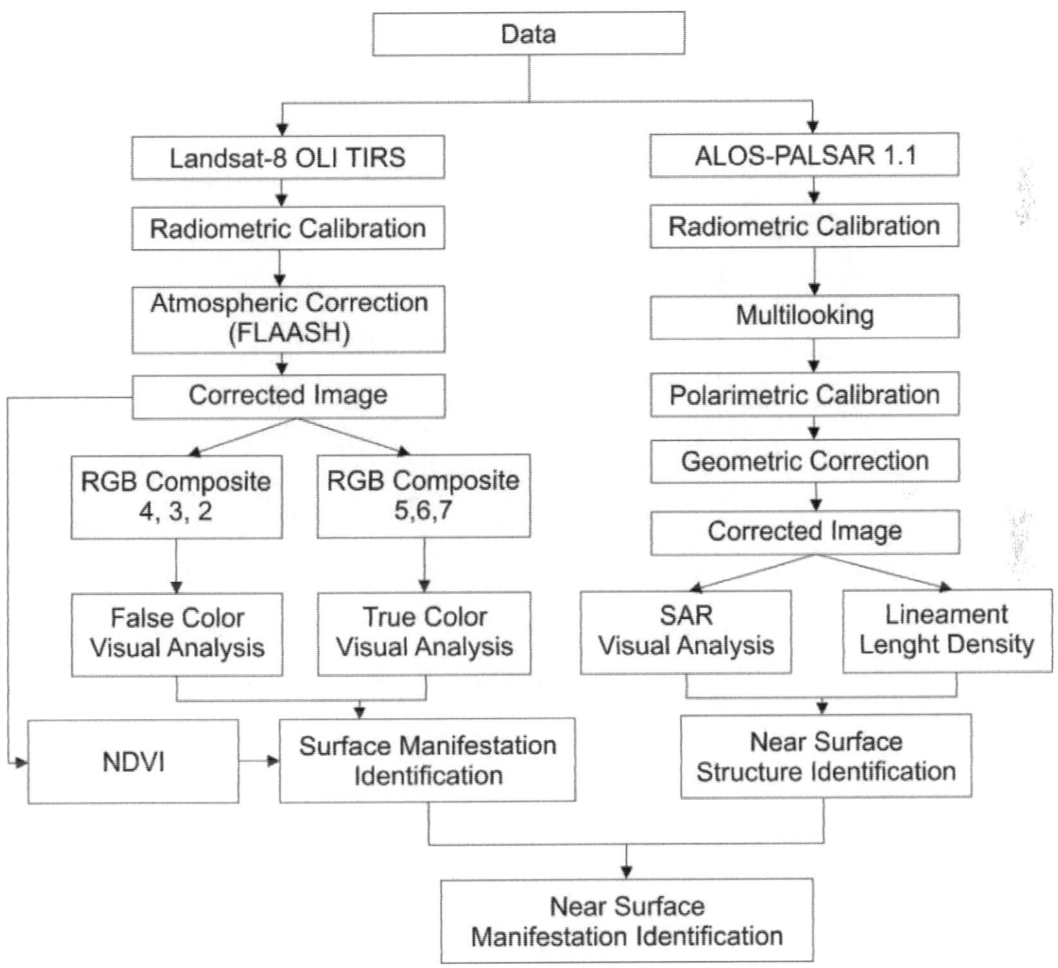

Figure 4 Work flow of optical and Synthetic Aperture Radar (SAR) remote sensing data processing and methodology.

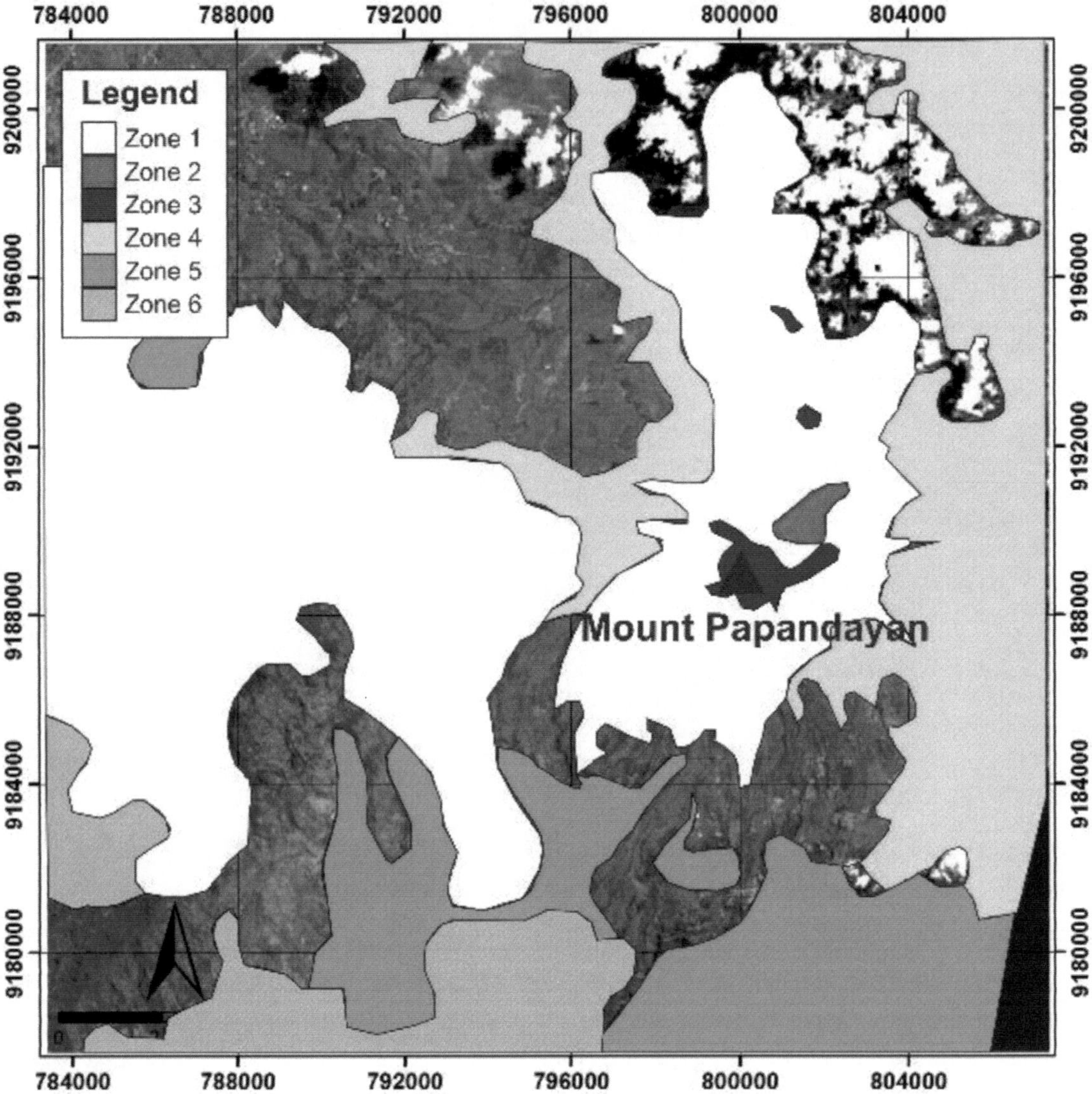

Figure 5 Surface feature classification based on TCC Landsat 8 OLI TIRS at Mt. Papandayan and surrounding area

Meanwhile detection of surface feature using FCC of Landsat 8 data shows 7 zones (figure 6) with detailed reflectance value mentioned in table 3. Based on delineation of FCC image, it shows different result with TCC image where FCC composite image has more zonation than TCC image. It proved that FCC shows texture, tone, and color better for visualization of surface feature zonation.

Table 3 Reflectance value of FCC for the classified geothermal surface feature zones.

Zone	Reflectance					
	R		G		B	
	Min	Max	Min	Max	Min	Max
1	0.008879	0.043441	0.004585	0.030404	0.002382	0.021748
2	0.008128	0.041081	0.005343	0.028651	0.003318	0.018798

3	0.010170	0.031580	0.005416	0.015168	0.003187	0.008098
4	0.010001	0.059358	0.004240	0.052905	0.002473	0.044049
5	0.008121	0.046428	0.004437	0.023461	0.002415	0.014150
6	0.012838	0.048101	0.005313	0.024359	0.002687	0.017803
7	0.011043	0.055535	0.005432	0.027421	0.003189	0.019138

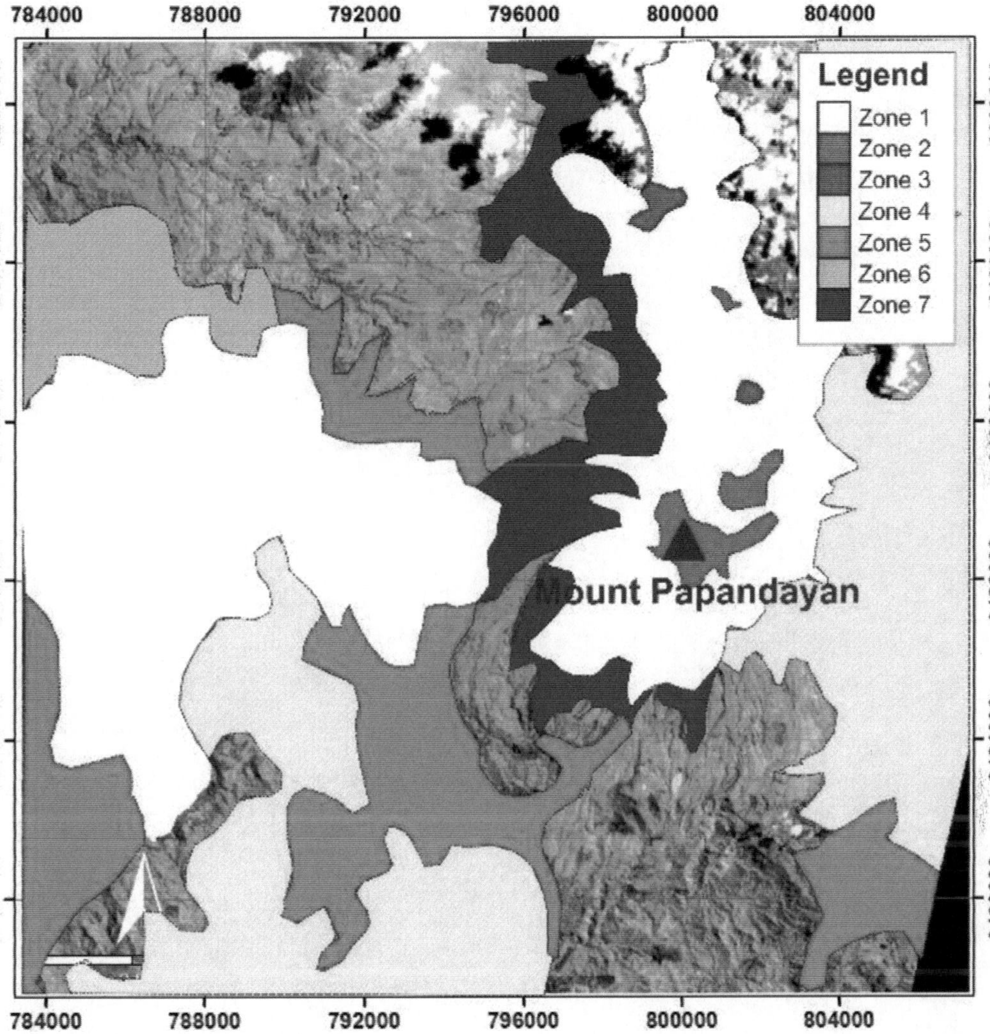

Figure 6 Surface feature classification based on FCC Landsat 8 OLI TIRS at Mt. Papandayan and surrounding area.

From optic image surface feature delineation, target of surface manifestation in Mount Papandayan geothermal field categorized as zone 1, 2 and 3 according to its color, tone, and texture. Zone 1 in TCC have a wider cover area than zone 1 in FCC. On the other hand, zone 2 and 3 both images show different shape and location because of different color and texture. Comparison of both composites shows that FCC has more precise delineation than TCC for detect surface manifestation.

Based on NDVI analysis compared with TCC, low vegetation zone assembled in the peak of Mt. Papandayan, particularly in Tegal Alun-alun Crater, Mas Creater, Manuk Creater and in the north-east side of NDVI map (figure 7). Rare vegetation zone was marked with black rectangle while cloud coverage marked with white rectangle. Rare vegetation could be caused by hydrothermal activity from geothermal system beneath the ground. According to earlier geological study [3] areas marked with black rectangles in NDVI map is the location of altered ground and fumaroles from Papandayan hydrothermal activity. Except in the Papandayan peak, none of surface manifestations were found in the other area.

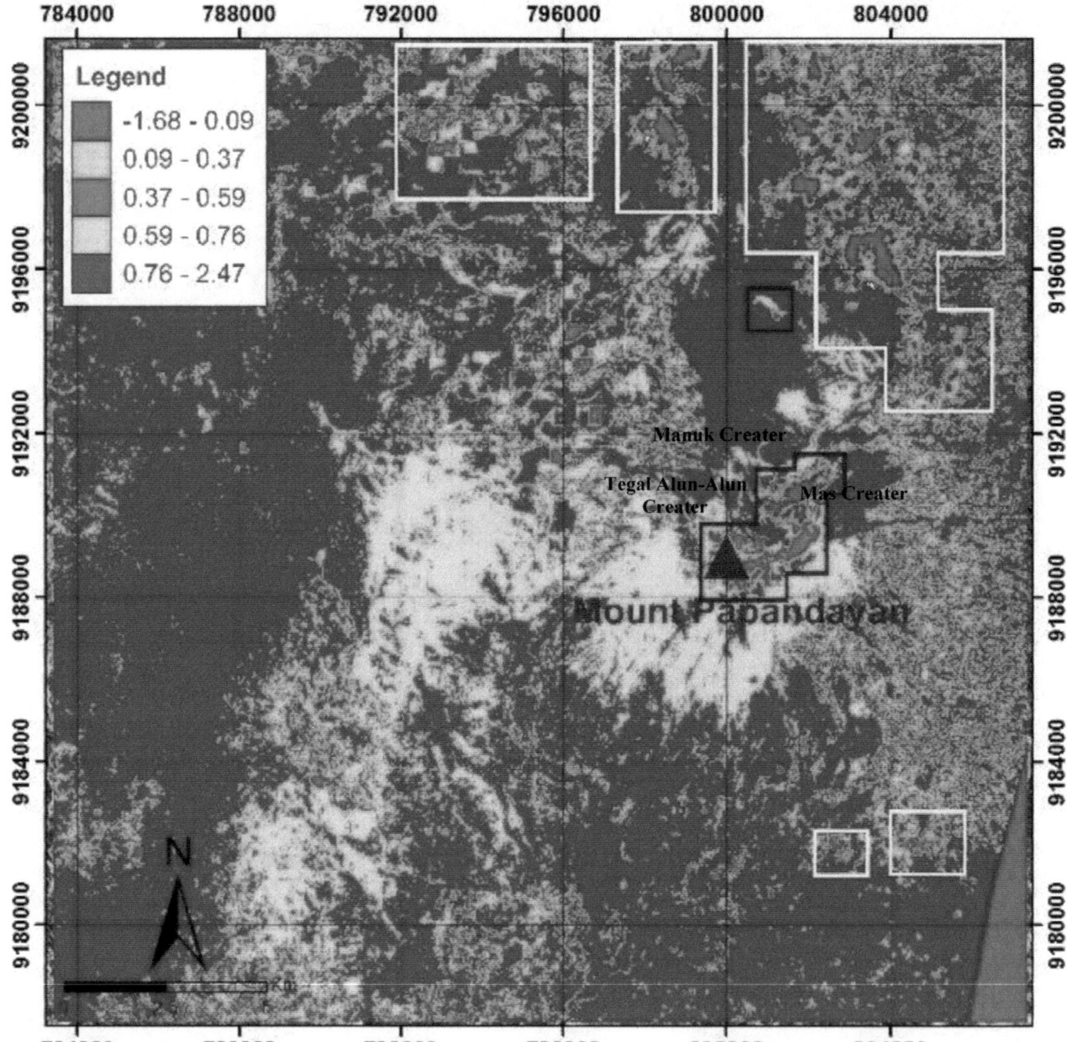

Figure 7 Image of TCC Landsat 8 OLI TIRS Band 4,3,2 of Mount Papandayan and surrounding area overlaid to NDVI map. Black rectangles are rare vegetation zone and white rectangle represents cloud coverage zone.

3.2 Near Surface Structure on ALOS PALSAR

Ridges morphology is one of geological structure feature. Geological structure makes the effected rock get weaker and create valley because of intensive erosion. In the other hand, ridges created from resistant rock that not effected by the erosion, or because of fold structure. Lineament study of ridges in Papandayan field (figure 8A) generate LLD map (figure 8B) overlaid with main structure according to geological map[12] reveal that geological structures happened to be located at low – high line density area. That area also be the place where of geothermal manifestation found. LLD analysis shows that high lineament density does not always show high structural activity.

Comparison between LLD map with optical visual study reveal that zone 1 in TCC and FCC composite overlaid with mostly high lineament density area, while zone 2 in both image overlaid with geological structure and surface manifestation.

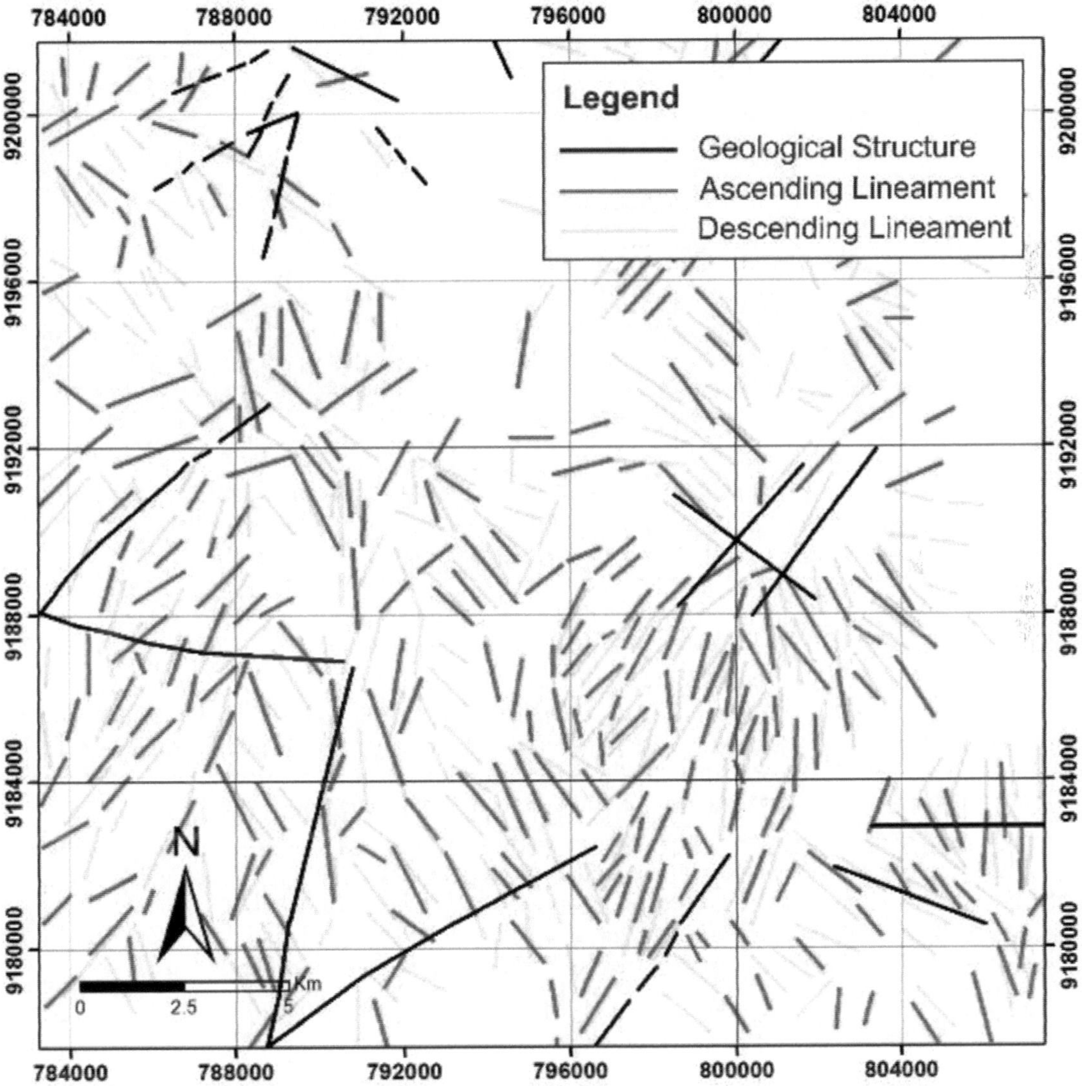

a

Figure 8 Detected lineament using ALOS PALSAR in ascending and descending orbit images (A) and lineament density map (B) overlaid with geological structures from regional geological map [7]

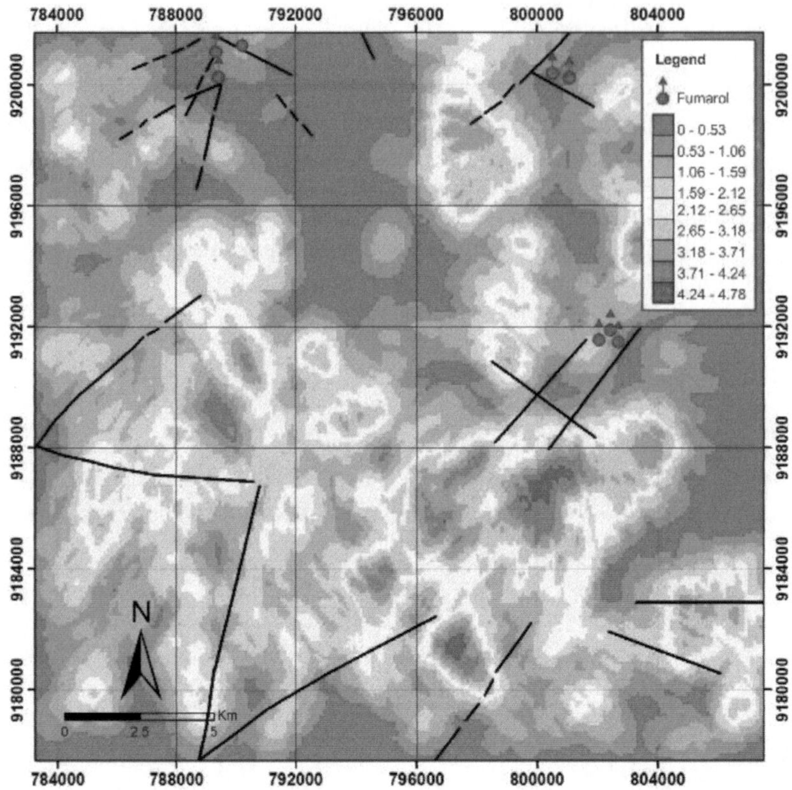

b

Figure 8 Detected lineament using ALOS PALSAR in ascending and descending orbit images (a) and lineament density map (b) overlaid with geological structures from regional geological map [7]

3.3 Detected Near Surface Geothermal Manifestation

According to visual analysis of Landsat-8, this remote sensing imagery effectively to detect geothermal surface manifestation with FCC image better than TCC. Comparison of visual analysis result of landsat-8 with lineament study of ALOS-PALSAR shows that high lineament density zone of Papandayan and surrounding area does not overlay with particular zone in FCC image delineation. We assumed that lineament density map detects near surface manifestation that cannot be detected with visual analysis. Near surface manifestation may cover with vegetation canopy or surface weathering. According to FCC visual analysis, NDVI and LLD detected near surface geothermal manifestation is located at Manuk Creater, Tegal Alun-Alun Creater, Mas Creater, and papandayan southeast and southwest flank (figure 9). To prove the existence of manifestation near the surface a ground check is necessary.

4. Conclusion

Visual analyses using optical image in Papandayan geothermal field shows that surface feature delineation using RGB composite 5,6,7 (False Color Composite, FCC) is better than RGB composite 4,3,2 (True Color Composite, TCC) presented by its 7 zonation while TCC only have 6 zonation. On the flip side for detecting vegetation coverage on the surface, TCC show a better result than FCC because of its ability to show real color of surface condition. Green represents vegetation canopy while the other color represents another surface condition.

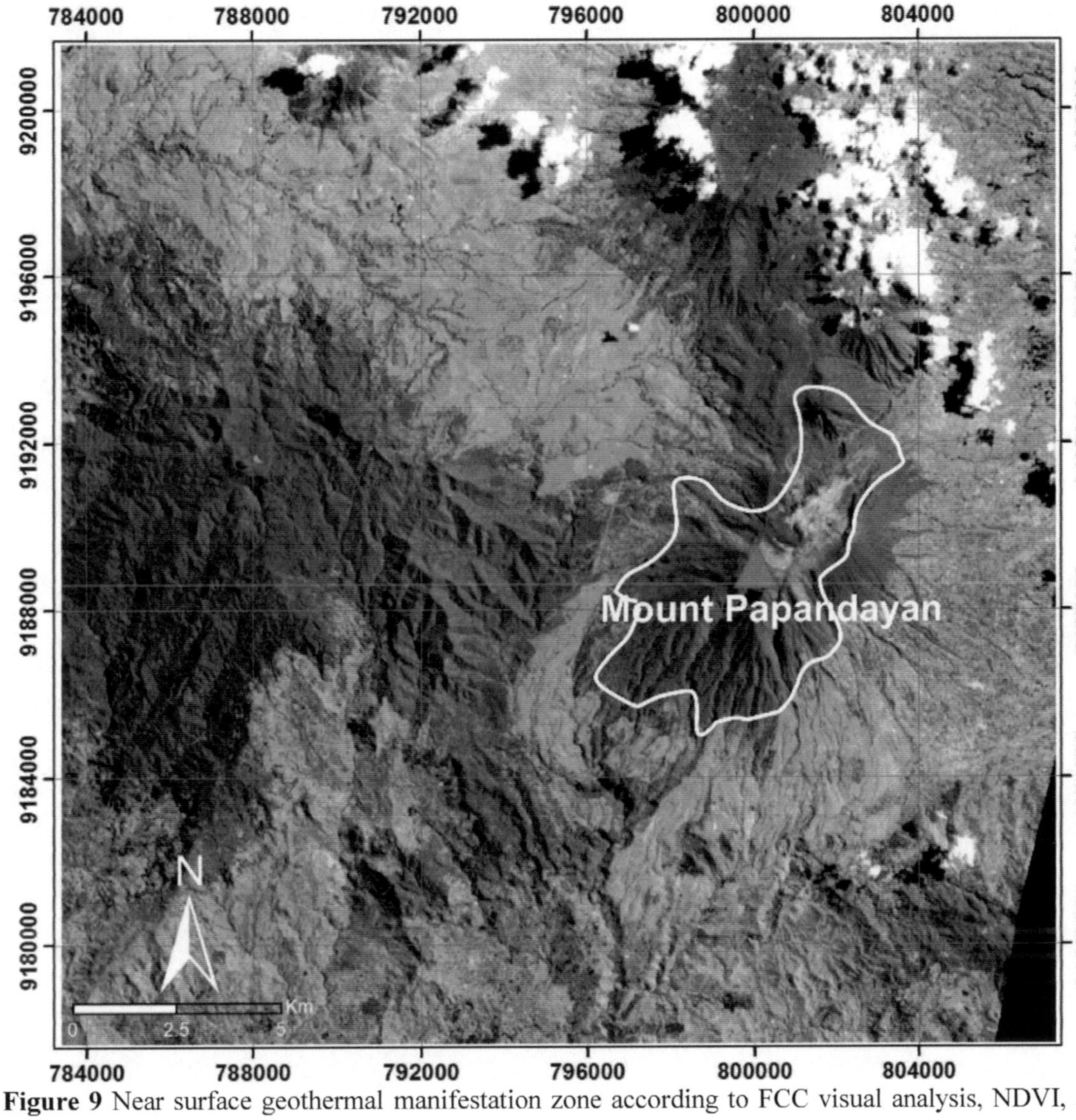

Figure 9 Near surface geothermal manifestation zone according to FCC visual analysis, NDVI, and LLD marked with white outline.

FCC image have better contrast that makes easier to delineate manifestation near the surface. A longer wavelength of FCC 0.85 – 2.29 μm than TCC 0.45 -0.67μm is predicted to have high penetration into surface layer. On the other hand, visible band with shorter wavelength than short infrared wavelength caused optimum surface layer detection such as vegetation coverage.

The Normalized Different Vegetation Index (NDVI) and visual analysis of TCC indicate low vegetation zone that might related with geothermal activity in mount Papandayan located at Tegal Alun-Alun Creater, Manuk Creater and Mas Creater.

The lineament length density analysis point that some high lineament density area associate with local structure however highest length density doesn't relate with any structure or geothermal manifestation in the flank of Mount Papandayan. Those high line density zone might come from near surface manifestation that covered with vegetation or weathering.

All things considered that the TCC, FCC, NDVI, and LLD were effective to recognize near surface manifestation of geothermal field in a preliminary state. TCC surface feature delineation that validate with NDVI detected surface manifestation area prospect that associate with low vegetation index. Combination of FCC and LLD provided accurate detection to near surface manifestation zone in geothermal field. The next step of this study is validation with ground check and available geological and geophysical data to validate the result.

References

[1] Sumintadiredja P and Saepuloh A 2013 *Significant Role of Remote Sensing Technology in Geology Exploration* Bali: 34th Asian Conference On Remote Sensing.
[2] Melosh Glenn 2016 *Simple Method To Use Spectral Remote Sensing Data In Geothermal Exploration* Stanford: 41St Workshop On Geothermal Reservoir Engineering
[3] Nasution Asnawir 2008 Geothermal Geophysical Study Of Mt. Papandayan, Garut District, West Java Indonesia *Proceedings of The 8Th Asian Geothermal Symposium*
[4] Van Padang N 1951 *Catalogue of Volcanic Activity and Solfatara Fields*
[5] Kusumadinata K 1979 *Data Dasar Gunungapi Indonesia* Bandung: Ditjen Vulkanologi
[6] Bernstein L S Adler-Golden S M Sundberg R L and Ratkowski A J 2006 Improved Reflectance Retrieval from Hyper- and Multispectral Imagery Without Prior Scene or Sensor Information *Proc. SPIE Remote Sensing of Clouds and the Atmosphere XI* vol 6362 pp 63622P
[7] Nugraha Bangkit Adhi Asep Saepuloh and Ketut Wiantika *Ground Surface Changes Detection At Geothermal Surface Manifestations Using Multi-Temporal Landsat Imageries* Bandung: Institut Teknologi Bandung
[8] Suryantini Wibowo H Rahman K R Woldai T 2013 Application of Normalized Different Vegetation Index (NDVI) Method to Identify Thermal Anomaly Area from Remote Sensing *Proceeding 13th Indonesia International Geothermal Convention & Exhibition 2013*
[9] Saepuloh A Koike K Omura M 2012 *Applying bayesian decision classification to Pi-SAR polarimetric data for detailed extraction of the geomorphologic and structural features of an active volcano* IEEE-GRSL
[10] Saepuloh A Susanto A Sumintadireja P Suparka E 2015 *Characterizing Surface Manifestation Of Geothermal System Under Torrid Zone Using Synthetic Aperture Radar(SAR) Data* Melbourne: World Geothermal Congress 2015
[11] S C Liew 2001 *Principles Of Remote Sensing – Microwave Frequency* Singapore: Centre for Remote Imaging, Sensing and Processing National University of Singapore
[12] Alzwar M Akbar N and Bachri S 1992 *Peta Geologi Lembar Garut dan Pamengpeuk* Bandung: Geological Research and Development Centre

Preliminary study of Songa-Wayaua geothermal prospect area using volcanostratigraphy and remote sensing analysis

Ribka Asokawaty[1], Indra Nugroho[1], Joshua Satriana[1], Muhamad Hafidz[1], Suryantini[1,2]

[1] Geothermal Engineering Master Program, Faculty of Mining and Petroleum, Bandung Institut of Technology (ITB)
[2] Faculty of Earth Sciences and Technology, Bandung Institut of Technology (ITB), Jalan Ganesha no. 10, Bandung 40132, West Java, Indonesia

email: ribkaasokawaty1410@students.itb.ac.id

Abstract. Songa-Wayaua geothermal prospect area is located on Bacan Island, Northern Molluca Province. Geothermal systems in this area associated with three Quartenary volcanoes, such as Mt. Pele-pele, Mt. Lansa, and Mt. Bibinoi. Based on literature study, five surface manifestations such as hot springs and alteration occurred within this area. The active manifestations indicate that Songa-Wayaua area has potential geothermal resource. This study objective is to evaluate Songa-Wayaua geothermal system on preliminary study stage by using volcanostratigraphy and remote sensing analysis to delineate the boundary of geothermal system area. The result of this study showed that Songa-Wayaua prospect area has four heat sources potential (e.g. Pele-pele Hummock, Lansa Hummock, Songa Hummock, and Bibinoi Hummock), controlled by geological structure presented by Pele-pele Normal Fault, and had three places as the recharge and discharge area which are very fulfilling as a geothermal system.

1. Introduction

The tectonic position of Indonesia which is on the Ring of Fire resulting a big potential of geothermal resources. This renewable energy is expected to replace the fossil energy later. Geothermal energy development on the eastern part of Indonesia is less developed than on the western part of Indonesia, whereas eastern part of Indonesia is still lack of electricity. Thus, it needs more geology explorations in order to find new potential area of geothermal resources. Determination of an area that has geothermal prospect usually depends on the geothermal surface manifestations. However, the occurrence of these geothermal manifestations is very limited and sometimes absent, so it can be obstructed in determining the geothermal system, especially in volcanic regions and green field area.

The preliminary studies which can be used before the field survey are volcanology study and remote sensing analysis [1]. Volcanostratigraphy study aims to classify the eruption products based on source. The classification based on the eruption center is important to localize the areas that most likely located near a heat source or associated with the magmatic intrusions that are still active or dormant. Then, remote sensing analysis aims to find thermal anomalies on the research area so we can obtain the conceptual model of permeable zone beneath the research area, also the lineaments study so knowing the trend and stress regime which developed at the research area [2, 3]. Furthermore, the study objective is going to identify each components of geothermal system in Songa-Wayaua prospect area (figure 1).

Songa-Wayaua is located on the eastern part in Indonesia, which needs more electricity, and has a prospect of geothermal potential. Tectonically, Songa Wayaua is very complex and the chemistry of the volcanic rocks different with Halmahera region [4]. Most of the basement rock formation of Songa Wayaua is from continental crust (Australia plate), while the volcanic rocks of Songa Wayaua was the result of magmatism related to movements along the sinistral Sorong fault [5].

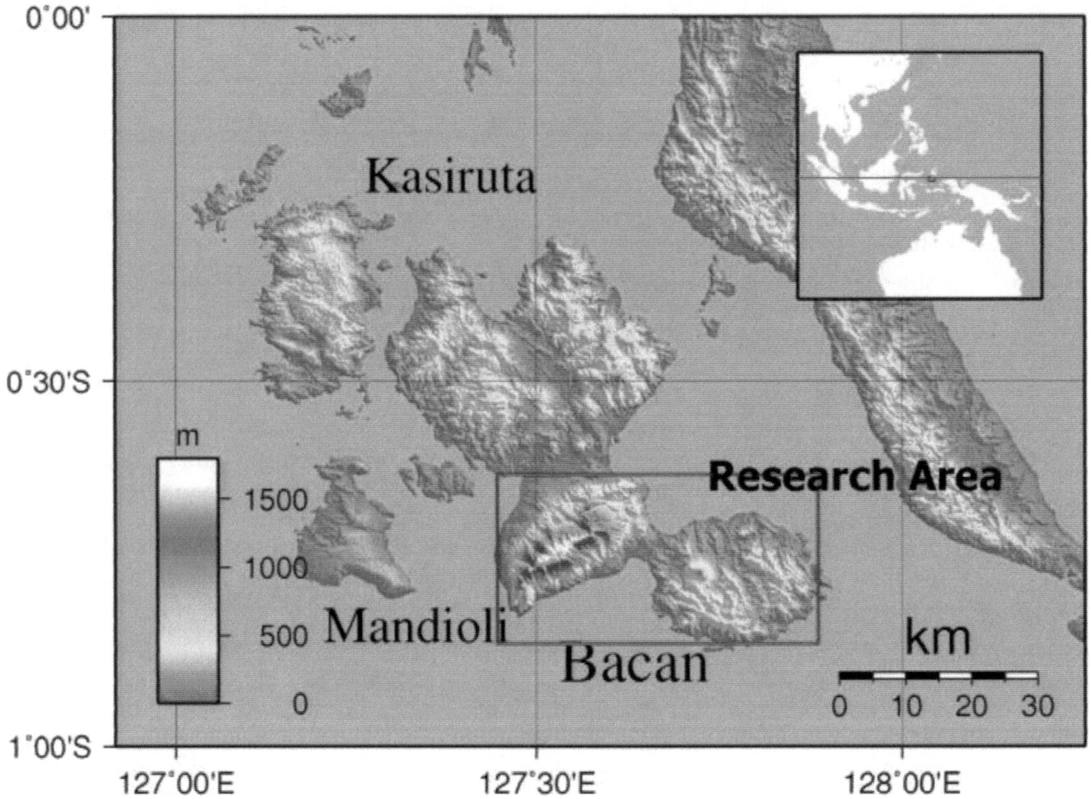

Figure 1. The research area at Songa Wayaua, Bacan Island, South Halmahera Regency, North Mollucas Province, presented by red rectangle.

2. Methods

2.1. Volcanostratigraphy Analysis

The basic method of geological mapping in volcanic region is volcanostratigraphy. This method is used because the volcanic facies rapidly changes both lateral and vertical succession. The volcanistratigraphy unit which similar to Formation is Crown, while smaller unit is called Hummock. The group consist of two or more Crowns are called Brigade, and Brigade series compose an Arc [6].

The initial stage of volcanostratigraphy unit determination is delineating non-volcanic rocks to define the base of volcano in research area, so that we become more focused on volcanic rocks mapping.

The first step is overlaying the same scale of topographic maps with the regional geological map. The next step is identifying of volcanostratigraphy unit using the topographic maps (scale 1:100,000). Volcanostratigraphy units are detemined based on the eruption centre, ridge lineaments, stream or drainage system, and also base of volcano boundaries.

IOP Conf. Series: Earth and Environmental Science **103** (2017) 012020 doi:10.1088/1755-1315/103/1/012020

The stage of volcanostratigraphy unit analysis begins with local highest peaks identification which commonly symbolized by the triangle symbols on the maps and with specific elevation. Furthermore, identification of circular features that is generally characterized by circular pattern elevation contours with concentric pattern elevation contours in the middle. These structures characterize the existence of circular depression, caldera, or crater as the central of major eruption.

In this research, volcanostratigraphy analysis was based on topographic maps: Silang sheet [7] scale 1:100,000, North Mollucas, which next is compared with geological map: Bacan sheet [8]. The research area administratively is located on Songa Village and Wayaua Village, East Bacan District, South Halmahera Regency, North Mollucas. Based on regional geological maps, the research area consists of three major lithologies [9], i.e. Holocene Volcanic Rocks and Bacan Formation, composed by volcanic rocks, on the centre and eastern part of research area, and Sibela Metamorphic Complex, on the western part of the research area as basement rocks.

Generally, ridges and streams surrounding the central eruption forms radial drainage system, even when the volcano has undergone advance erosion [10]. The assumption was used to withdraw base of volcano border is when the ridges and the drainage systems form concentric radial pattern, then it is considered has same source [11]. Crown unit is characterized by concentric contour pattern of a big size volcano cone or group of volcano cones that are related genetically, while hummock unit is characterized by the small volcano cone contour pattern around the crown. Each crown or hummock that had been identified subsequently named according to its local geography.

2.2. Remote Sensing Analysis

Remote sensing is one of typical study to identify the geological structures by observing the lineaments. The identification of lineaments could be done by using topography maps and available remote sensing data [12]. The lineaments can be formed by the erotional process on the ridge pattern, specific topography forms, and drainage system. In this research, the lineaments were indentified using synthetic aperture radar (SAR) image using Sentinel 1-A satellite: dual orbits, consists of ascending and descending with single polarization.

Ascending and descending images can be distinguished by the acquisition process. Ascending images were acquired from west to east side-looking with the satellite led from the south to north. , while the descending image were acquired from east to west side-looking and satellite led from north to south. This acquisition would display different results, so the combination of both lineaments of these images will obtain a comprehensive lineament patterns.

All lineament patterns anaysis might indicate that the research area has faults and fractures as geothermal permeable zone. To support the classification based on the area, we calculate the quantity and the length of each lineament which overlay on 1x1 km grid. The lineaments length of each grid was mapped. This analysis was known as Fault Fracture Density (FFD) Analysis [13].

The area of geothermal system is usually characterized by manifestations with high temperature relatively, resulting plants or vegetation on the surface become damaged, usually called vegetation stress [5]. We also do NDVI analysis to detect the appearance of vegetation stress from remote sensing by calculating the surface reflectance values using Normalize Different Vegetation Index (NDVI) formula (equation 1). The calculation of NDVI index is using satellite imagery that has visible red and near infrared band.

$$NDVI = \frac{(Near\ Infrared - Visible\ Red)}{(Near\ Infrared + Visible\ Red)} \quad (1)$$

Index of NDVI calculation will show the distribution of vegetation stress or healthy vegetation location [14]. However, it may need further analysis, because small NDVI index also may show there is no vegetation on the location for the residential or manmade, and also the presence of cloud covering the research area.

IOP Conf. Series: Earth and Environmental Science **103** (2017) 012020 doi:10.1088/1755-1315/103/1/012020

3. Results

3.1. Volcanostratigraphy Study Result

Based on the analysis of pattern and population of ridge and valleys (stream) at topographic maps, scale 1:100,000 Silang sheet, and then correlated with the geological maps, Bacan sheet, the volcanostratigraphy units consists of Songa Crown, Pele-pele Hummock, Lansa Hummock, Songa Hummock, and Bibinoi Hummock (figure 2) [8]. This result indicated that there was correlation among the distribution of volcanic rock units and the border of volcanostratigraphy units, so that we could know the distribution of eruption products controlled the morphology of Songa-Wayaua. If there was a difference between the borders of volcanostratigraphy units with the volcanic rocks units in this study, it is due to the border only withdrawn based on morphology, where geological maps used only as a comparison. We also interpreted possible heat source for geothermal system, i.e. of Songa Crown, Pele-pele Hummock, Lansa Hummock, Songa Hummock, and Bibinoi Hummock. The crown on this research area indicated that Songa crown was once an active volcano, from the radial pattern of eroded ridge and drainage pattern. While the hummocks in this area indicated there were still volcanic activity developed below the surface that allowed it to be the heat source.

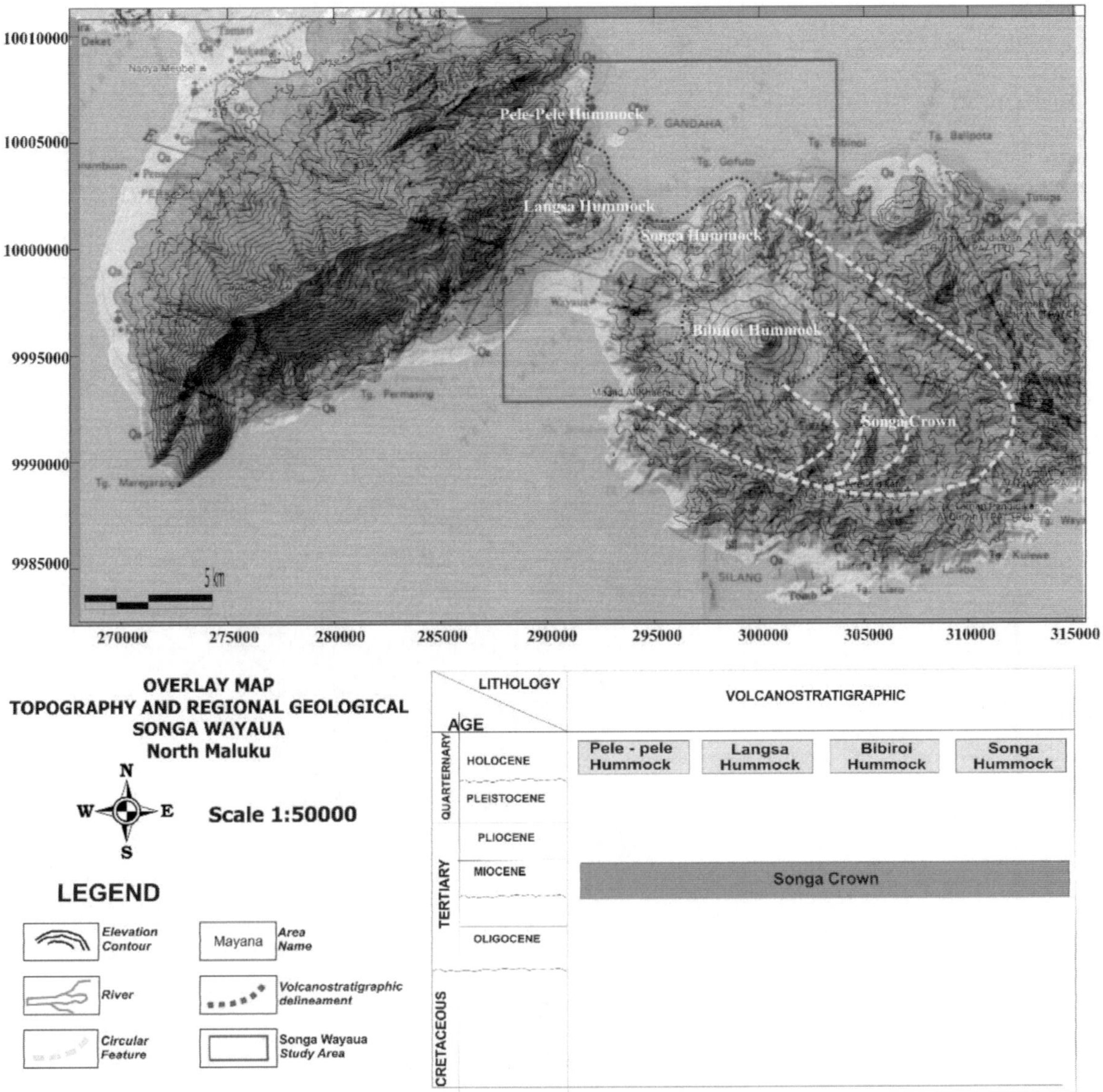

Figure 2. Distribution of volcanostratigraphy border on the overlayed of topography and regional geological maps of Songa-Wayaua and surrounding area (modified from Bacan Geological Map [8]).

3.2. Remote Sensing Study Result

The result of lineaments study base on SAR images lineaments show in the figure 3.

Rossette diagram (figure 4) was used to determine dominant trend of lineaments. The result showed that the major lineament trend of the research area was northwest – southeast. This was similar to the regional fault based on geology where Pele-pele Normal Fault, and Bibinoi Fault has some direction. It showed the west part of the research area had northwest-southeast trending direction that became the major structural control thus allowing the surface manifestations existence.

From FFD analysis acquired, the areas that had highest value were on the northwest, southeast, and northeast area of the research. Those areas were interpreted as permeable zone which were estimated as the recharge area of a geothermal system on the research area (figure 5).

IOP Conf. Series: Earth and Environmental Science **103** (2017) 012020 doi:10.1088/1755-1315/103/1/012020

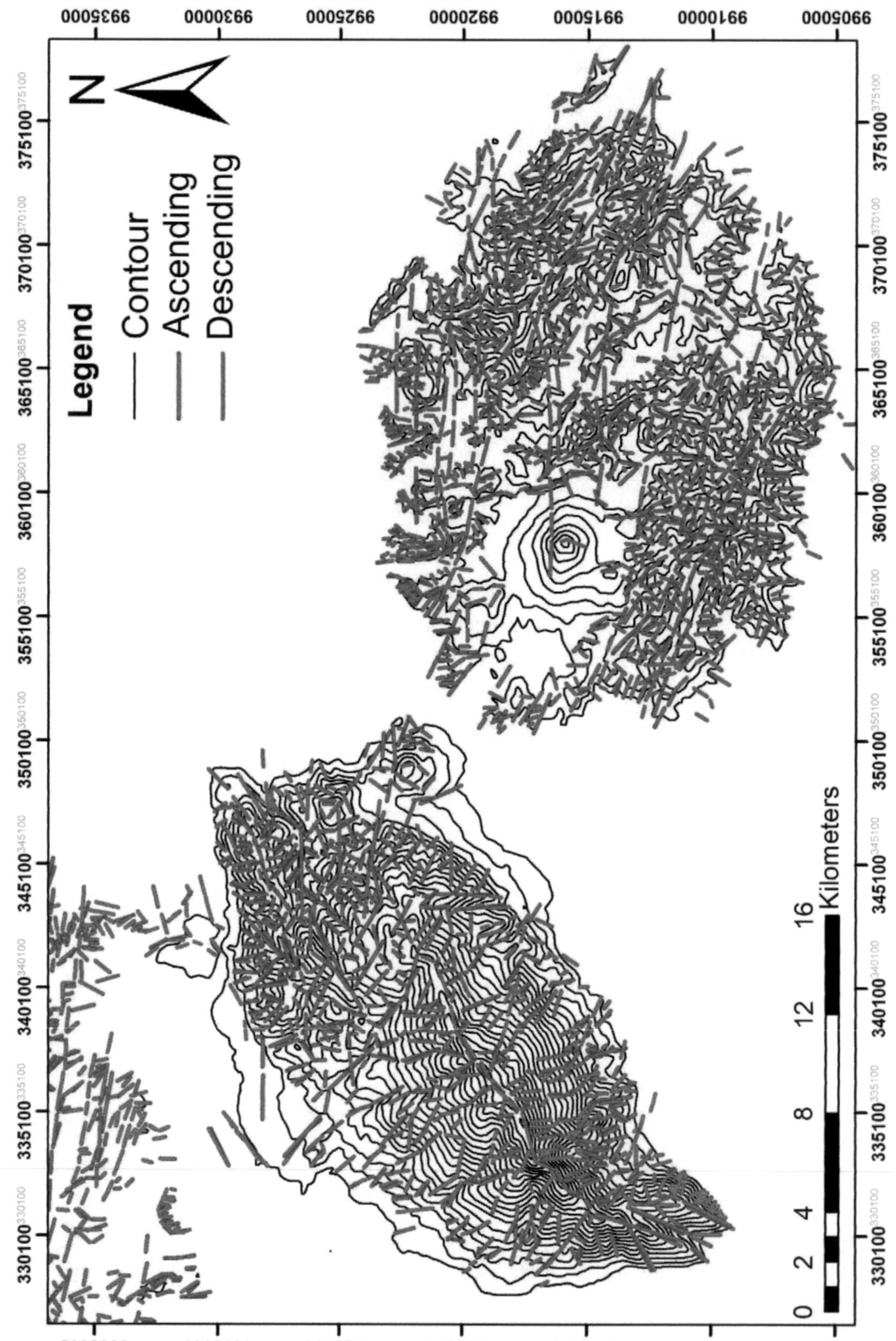

Figure 3. Lineaments pattern based on visual observation of Ascending (red) and Descending (blue) images combination overlain on contour lines of topography.

Geological structures which exist on the northwest of the research area indicated the intensity of the tectonic and lithology effect (metamorphic rock). Thus enable the fluid from surface enter the rocks and become the recharge area. But it needs to be studied more with further research to gain more informations. While on the volcanic rocks, the stuructures enables the fluids from the subsurface to release on the surface and becomes surface manifestations (discharge area).

The calculation of NDVI analysis should be compared to the satellite imagery using natural colour composite to represent actual appearance of the earth's surface. This step aims to ensure if NDVI index value became low, it was caused by the clouds or manmade. Processing satellite imagery data was done using ILWIS 3.3 and the image was acquired from Landsat 8 OLI satellite which had 11 bands where for the visible red band using band 4 and near infrared band using band 5.

Natural colour composite map displayed that residential area are located around the middle of the island up to the coast towards the northeast and southwest. In addition, at the time of image acquisition, there were clouds on the north, so it was estimated to result low NDVI index value originated from the clouds (figure 6).

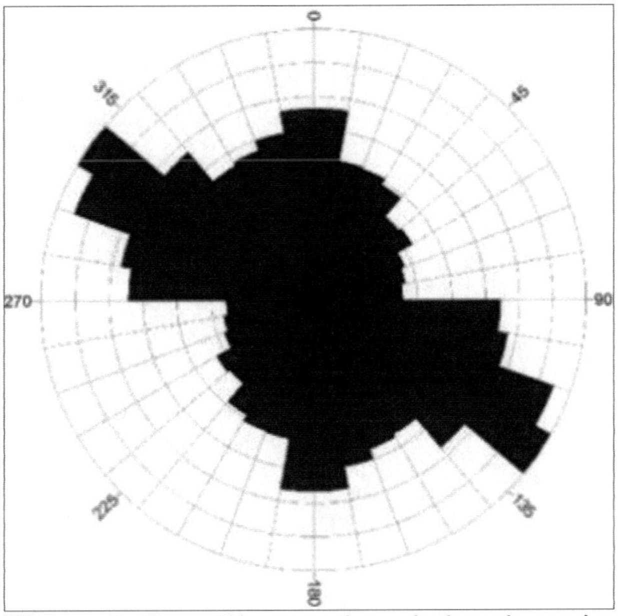

Figure 4. Rose Diagram showed that the major lineament trend of the Songa Wayaua area was northwest – southeast.

To support the analysis, NDVI value of Songa-Wayaua divided into three classes. Low NDVI with value -1.00 to 0.11 was poor vegetation shows in red colour, medium NDVI with value 0.11 – 0.47 shows in yellow colour, and good NDVI with value 0.47 – 1.00 was healthy vegetation and shows in green colour. Furthermore, there was value that could not be determined, valued in white colour, it was alleged because of the limitations of software in calculating NDVI value. Songa-Wayaua area had low NDVI values around the east coast, which was also the residential area. Beside the eastern area, there were also low NDVI values on the southern area of Songa, which was still Geothermal Working Area Songa-Wayaua.

IOP Conf. Series: Earth and Environmental Science **103** (2017) 012020 doi:10.1088/1755-1315/103/1/012020

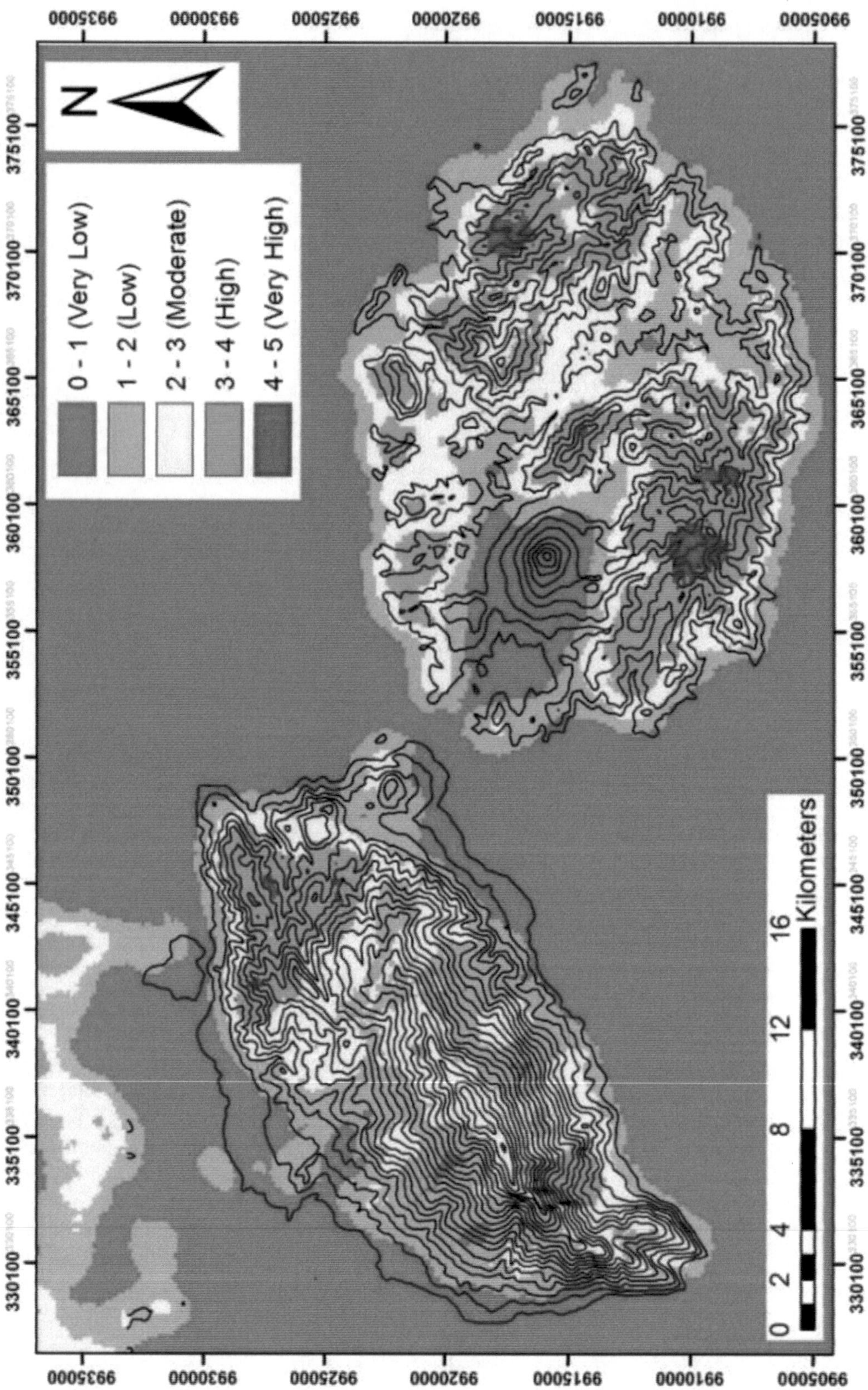

Figure 5. Fault Fracture Density of Songa-Wayaua area presented by color scale overlain on contour lines of topography

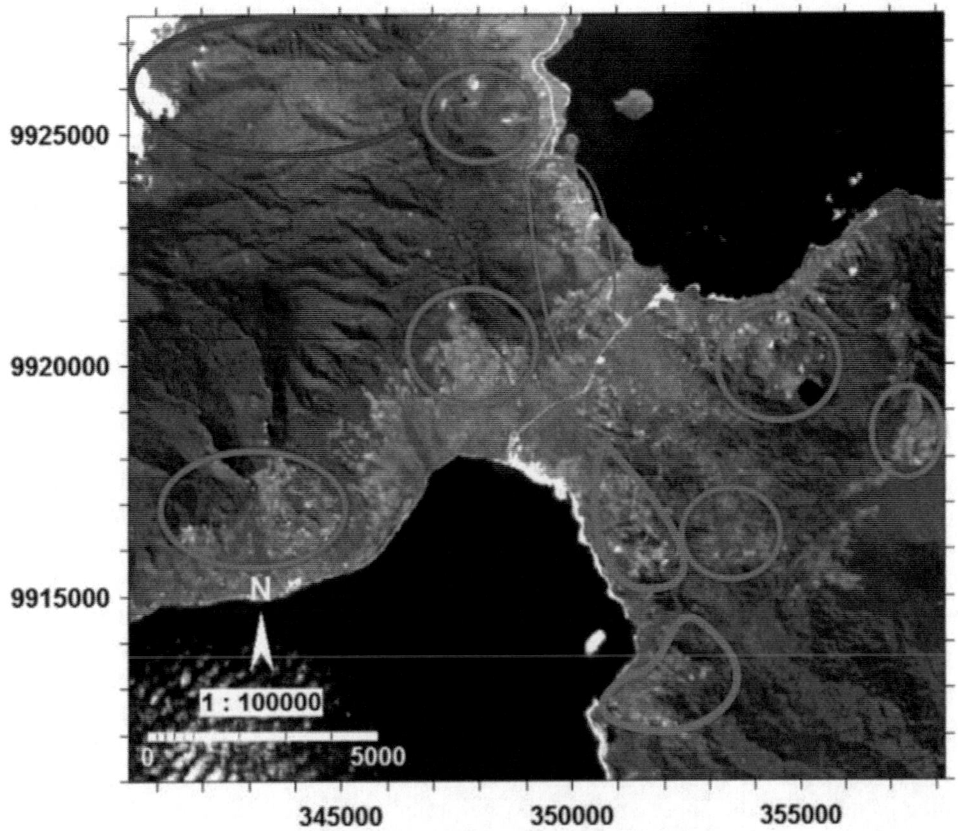

Figure 6. Natural color composite map

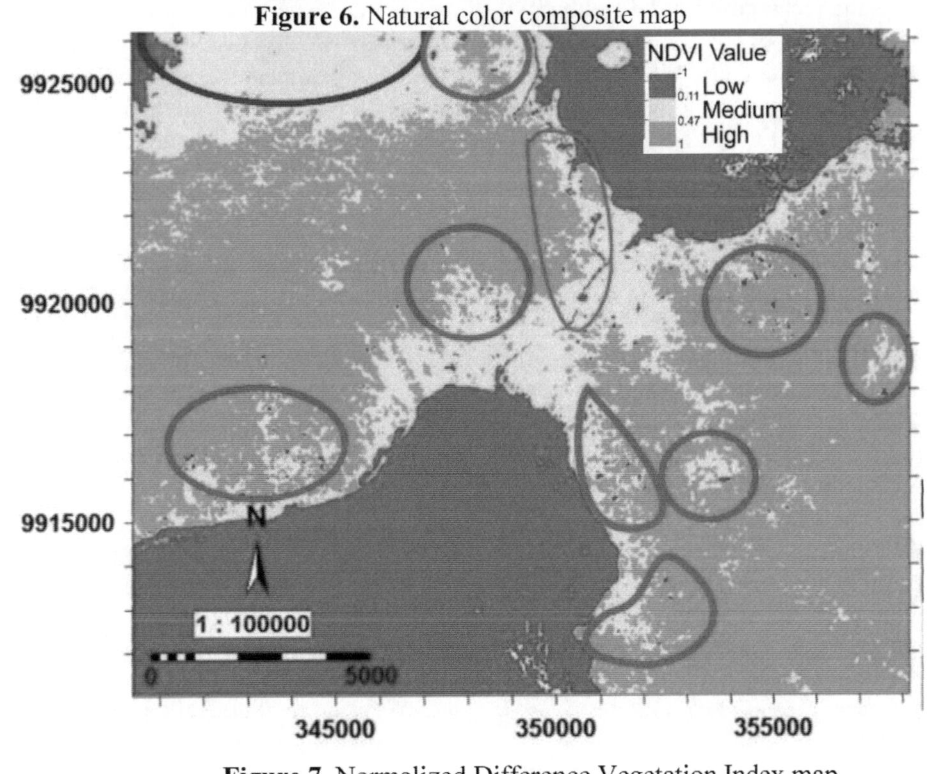

Figure 7. Normalized Difference Vegetation Index map

From the analysis, it was estimated that geothermal manifestations were on the purple-circled area in the figure 7, while the blue circles were not indicate the manifestations, but the thin clouds covering the vegetation underneath. Geothermal manifestations on the north and southeast were warm springs [15], while those on the south and southwest were still not carried out further exploration yet. Geothermal manifestations in the form of hot spring could not distinguish by this method, because the manifestations were adjacent to the residential area, which already had low vegetation values.

From volcanostratigraphy and remote sensing analysis, were concluded that there is one geothermal system, where Mount Lansa and Mount Songa area had more potential resource [15].

4. Conclusion

Based on the analysis of topographic maps, remote sensing, and regional geology for the geothermal system of Songa-Wayaua area, there are some conclusions as follows:

- The volcanostratigraphy of the research area are Songa Crown, Pele-Pele Hummock, Lansa Hummock, Songa Hummock, and Bibinoi Hummock. The Crown on this research area indicated the area was once an active volcano. While the hummocks in this area indicated there were still volcanic activity developed below the surface that allowed it to be the heat source in the research area.
- Geological structures of Songa-Wayaua were dominated northwest-southwest trending, consisting of Pele-Pele Normal Fault. This geological structure was likely to be primary control of Songa-Wayaua geothermal system area thus allowing the surface manifestations existence. These structures developed at areas with low elevation, which were the central part of the research area, and some along the eastern coastline of the research area. These structures enabled these locations as the discharge area
- From FFD analysis acquired, the areas that had highest value were on the northwest, southeast, and northeast area of the research. Those areas were interpreted as permeable zone. Those areas were estimated as the recharge area of a geothermal system on the research area
- From the NDVI, could be found several vegetation stress locations on research area. It was correlated with the field data, which were several manifestations of warm springs occurred.
- From volcanostratigraphy and remote sensing analysis, we concluded that there is one geothermal system, where Mount Lansa and Mount Songa area had more potential resource than other areas.

References

[1] Saepuloh A and Bakker E 2016 The Significance of SAR Remote Sensing in Volcano-Geology for Hazard and Resource Potential Mapping *Proceeding of the 6th International Symposium on Earth-Hazards and Disaster Mitigation (ISEDM)* Bandung

[2] Akbari D and Saepuloh A 2016 Identification of Surface Manifestation at Geothermal Field Using SAR Dual Orbit Data *IOP Conference Series: Earth and Environmental Science* vol 42 no 1 pp 012001

[3] Haeruddin Saepuloh A Heriawan M N Kubo T 2016 Identification of Linear Features at Geothermal Field Based on Segment Tracing Algorithm (STA) of the ALOS PALSAR data *IOP Conference Series: Earth and Environmental Science* vol 42 no 1 pp 012003

[4] Saepuloh A, Urai M, Sumintadireja P, Suryantini 2012 Spatial priority assessment of geothermal potentials using multi-sensor remote sensing data and applications, *Proceeding of the 1st ITB International Geothermal Workshop* 2012 Bandung

[5] Malaiholo J F A and Hall R 1996 The geology and tectonic evolution of the Bacan region, east Indonesia *Geological Society, Special Publications* London vol 106 p 483-497

[6] Komisi Sandi Stratigrafi Indonesia 1996 *Sandi Stratigrafi Indonesia* (Jakarta: Ikatan Ahli Geologi Indonesia)

[7] Army Maps Service 1944 *Silang, Molukken Islands, Gouvernement der Molukken, Residentie*

Ternate (Washington D C: Army Maps Service (QU) U S Army)
[8] Yasin A 1980 *Peta Geologi Lembar Bacan, Maluku Utara* (Bandung: Pusat Penelitian dan Pengembangan Geologi)
[9] Darman H and Sidi F H 2000 *An Outline of The Geology of Indonesia* (Jakarta: Ikatan Ahli Geologi Indonesia)
[10] Huggett R J 2007 *Fundamentals of Geomorphology* 2nd edition Routledge pp 458
[11] Van Zuidam R A 1985 *Aerial Photo Interpretation in Terrain Analysis and Geomorphological Mapping* (Amsterdam: Smits Publisher, The Hague)
[12] Saepuloh A 2015 Detecting surface structures after large eruption of Mt. Merapi in 2010 using ALOS/PALSAR data *Procedia: Earth and Planetary Science* vol 12 pp 84–92
[13] Soengkono S 1999 Te Kopia geothermal system (New Zealand) - The relationship between its structure and extent *Geothermics* vol 28 no 6 pp 767-784
[14] Suryantini Wibowo H Rahman K R and Woldai T 2013 Application of Normalized Different Vegetation Index (NDVI) Method To Identify Thermal Anomaly Area From Remote Sensing Proceedings 13th Indonesia International Geothermal Convention & Exhibition (Jakarta: Asosiasi Panasbumi Indonesia)
[15] LVG ITB 2012 Peningkatan Kualitas/Kuantitas dan Validasi Data Dalam Rangka Penentuan Lokasi Sumur Eksplorasi di Lapangan Songa Wayua Maluku Utara *Laporan Penelitian Panas Bumi* (Bandung: Institut Teknologi Bandung)

Review of subduction and its association with geothermal system in Sumatera-Java

A F Ladiba[1], L Putriyana[1], B br. Sibarani[2], H Soekarno[1]
[1] Research and Development Center of Electricity, New and Renewable Energy, and Energy Conservation, Ministry of Energy and Mineral Resources
[2] Geothermal Master Study Program, Bandung Institute of Technology

ayuta.fauzia@esdm.go.id

Abstract. Java and Sumatera have the largest geothermal resources in Indonesia, in which mostly are spatially associated with volcanoes of subduction zones. However, those volcanoes are not distributed in a regular pattern due to the difference of subduction position. Subduction position in java is relatively more perpendicular to the trench than in Sumatera. In addition, Java has a concentration of large productive geothermal field with vapour dominated system in the western part of Java, which may be caused by the various subduction dip along the island. In order to understand the relationship between the subduction process and geothermal system in the subduction zone volcanoes, we examined several kinematic parameters of subduction that potentially relevant to the formation of geothermal system in overriding plate such as slab dip, subduction rate, and direction of subduction. Data and information regarding tectonic setting of Sumatera and Java and productive geothermal field in Sumatera and Java have been collected and evaluated. In conclusion, there are three condition that caused the geothermal fluid to be more likely being in vapour phase, which are: the subduction is in an orthogonal position, the slab dip is high, and rate of subduction is high. Although there are plenty researches of subduction zone volcanoes, only a few of them present information about its formation and implication to the geothermal system. The result of this study may be used as reference in exploration of geothermal field in mutual geologic environment.
Keywords: subduction zone, slab dip, volcanic arc, vapour phase

1. Introduction

Almost 75% of total productive geothermal power plant in the world are associated with subduction zone volcanoes [1]. Indonesia which is located in Indo-Australian and Eurasian subduction zone, has a line of volcanoes along the subduction zone. Those volcanoes make a number of potential geothermal energy resources, where Sumatera and Java Island have the greatest number of them.

There are two categories of volcano; volcano at the volcanic arc which is genetically associated with subduction, and back arc volcano which its magmatism does not directly relate to the subduction [2]. In a study about relation between volcanic arc and geothermal field in Java Island, Setijadji [3] concluded that the highest prospectivity of geothermal energy is occurred in volcanic arc with genetic association with subduction process.

There is difference density of volcano distribution ratio between Sumatera-Java with 3: 3.5 ratio, which is caused by the difference of the slab dip in each part of the subduction. In spite of having a same subduction zone, there is irregularity trend formed in Java. Western part of Java (referred hereafter as

West Java) is the most seismically active region and has the highest volcanic density in the whole island [3]. It is also the one and only region where we can find vapour dominated geothermal sytems in Indonesia. It has five vapour dominated geothermal fields; Kamojang, Darajat, Patuha, Telaga Bodas, and Wayang Windu. West Java also has the tighest contour map of WBF depth which means it has the steepest angle of subduction (referred hereafter as slab dip) below it [4].

There are many researches about subduction and volcanic arc have been published, especially the one in Indo-Australian and Eurasian subduction zone. For instance, there are study about 3D geometry and evolution of the Indo-Australian plate by Richards, S., et al. [5], structural control of arc volcanism by Acocella, V., and Funiciello, F., [6], and geothermal systems in volcanic arcs by Stelling, P., et al., [1], but it is hard to find the connection between subduction process and its implication to the geothermal system

2. Methods

To understand the association between subduction with geothermal system in volcano subduction zone, we have collected data from various scientific literature about tectonic setting include geothermal system of some productive geothermal fields in Sumatera-Java. We examined some kinematic parameters of subduction which appear to be relevant with geothermal system formed at the overriding plate; slab dip, rate of subduction, and direction of subduction. Providing a simple framework in which to investigate its implications for the geothermal systems, the result of this study could be helpful for better understanding about the nature of volcanic arc and might be used as reference in exploration of geothermal systems in mutual geologic environment.

3. Regional Geology

The island of Sumatera and Java sit atop the Eurasian plate, which overrides the subducting Indo-Australian oceanic plates that converges obliquely at about 50 to 70 mm/yr [7]. The subduction along Sunda trench contributes to the development of Sunda Orogeny in Java and Barisan Orogeny in Sumatera [8]. The orogenies formed in parallel with the subduction is known as Sunda Arc which spreads along northwestern Sumatera to Flores. This northward subduction nearly perpendicular to the arc front in Java and increasingly oblique towards Sumatera [9], which affects the structure formed in Sumatera and Java.

There are several differences in geological setting of Java and Sumatera which affect the styles of geothermal fields there. Sumatera has a continental basement and its subduction style is an active continental margin. Meanwhile, Java is underlain by different basement compositions, from a continental basement in West and Central Java, to an island arc crust in East Java [10].

The potential geothermal field along Sunda arc, especially in Sumatera and Java island mostly has high-enthalpy geothermal system. The densely populated of potential geothermal system in Sumatera and Java island dominantly affected by typical of subduction zone.

Stresses resulting from the oblique subduction have been released periodically by dextral fault movements parallel to the plate margin [11]; these have produced the major Sumatera Fault System (SFS), which runs the entire length of the island [12]. The regional segmentation of West, Central and East Java volcanic arc has directly affected different geothermal prospectivity of each segment. In this case, prospectivity increases from the east to the west. West Java which is the most prospective at a regional scale is spatially associated with the highest crustal heatflow and the highest crustal seismic activities [3].

The geothermal fields in West Java are a major contributor of geothermal energy in Indonesia (figure 1). From 1513.5 MWe installed capacity in Indonesia, which is the number 3 installed national capacity in the world, West Java supplied 77 % of it, around 1164 MWe [13].

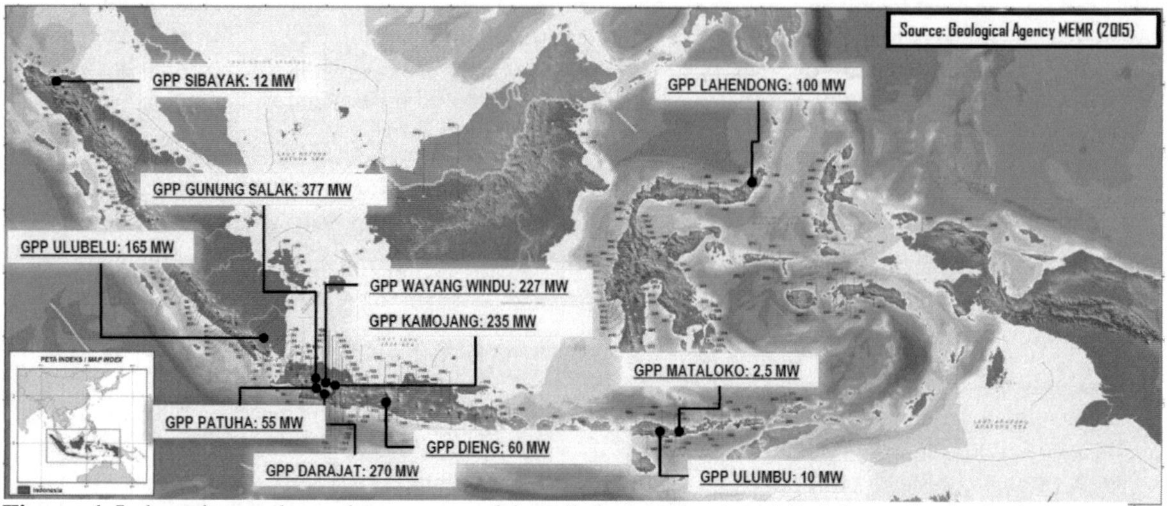

Figure 1 Indonesia geothermal potency and installed capacity (modified from Geological Agency, Ministry of Energy and Mineral Resources of Indonesia, in Directorate General of New, Renewable Energy, and Energy Conservation, 2016)

4. Discussion

Formation of volcanic arc is strongly correlated with the subduction process. With examining some kinematic parameters of subduction, we can interpret its correlation to the geothermal system which is associated with volcanic arc at the overriding plate.

The oldest and the most productive geothermal field in Indonesia are located in West Java. This region have five of eight vapour dominated geothermal fields of the earth. The factor contributing to the occurrence of the five vapour-dominated reservoirs in West Java are intense heating due to prolonged active volcanism, an absence of shear faulting, and the restrictive permeability range of the host and caprocks surrounding a relatively permeable reservoir [14]. Since vapour dominated geothermal field tends to be more stable and productive, this study is also conducted to review whether there is subduction factor interfere in controlling the formation of vapour dominated system.

4.1. Slab dip

Tatsumi [15] emphasized that pressure controls the magma in subduction zones. Precise earthquake hypocentral locations reveal that the depth to the top of the slab beneath the fronts of volcanic arcs is constant, to within a few kilometres, along individual segments of arc, but varies from 80 km to 160 km between different segments [16]. When the subducting plate reach the certain depth with certain pressure, it could produce magma which would rising and implied to formation of volcanoes at the above overriding plate. The shallower the dip of the slab, the greater the distance of the volcano arc formed from the trench.

We examined the distance of the volcanoes to the trench in Sunda Arc, and found insignificant difference of them (figure 2). In the other hand, from the contour depth of WBF zone, it can be seen that West Java has the steepest slab dip among the Sunda Arc [4]. From this sight, the relation between slab dip and the volcanic arc is not clearly seen.

England and Katz [16] have studied about location of volcanic arc formation which said that the mechanism whereby the melt reaches the Earth's surface are controlled by the dynamics of the wedge. Hence, the arc front is located somewhere above the region in which the temperature in the wedge exceeds a condition called as anhydrous solidus (the state where water contents are about ten times lower than the concentrations estimated for wet melting), because the degree of hydrous melting increases rapidly there. It supports the previous study by England and Wilkins [17] which said that the locations of the arc controlled by a strongly temperature-dependent process either at the top of the slab or in the wedge of mantle beneath the arcs, while the temperature depends on the convergence rate, V, and dip, δ, of the slab.

Since they are counted as inseparable parameters, we would review the rate of subduction and its influence to the subduction process and volcanic arc to understand the relation between two of them to the temperature of the wedge.

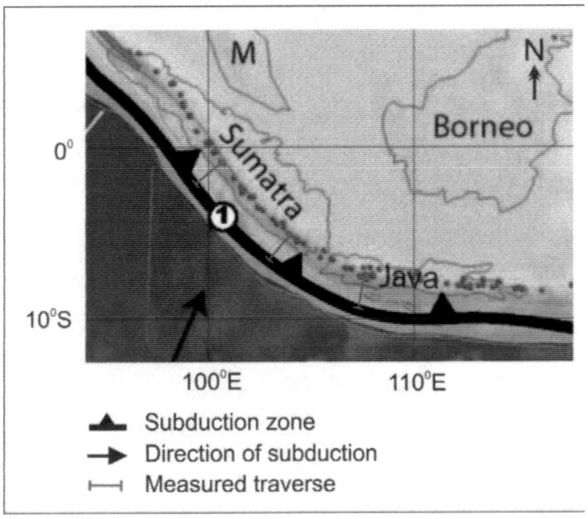

Figure 2 Subduction zone in Sumatra and Java (modified from Richards, et al., 2007) [5].

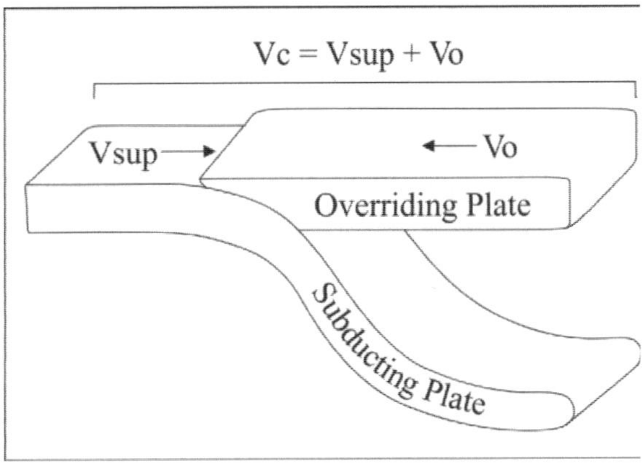

Figure 3 Definition of rate of subduction in scheme. Convergent velocity (Vc) is sum of subducting plate velocity (Vsup) and overriding plate velocity (Vo) (modified from Acocella and Funiciello, 2009) [6].

4.2. Rate of subduction

The rate of subduction could be defined as convergent velocity (Vc) which is sum of subducting plate velocity (Vsup) and overriding plate velocity (Vo) [6] (figure 3).

The subducting plate velocity (Vsup) which is included in rate of subduction (Vc) has been observed together with depth beneath volcanic arc to study about factor controlling the location of volcanic arc, using a simple mathematical model [16]. It was a further study of previous research about temperature structure in subduction zones which was concluded that temperature of the slab and the mantle wedge which controlled the location of the arc, depend on a single dimensionless distance $Vr\,\delta^2/\kappa$, where V is the rate of plate convergence (or Vc based on previous description), r is distance from the corner of the wedge, δ is the dip of the slab and κ is thermal diffusivity (figure 4). In this scaling, the maximum temperature in the mantle wedge T_r is given by equation 1:

$$T_r \approx T_0 exp\left[-B\left(\frac{Vr\delta^2}{\kappa}\right)^{\beta}\right] \quad (1)$$

where T_0 is a scale temperature, and B, β and γ are constants, the values of which depend on the details of the flow near the top of the slab.

The study has calculated the steady state wedge, T_r, plotted against $Vr\,\delta^2/\kappa$, and it is confirmed that T_r is proportional with the dimensionless distance (figure 5).

The greater the number of Vδ, the more the temperatures it needs to reach the condition of melting slab to produce magma and form the volcanic arc.

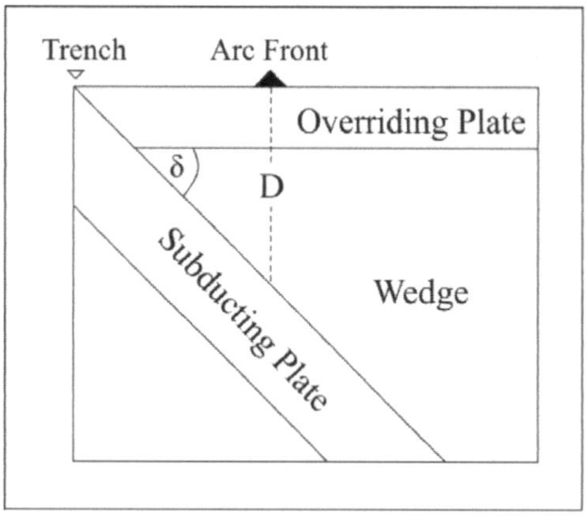

Figure 4. Schematic of subduction zone (modified from England and Katz, 2010)

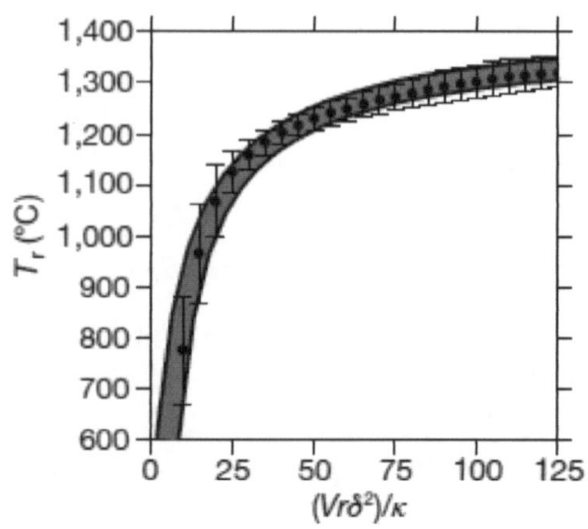

Figure 5. Maximum temperature in the wedge, Tr, as a function of dimensionless distance from the wedge corner, Vrd2/k, where V is convergence speed, δ is dip of the slab, and κ is thermal diffusivity [6].

4.3. Direction of subduction

Plate motion vectors can be used to predict deformation which produced as subduction outcome. Where subduction is strongly oblique and steep, arc-parallel strike-slip faults can form. These frequently run along the volcanic arc and have associated with permeable secondary structures whose orientations can be determined similarly [18].

According to Hochstein and Browne [19], reservoir permeability in vapour-dominated system is >10 mD and surrounded by impermeable rock which has permeability <1-3 mD to prevent flooding from the outside. Generally, fault has strong impact in creating secondary permeability in reservoir. In Kamojang, the outer zone of reservoir has low permeability, 0,02 mD and in the inner zone is 150 mD [20]. Meanwhile, the outer zone of Darajat reservoir is about 0,04-0,16 mD [21]. The occurence of strike-slip which is affected by shear stress will generate transtension which could enrich the permeability of reservoir. Therefore, the absence of shear faulting could support the vapour dominated system to be occurred.

Great Sumatera dextral strike-slip was formed parallel with the arc. Gunderson, et al., [22] and Santoso, et al., [23], have mapped north striking faults and identified half graben and graben structures. Intensive structure in Sumatera caused the high permeability of reservoir and surrounded rock. It generates the high rates of water inflow and produce liquid phase in reservoir. Study about exclusivity of the vapour dominated geothermal system in West Java concluded that shear stress on faults that cross through the reservoir and associated enhanced permeability possibly play an important role in preventing a vapour-dominated system from developing [14].

Moving south from Sumatera to Java there is a change from oblique to perpendicular subduction. While geothermal system in Sumatera are highly affected by Great Sumatera Fault, West Java geothermal system have not been associated with regional faults like Sumatera's. West Java can be considered to be located at the transitional zone between oblique subduction in Sumatera and frontal subduction at the eastern part of West Java [24]. In West Java, oblique subduction occured during Cretaceous while orthogonal subduction during Tertiary periods (figure 6). Each of subduction activity has formed different structural pattern, depending on the position of subduction track at that time [25]. These two kinds of subductions in Java tectonic were accommodated by wrench faults in the upper crust in NW direction.

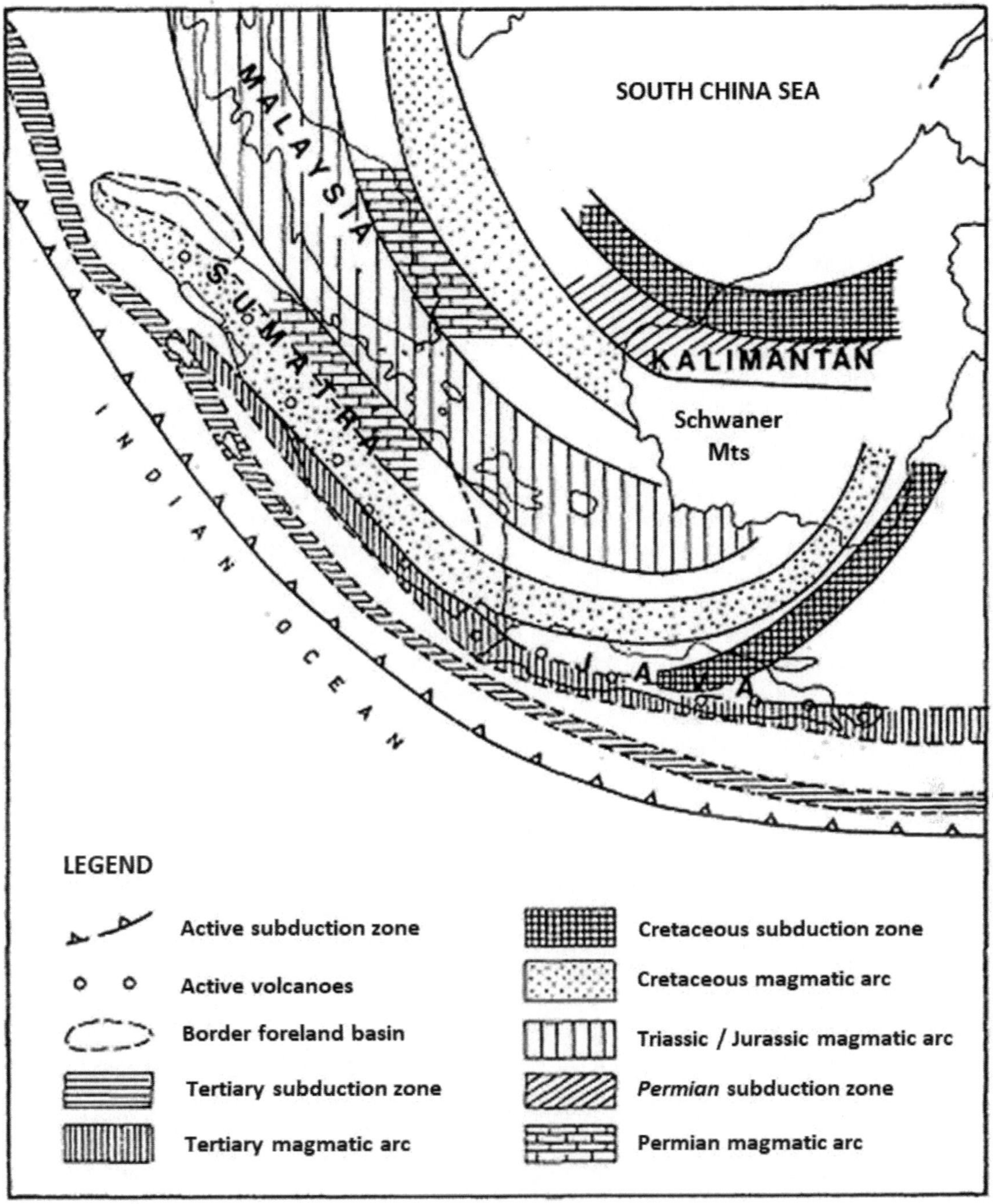

Figure 6. Evolution of subduction zones and magmatic arc of West Java [26].

The West Java Fault as a major structure has two main NW trending of fault splays that formed Garut circular structure or Garut Basin by negative flower structure. A research about regional structure in West Java by Corbett and Leach [27], was conducted to know about deep seated structure and surface structure relationship and resulted that the N-S to NNE-SSW trending surface structure is the surface manifestation of NW-SE to WNW-ESE deep structures extensional system in West Java is N-S.

Movements along laterally offset basement faults generate popup structures and pull-apart grabens in restraining and releasing bends, respectively [28]. If the vertical component is normal, faults tend to be listric and to form a negative flower structure, which forms a depressed area [28]. The upper part of negative flower structure is a pull-apart basin which is bounded by faults. This regional major deep seated structure's role in localizing magmatic heat source in an arc-volcanism and secondly, extensional setting of structure that provide the favourable condition for secondary permeability in fluids interaction and accumulation [27]. It can be argued that the oblique subduction has a important role in generating the pathway for intrusion and enhancing the permeability.

On the other hand, the Indo-Australian plate is subducted under the Eurasian plate northwardly at N20^{0}E and at rate of 6-7 cm/yr in West Java area [30]. This orthogonal subduction since Tertiary period accomodate the appearance of 4 structure pattern in West Java, N-S, N-E, E-W, and N-W trending. The fault structures of NE-SW trending are associated with oblique subduction activity during Cretaceous period and has been reactivated during the Early Tertiary tectonic period with compression forces of N-S trending. This trend has generated fault structures of NS striking as an extensional fault and fault pattern with NNE-SSW direction as Synthetic Riedel Shear. It can be argued here that West Java Fault probably the remnant of previous structure style when West Java was in oblique subduction.

There are NE and N-S striking faults found in 5 vapour dominated fields which act as favourable fault in providing permeability zone. It is interpreted that these structures control the permeability in reservoir while not intensify the surrounded permeability. That fact is supported by the absence of shearing faulting which occured during Createaceous period. It caused the permeability in reservoir is much higher than surrounding so they are able to produce vapour phase fluid. The transtension zone which controlled the localizing of heat source also support the occurence of intense heating.

It is confirmed that direction of the subduction affected the structure formed at the overriding plate which can support the act of the geothermal fluid in a geothermal system. The oblique subduction in Sumatra generates NW-SE striking strike slip which enhance the permeability and caused high rates water inflow. Its condition is suitable to create the liquid phase reservoir. The other way, transition subduction that occured in West Java caused the lacking of shearing fault and enhance the higher permeability in reservoir than surrounding. It support the vapour dominated reservoir to develop.

5. Conclusion

If the multiplicative value of slab dip and rate of subduction of Java is higher than Sumatera, the heat source in Java tends to be higher, although there is uncertainty in this deduction since they are not the only factors to be considered in this correlation.

The higher the obliqueness of direction of subduction, the higher the permeability formed in overriding plate. It is caused by oblique subduction which generally forms extension in transtension zone with permeable character. The permeability in Java is generally lower than in Sumatera due to the absence of shearing fault in West Java. It has generated certain geological setting which makes some area in West Java have significant differences between permeability of reservoir and surrounding rocks. Coupled with intensive heating from the heat source, it leads vapour dominated geothermal system to developed in West Java.

However, the precise correlation between kinematic of subduction process and geothermal system at overriding plate remains unclear. This may be caused by lack of data on exact value of slab dip, rate of subduction, and distribution of permeability of geothermal system. In order to understand more about implication of subduction to the geothermal prospectivity, research about act of slab dip and rate of subduction in structure formation at overriding plate, and further study about localization of vapour dominated geothermal system in West Java are needed.

References

[1] Stelling P Shevenell L Hinz N Coolbaugh M Melosh G Cumming W 2016 Geothermal systems in volcanic arcs: Volcanic Characteristics and Surface Manifestations as Indicators of Geothermal Potential and Favorability Worldwide *Journal of Volcanology and Geothermal Research*

[2] Setijadji L D 2005 *Geoinformation of Island Arc Magmatism and Associated Earth Resources: A Case Study of Java Island, Sunda Arc, Indonesia* PhD thesis Kyushu University Japan

[3] Setijadji L D 2010 Segmented Volcanic Arc and its Association with Geothermal Fields in Java Island, Indonesia *Proceedings World Geothermal Congress Indonesia* Bali

[4] Abdurrachman M Widiyantoro S Priadi B Alim M Z A Dewangga A H 2015 Proposed New Wadati-Benioff Zone Model in Java-Sumatera Subduction Zone and Its Tectonic Implication *Proceedings: Joint Convention Balikpapan 2015, HAGI-IAGI-IAFMI-IATMI* Balikpapan Indonesia

[5] Richards S G Lister and Kennett B 2007 A slab in depth: Three-dimensional geometry and evolution of the Indo-Australian Plate *Geochem. Geophys. Geosyst., 8* Q12003 doi:10.1029/2007GC001657

[6] Acocella V Funiciello F 2009 Kinematic Setting and Structural Control of Arc Volcanism *Earth Planet. Sci. Lett. 2009* doi:10.1016/j.epsl.2009.10.027

[7] Prawirodirdjo L Bock Y Genrich J F Puntodewo S S O Rais J Subarya C Sutisna S 2000 One Century of Tectonic Deformation Along the Sumateran Fault from Triangulation and Global Positioning System Survey *Journal of Geophysical Research vol 105 no B12*

[8] Simandjuntak R O Barber A J 1996 Contrasting Tectonic Styles in the Neogene Orogenic Belts of Indonesia *Tectonic Evolution of Southeast Asia, Geological Society Special Publication no 106* pp 185-201

[9] Susanto A Suparka E Tsuchiya N Hirano N Kishita A Kusumah Y I 2011 Hydrothermal Alteration Study in Malabar Area, Northern part of the wayang windu geothermal Field, Indonesia *Proceedings of the 9th Asian Geothermal Symposium* pp 76-81

[10] Setijadji L D 2015 Effective Field Geological Mapping Techniques for Sumateran Geothermal Fields, Indonesia *Proceedings World Geothermal Congress 2015*

[11] Fitch F J 1972 Plate Convergence, Transcurrent Faults and Internal Deformation Adjacent to S.E. Asia and the Western Pacific *J. Geophys Res 77* p 4432-60

[12] Rocks N M S Syah H H 1982 Permian to recent volcanism in Northern Sumatera, Indonesia: A Preliminary study of its distribution, chemistry and Peculiarities *Bull Volcanol vol 45-2*

[13] Directorate General of New, Renewable Energy, and Energy Conservation of Ministry of Energy and Mineral Resources 2016 *Buku Pintar Panas Bumi 2016* Indonesia

[14] Raharjo I B Allis R G and Chapman D S 2012 Why Are the Only Volcano-Hosted Vapour-Dominated Geothermal Systems in West Java, Indonesia? *GRC Transaction vol 36*

[15] Tatsumi Y 2003 Some Contraints on Arc Magma Genesis *Geophysical Monograph 138* American Geophysical Union

[16] England P C Katz R F 2010 Melting Above the Anhydrous Solidus Controls the Location of Volcanic Arcs *Nature vol 467* Macmillan Publishers Ltd

[17] England P and Wilkins C 2004 A simple Analytical Approximation to the Temperature Structure in subduction zones *Geophys. J. Int. 159,* 1138–1154. doi: 10.1111/j.1365-246X.2004.02419.x

[18] Bogie I 2000 *The Use of Plate Motion Vectors to Predict the Orientation of Highly Permeable Structures in Active Geothermal Fields Related to Subduction*

[19] Hochstein M P and Browne P R L 2000 Surface manifestation of geothermal systems with volcanic heat sources *Encyclopedia of Volcanoes* editor H Sigurdsson B F Houghton S R McNutt H Rymer and J Stix eds, Academic Press p 835-855

[20] O'Sullivan M J Barnett B G Razali M Y 1990 Numerical Simulation of the Kamojang Geothermal Field, Indonesia *Geothermal Resources Council Transactions* vol 14 part II

IOP Conf. Series: Earth and Environmental Science **103** (2017) 012021 doi:10.1088/1755-1315/103/1/012021

[21] Kaya E and O'Sullivan M J 2010 Modelling of Injection into Vapour Dominated Geothermal System *Proceedings World Geothermal Congress* Bali

[22] Gunderson R P Dobson P F Sharp W D Pudjinato R Hasibuan A 1995 Geology and thermal features of the Sarulla contract area, North Sumatra, Indonesia *Proceedings WGC* 687-692

[23] Santoso D Suparka M E Sudarman S Suari S 1995 The geothermal fields in central part of Sumatra Fault Zone as derived from geophysical data *Proceedings WGC 1995* 1363-1366.

[24] Fauzi A Permana H Indarto S Gaffar E Z 2015 Regional Structure Control of Geothermal Fields in West Java *Proceedings World Geothermal Congress 2015*

[25] Haryanto I Sunardi E Sudrajat A Abdurrokhim Jamal 2014 Plate Tectonic and Regional Structural Geology in West Java *Geoscience for Energy, Mineral Resources, and Enviroment*

[26] Katili J A 2006 Plate Tectonics of Indonesia with Special Reference to The Sundaland Area *Proceedings Indonesian Petroleum Association.*

[27] Corbett G and Leach T 1997 Southwest Pacific Rim Gold-Copper Systems: Structure, Alteration, and Mineralization Short course Manual.

[28] Mann P Hempton R Bradley D C Burke K 1983 Development of Pull-Apart Basins *The Journal of Geology vol. 91 no 5* 529-554

[29] Harding T P 1985 Seismic characteristics and identification of negative flower structures, positive flower structures, and positive structural inversion *Bulletin of the American Association of Petroleum Geologists* 69 582-600

[30] Hall R 2012 Late Jurassic-Cenozoic Reconstructions of the Indonesian Region and the Indian Ocean *Tectonophysics* 570-571 1-41

IOP Conf. Series: Earth and Environmental Science **103** (2017) 012022 doi:10.1088/1755-1315/103/1/012022

Identification of natural fractures and in situ stress at Rantau Dedap geothermal field

Andika Artyanto[1], Benyamin Sapiie[1], Chalid Idham Abdullah[1], Ridwan Permana Sidik[2]

[1] Bandung Institut of Technology (ITB), Ganesha 10, Bandung, Indonesia
[2] PT Supreme Energy, Menara Sentraya, 23AA floor, Kebayoran Baru, Jakarta, Indonesia

email: andika.artyanto@gmail.com

Abstract. Rantau Dedap Area is a geothermal field which is located in Great Sumatra Fault (GSF). The fault and fracture are main factor in the permeability of the geothermal system. However, not all faults and fractures have capability of to flow the fluids. Borehole image log is depiction of the borehole conditions, it is used to identify the natural fractures and drilling induced fracture. Both of them are used to identify the direction of the fracture, direction of maximum horizontal stress (S_{Hmax}), and geomechanics parameters. The natural fractures are the results of responses to stress on a rock and permeability which controlling factor in research area. Breakouts is found in this field as a trace of drilling induced fracture due to *in situ* stress work. Natural fractures are strongly clustered with true strike trending which first, second, and third major direction are N170°E – N180°E (N-S), N60°E – N70°E (NE-SW), and N310°E – N320°E (NW-SE), while the dominant dip is 80° –90°. Based on borehole breakout analysis, maximum horizontal stress orientation is identified in N162°E – N204°E (N-S) and N242°E (NE-SW) direction. It's constantly similar with regional stress which is affected by GSF. Several parameters have been identified and analyzed are S_{Hmax}, S_{Hmin}, and S_y. It can be concluded that Rantau Dedap Geothermal Field is affected by strike-slip regime. The determination of in situ stress and natural fractures are important to study the pattern of permeability which is related to the fault in reservoir of this field.

1. Introduction

In situ stress, active fault, and natural fractures are known as the determinant factors for permeability. Fault is the manifestation of active deformation that caused by changes in regional stress pattern. Active deformation associated with the fault movement lead to the formation of fracture and disruption to *in situ* stress [1]. Distribution of permeability around the fault affected the formation of fracture due to *in situ* stress. This causes a difficulty in determining fluid flow system of fracture.

Rantau Dedap Field has fractured reservoir since it is a fault controlled system. Fracture reservoir is located on areas with complex geological fault structure. The fault affects distribution of permeability on the area. Reservoir that has low permeability of matrix, would not have no significant fluid flow if they are not fractured [2].

2. Geological and Structural Setting

Rantau Dedap Area is a geothermal field located in Great Sumatra Fault (GSF). The oblique convergent is resulted by N-S regional compression and possible dextral strike-slip faulting. The dextral transcurrent Sumatran Fault runs in the entire length of Sumatra, from Banda Aceh to the Sunda Strait. The fault is coincident with the Sumatran volcanic arc and parallel to the offshore trench [3]. The Island of Sumatra is an island arc produced by subduction of the Indian-Australian Plate beneath the Southeast Asia Plate. The relative 7.7 cm a NNE-directed motion of the Indian Ocean results in oblique (c. 45°) subduction at the Sunda Trench (see figure 1) [4].

The tectonic phase from the field data suggest the extensional tectonic regime work at Rantau Dedap field from central area through Bukit Besar area, while a compressive phase located more to the south seemed to be the limit of opening zone [5].

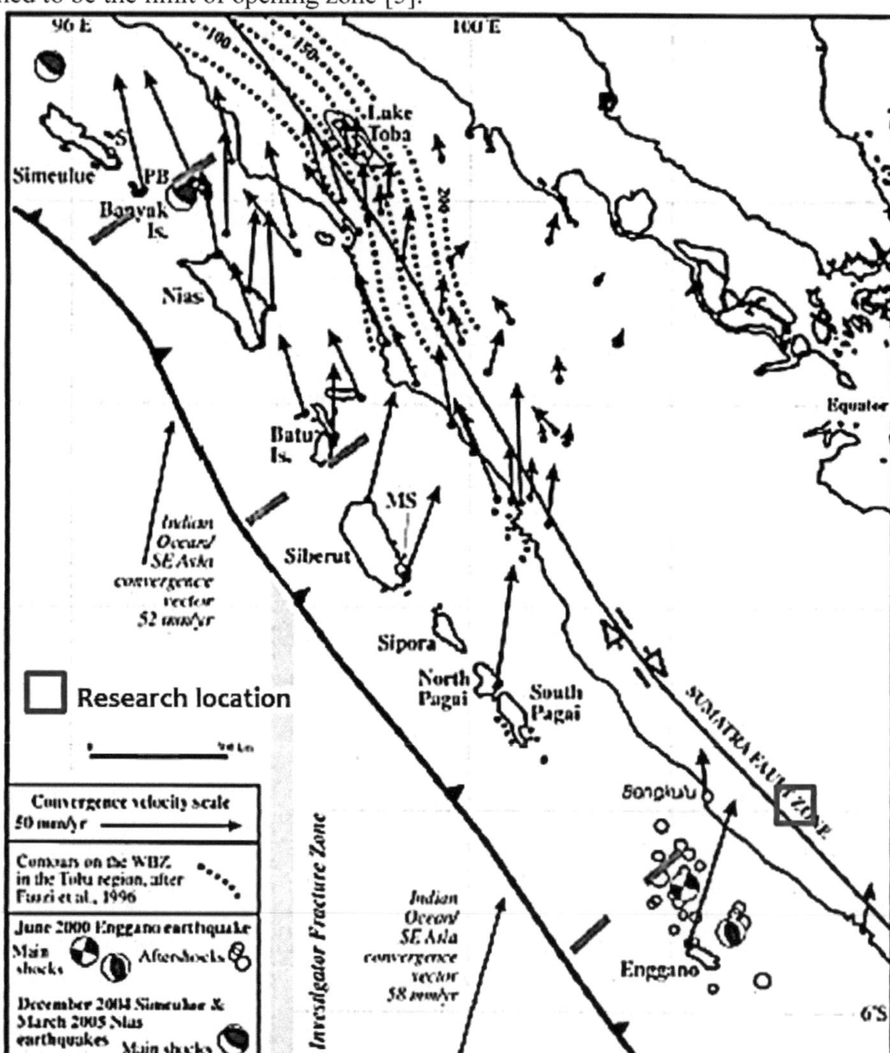

Figure 1. Regional Rantau Dedap structure frame work [1]. Vectors show rates of movement relative to a stable Southeast Asia. They imply stress accumulation in parts of the subduction region.

Bukit Besar area was inferred as a morpho-tectonic that has main effect to Rantau Dedap structure since it is denote as part of a sinistral fault's evident trending NE-SW toward higher elevation of Bukit Besar itself. Rantau Dedap is likely under influence of two major structures, i.e. Manna Fault (GSF) and Kikim Fault as shown in figure 2. Kikim Fault as normal fault with striking NE-SW to Bukit Besar area. Even though, no clear expression at surface (see figure 1). These structures control the locations with respect of thermal manifestation distribution more likely bound by such faults configuration [4].

The structural map of study area is shown in figure 2. The oldest tectonic phase of WNW-ESE compressional stress probably produced the Manna Fault and mostly recorded in the pre-Besar volcanic in the southeastern area [5]. The next phase was sinistral movement of Kikim Fault that had produced the N-S compressional stress and generated Cawang Fault. Subsequent phase of NW-SE extensional stress that possibly acted as releasing period of previous compressional stress have opened most of the Cawang Rift structures. As the Cawang Fault was cut and superimposed by a series of lateral collapse structures striking NW-SE and sliding toward WSW direction (see figure 2). The distribution of thermal manifestations in this field is controlled by strike slip faults and their associated fractures particularly in NE-SW, NW-SE and N-S trends [6].

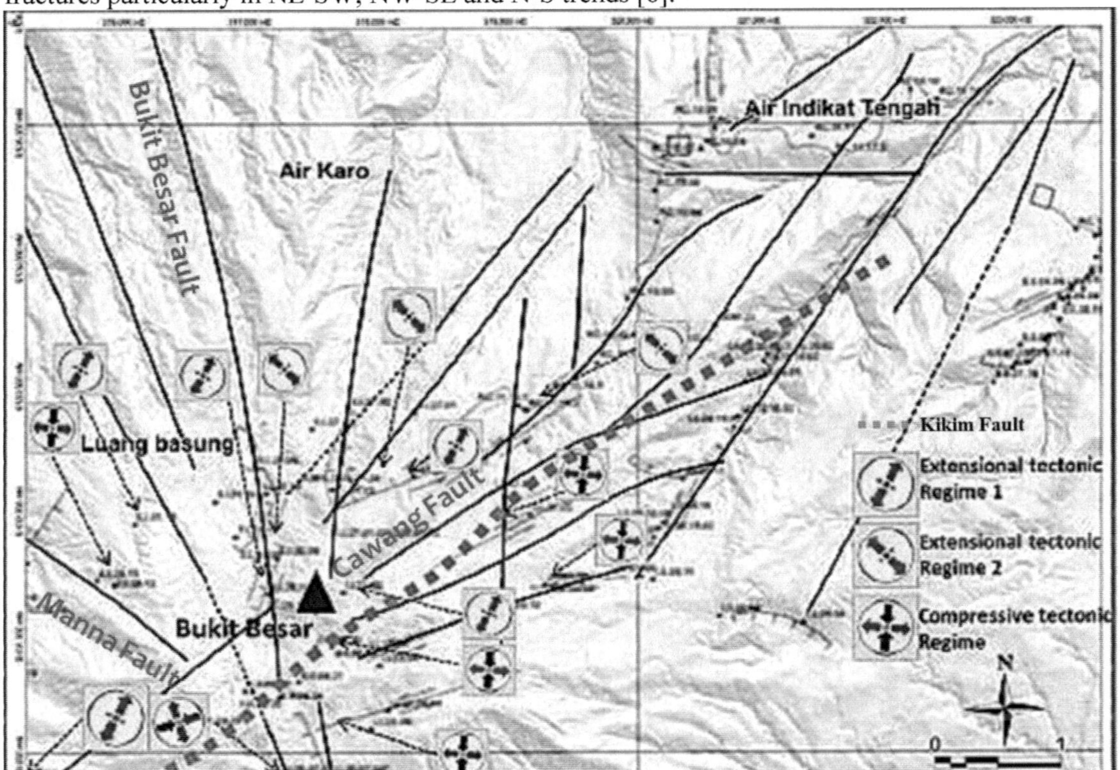

Figure 2. Structural Map of Rantau Dedap project. The circle data (stress orientation) and the red lines shows distribution of structural data measurement [5].

3. Geological and Structural Setting

3.1. Natural Fracture

The natural fracture in the borehole image consists of conductive and resistive fracture. Conductive fractures have a low resistivity because it is filled by drilling mud which gave a conductive respond. Partial fracture is part of conductive fracture which is formed partially sinusoidal in wellbore. A resistive fracture will have high resistivity due to the minerals, while the surrounding of wall formation has a low resistivity value. Conductive natural fractures which as basis for determining the direction of current working stress at the moment. Sometimes these fractures are filled with conductive minerals and would appear darker even though they are closed (see figure 3).

Based on data analysis, the two types of natural fractures were picked independently: 1290 strike-dips of partial fracture and 343 strike-dips of conductive fracture were measured in AA-1, AA-2 and AA-3. The data were statistically obtaining from the clustered direction of natural fracture. The first dominant direction is N170°E – N180°E (N-S), the second direction is N60°E – N70°E (NE-SW) and the third direction is N310°E – N320°E (NW-SE), while the dominant dip are 80° –90° (see figure 4). This result is similar to surface fault of Rantau Dedap Geothermal Field (see figure 6) [5].

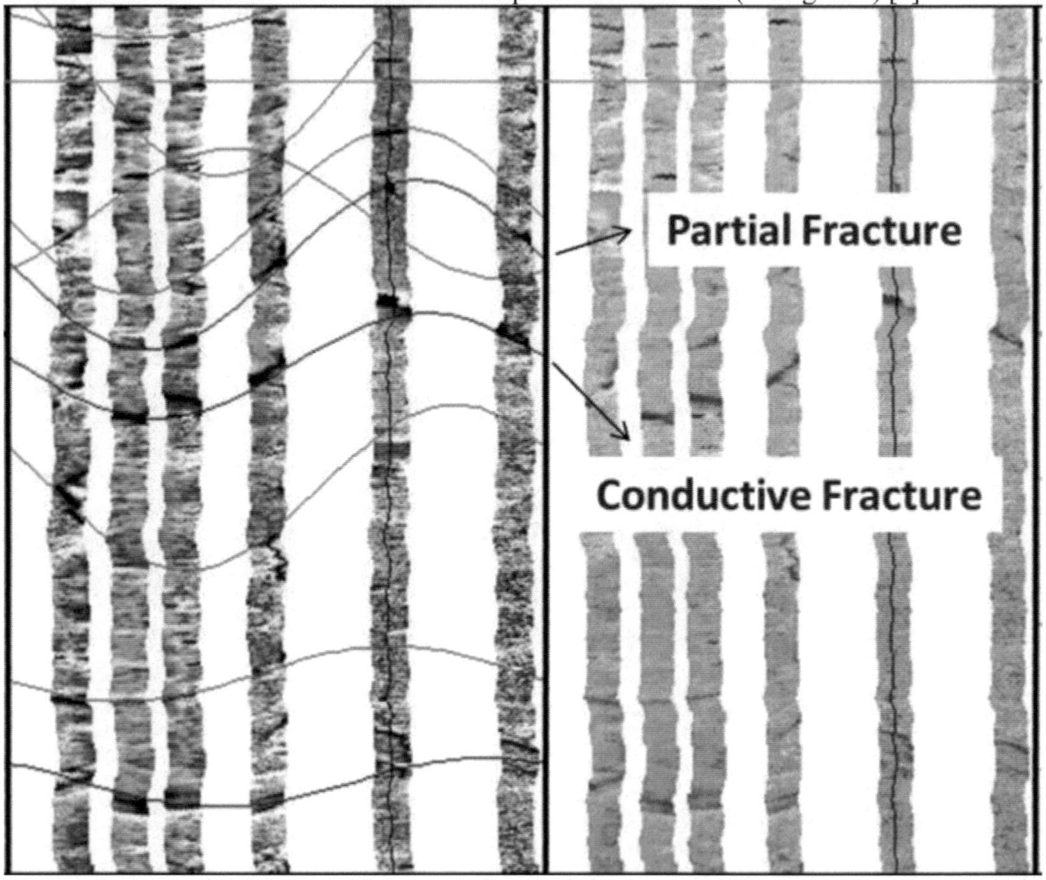

Figure 3. Examples of sinusoidal trace of natural fractures (NF) intersecting the borehole wall.

3.2. S_{Hmax} Azimuth Modelling

Determination of the orientation S_{Hmax} conducted by applying the directions orientation from induced tensile and breakouts. Unfortunately the method could not be directly applied since it was obtained from an inclined or directional well drilling. A formulation of specific equation must be constructed to resolved S_{Hmax} direction (see figure 5).

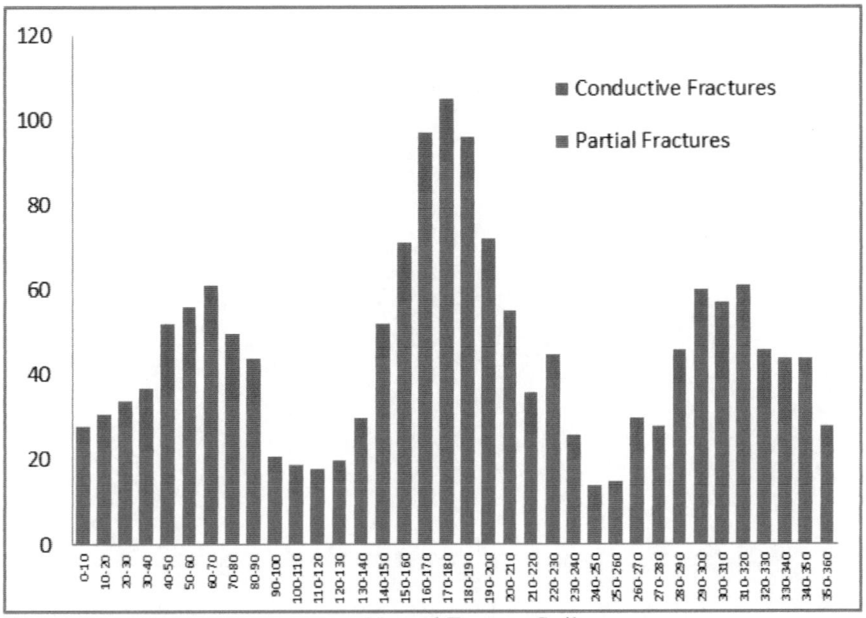

a. Natural Fracture Strike

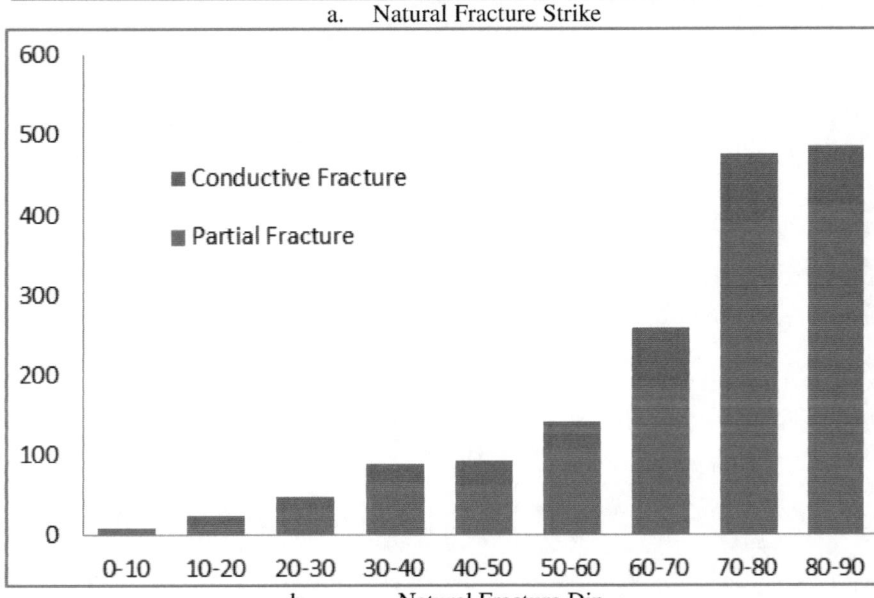

b. Natural Fracture Dip

Figure 4. Histogram of natural fracture strikes (a) and dips (b) in Rantau Dedap well.

The modelling of S_{Hmax} magnitudes and the azimuth are necessary to produce the breakouts of the observed width and position in the highly inclined well which was conducted using the software Stress and Failure of Inclined Boreholes (GMI·SFIB) designed by GeoMechanics International (GMI). The parameters used are the remote stress tensor onto the borehole surface, mud pressure in the borehole, thermal stress which accompanying temperature change through the coefficient of linear expansion, and the poroelastic distortion adjacent to the borehole wall through Biot's coefficient and Poisson's ratio [8]. Based on breakout on each well, the direction of maximum horizontal stress at each well is determined. The results indicate that the direction of maximum horizontal stress at AA-1 has N242°E azimuth direction (see figure 6), AA-2 has N204°E azimuth direction (see figure 6), and AA-3 has N162°E azimuth direction (see figure 7).

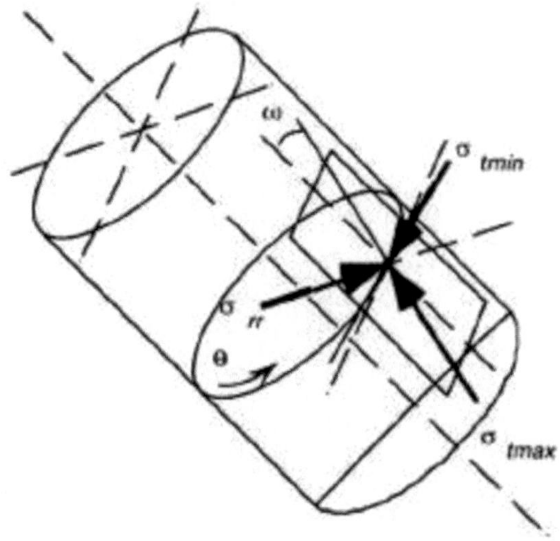

Figure 5. Stresses acting in the wall of a directional well [7].

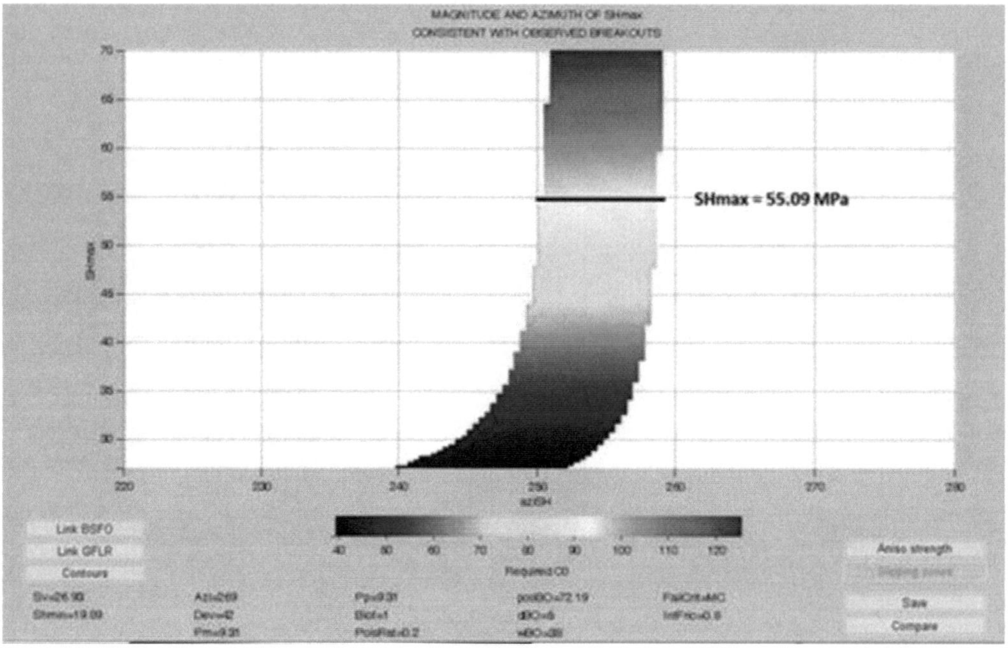

a. AA-1.

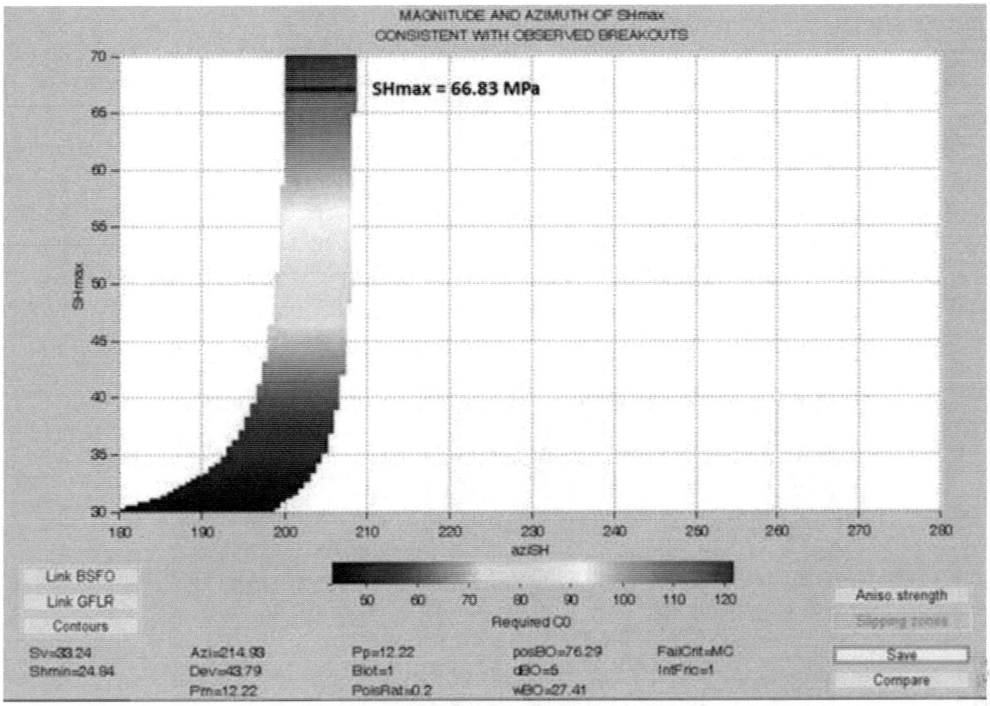

b. AA-2

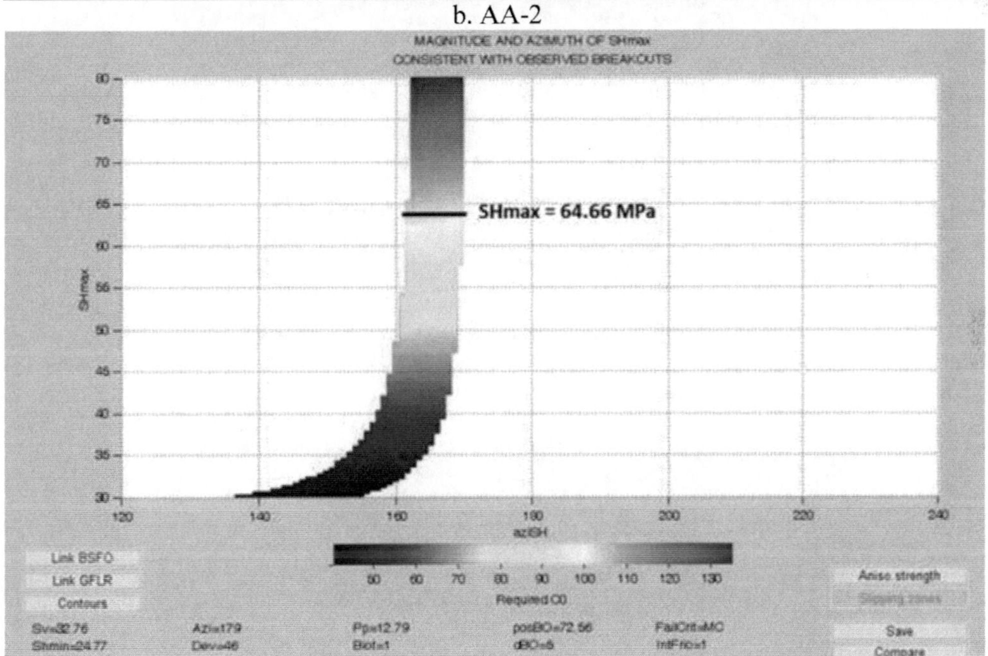

c. AA-2

Figure 6. The orientation S_{Hmax} on AA-1 (a), AA-2 (b) and AA-3 (c) are calculated based on the parameter magnitudes Geomechanics using GMI-SFIB module. The value obtained on the basis of the equation S_{Hmax} [7].

3.3. Principal Horizontal Stress Magnitudes Modelling

The maximum horizontal stress magnitudes are the most difficult parameter to be calculated. This parameter cannot be obtained specially using direct measurement. However, this parameter is adjusted by using specific calculation or using borehole breakout and frictional condition to constraint the data.

First step, S_{hmin} is evaluated magnitudes ranging from critically stressed for normal faulting at a coefficient of friction of 0.8 to 1.0 in which S_{hmin} approaches S_V, thus spanning the range of stresses consistent with normal faulting to strike-slip faulting. The 0.8 coefficient of friction is a conservative estimation based on frictional strength studies of the crust, whereas volcanic andesite is expected to have a coefficient of friction of 0.6 [9]. For this range of S_{hmin} magnitudes we derived corresponding S_{Hmax} azimuths consistent with breakout position and width for each pair of breakouts in the borehole as a function of rock strengths (UCS).

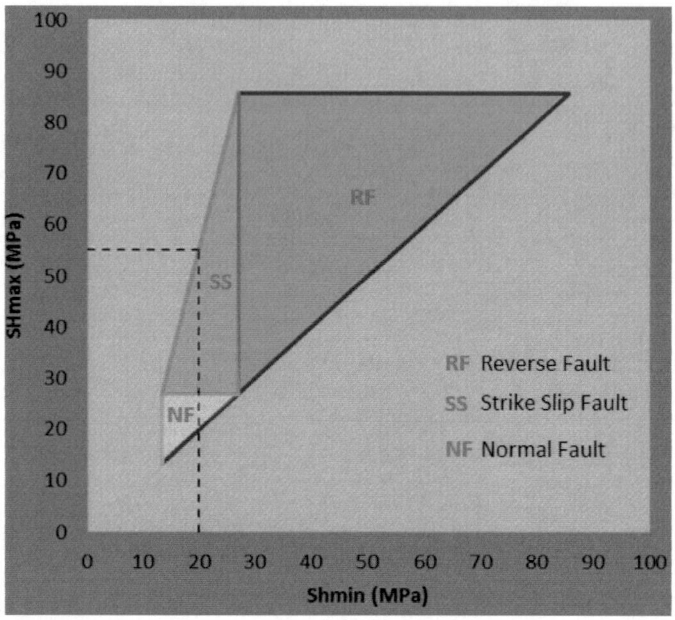

a. AA-1.

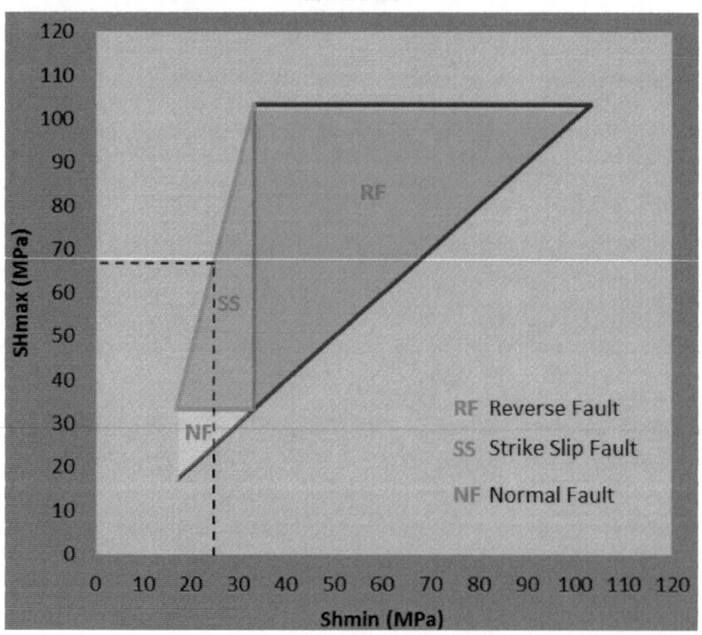

b. AA-2.

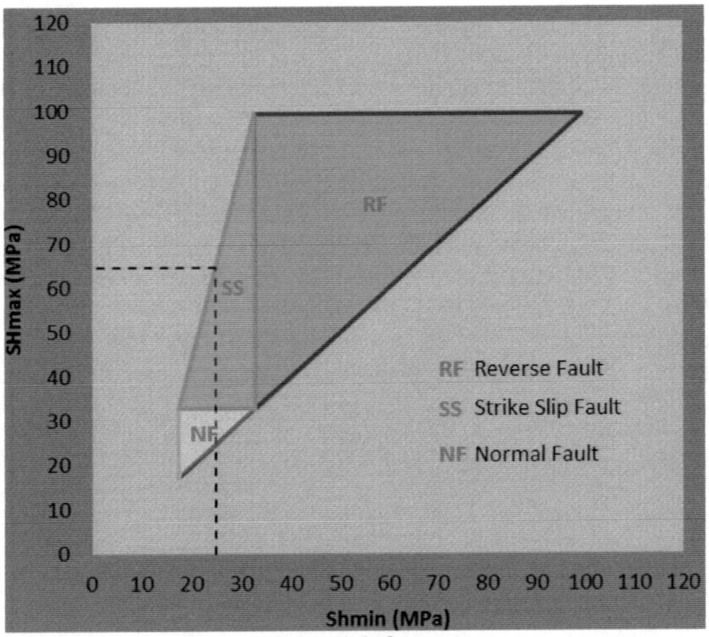

c. AA-3.

Figure 7. Analysis of in-situ stress using stress polygon on the well AA-1(a), well AA-2 (b) and well AA-3 (c).

Next step is to combine of S_{hmin} and S_{Hmax} magnitudes together with S_V, P_p and the wellbore conditions that could reproduce the breakout positions and widths (see figure 7). The polygon in figure 7 defines the range of principal horizontal stress magnitudes scaled to the vertical stress that can be supported by the frictional strength of the crust, conservatively estimated to be 0,8. The UCS necessary to allow the modelled breakout to form it's contoured as a function of these principal horizontal stresses; only the ranges of S_{hmin} and S_{Hmax} that producing breakouts are contoured. For any given breakout, horizontal stress magnitudes scaled to S_{Hmax} that are consistent with the breakout occurrence and thus is a measurement of the uncertainty in the stress magnitude model.

The analysis shows that the AA-1 has the value S_{Hmax} 55.1 MPa, formed by strike slip fault regime, AA-2 has a value of S_{Hmax} 66.8 MPa, formed by the strike slip fault regime, and AA-3 has the value S_{Hmax} 64.6 MPa, formed by the strike slip fault regime.

4. Conclusions

This detailed analysis of borehole image data recorded in AA-1, AA-2 and AA-3 well have provided a well resolved orientation of the horizontal principal stresses that agrees with other observation of the stress state at the Rantau Dedap site. Natural fracture are strongly clustered with true strike trending which the first, second, and third dominant direction is N170°E – N180°E (N-S), N60°E – N70°E (NE-SW), and N310°E – N320°E (NW-SE) respectively, while the true dip dominant is 80° –90°. The magnitude of the maximum horizontal compressive stress constrained by these data is consistent with the occurrence both strike-slip mechanisms at this site.

References

[1] Hennings P, Allwardt P, Paul C, Zahm R, Reid Jr H, Alley R, Kirschner B, Lee and Hough E 2012 Relationship between fractures, fault zones, stress, and productivity in The Suban gas field, Sumatra, Indonesia *AAPG Bulletin.* **96** p 753 – 772

[2] Zoback M D 2007 *Reservoir Geomechanics* (New York: Cambridge University Press)

[3] Sieh K.and Natawijaya D 2000 Neotectonics of the Sumatran Fault, Indonesia *Journal of Geophysical Reserch* **105** p 22 – 295

[4] Barber A J and Crow M J 2005 *Sumatra: geology, resources and tectonic Evolution* (London: Geological Society Memoirs) p 175-233

[5] Sidik R P, Mussofan W, Santana S and Azis H 2016 Structure geology of southwestern sector Rantau Dedap Geothermal Field Jakarta *Proceedings Annual 4th Indonesia International Geothermal Convention & Exhibition* **4**

[6] Santana S, Abiyudo R, Hadi J and Sapiie B 2013 Structural concept (play) to reduce well targetting risk: Rantau Dedap Case Study *Jakarta Proceedings The 13th Annual Indonesian Geothermal Association Meeting & Conference* **13**

[7] Zoback M D, Barton C A, Brudy M, Castillo D A, Finkbeiner T, Grollimund B R, Moos D B, Peska P, Ward C D and Wiprut D J 2003 Determination of stress orientation and magnitude in deep wells" *International Journal of Rock Mechanics and Mining Sciences* **40** p 1049

[8] Batir J, Davatzes N C and Asmundsson R 2012 Preliminary model of fracture and stress state in The Hellisgeidi Geothermal Field, Hengil Volcanic System, Iceland *Proceedings Thirty-Seventh Workshop on Geothermal Reservoir Engineering Stanford University* **37**

[9] Byerlee J D 1978 Friction of rock *Pure Appl Geophys* **11** p 615 – 26

Institute of Physics
Dirac House, Temple Back
Bristol BS1 6BE UK

ISSN: 1755-1307
ISBN 978-1-5108-5554-0

CURRAN ASSOCIATES INC.
proceedings.com

IV International Conference on Modern Technologies for Non-Destructive Testing 2015

IOP Conference Series: Materials Science and Engineering Volume 132

Tomsk, Russia
5 – 10 October 2015

IOP
Institute of Physics